OIL SPILL DISPERSANTS

EFFICACY AND EFFECTS

Committee on Understanding Oil Spill Dispersants: Efficacy and Effects
Ocean Studies Board
Division on Earth and Life Studies

NATIONAL RESEARCH COUNCIL
OF THE NATIONAL ACADEMIES

THE NATIONAL ACADEMIES PRESS
Washington, D.C.
www.nap.edu

THE NATIONAL ACADEMIES PRESS 500 Fifth Street, N.W. Washington, DC 20001

NOTICE: The project that is the subject of this report was approved by the Governing Board of the National Research Council, whose members are drawn from the councils of the National Academy of Sciences, the National Academy of Engineering, and the Institute of Medicine. The members of the committee responsible for the report were chosen for their special competences and with regard for appropriate balance.

This study was supported by a contract between the National Academies and the following entities: Contract/Grant No. 50-DGNA-1-90024 from the National Oceanic and Atmospheric Administration, Contract/Grant No. 0103PO73652 from the Minerals Management Service and the U.S. Coast Guard, and Contract/Grant No. 2003-100861 from the American Petroleum Institute. Any opinions, findings, conclusions, or recommendations expressed in this publication are those of the author(s) and do not necessarily reflect the views of the organizations or agencies that provided support for the project.

International Standard Book Number 0-309-09562-X
Library of Congress Catalog Card Number 2005931459

Additional copies of this report are available from the National Academies Press, 500 Fifth Street, N.W., Lockbox 285, Washington, DC 20055; (800) 624-6242 or (202) 334-3313 (in the Washington metropolitan area); Internet, http://www.nap.edu

Printed in the United States of America

THE NATIONAL ACADEMIES

Advisers to the Nation on Science, Engineering, and Medicine

The **National Academy of Sciences** is a private, nonprofit, self-perpetuating society of distinguished scholars engaged in scientific and engineering research, dedicated to the furtherance of science and technology and to their use for the general welfare. Upon the authority of the charter granted to it by the Congress in 1863, the Academy has a mandate that requires it to advise the federal government on scientific and technical matters. Dr. Ralph J. Cicerone is president of the National Academy of Sciences.

The **National Academy of Engineering** was established in 1964, under the charter of the National Academy of Sciences, as a parallel organization of outstanding engineers. It is autonomous in its administration and in the selection of its members, sharing with the National Academy of Sciences the responsibility for advising the federal government. The National Academy of Engineering also sponsors engineering programs aimed at meeting national needs, encourages education and research, and recognizes the superior achievements of engineers. Dr. Wm. A. Wulf is president of the National Academy of Engineering.

The **Institute of Medicine** was established in 1970 by the National Academy of Sciences to secure the services of eminent members of appropriate professions in the examination of policy matters pertaining to the health of the public. The Institute acts under the responsibility given to the National Academy of Sciences by its congressional charter to be an adviser to the federal government and, upon its own initiative, to identify issues of medical care, research, and education. Dr. Harvey V. Fineberg is president of the Institute of Medicine.

The **National Research Council** was organized by the National Academy of Sciences in 1916 to associate the broad community of science and technology with the Academy's purposes of furthering knowledge and advising the federal government. Functioning in accordance with general policies determined by the Academy, the Council has become the principal operating agency of both the National Academy of Sciences and the National Academy of Engineering in providing services to the government, the public, and the scientific and engineering communities. The Council is administered jointly by both Academies and the Institute of Medicine. Dr. Ralph J. Cicerone and Dr. Wm. A. Wulf are chair and vice chair, respectively, of the National Research Council

www.national-academies.org

COMMITTEE ON UNDERSTANDING OIL SPILL DISPERSANTS: EFFICACY AND EFFECTS

JACQUELINE MICHEL, *Chair*, Research Planning, Inc., New Orleans, Louisiana
E. ERIC ADAMS, Massachusetts Institute of Technology, Cambridge
YVONNE ADDASSI, California Department of Fish and Game, Sacramento
TOM COPELAND, Commercial Fisherman (retired), Everson, Washington
MARK GREELEY, Oak Ridge National Laboratory, Oak Ridge, Tennessee
BELA JAMES, Shell Global Solutions, Houston, Texas
BETH MCGEE, Chesapeake Bay Foundation, Annapolis, Maryland
CARYS MITCHELMORE, University of Maryland Center for Environmental Science, Solomons
YASUO ONISHI, Pacific Northwest National Laboratory, Richland, Washington
JAMES PAYNE, Payne Environmental Consultants, Inc., Encinitas, California
DAVID SALT, Oil Spill Response Limited, Southampton, Hampshire, United Kingdom
BRIAN WRENN, Washington University, St. Louis, Missouri

Staff

DAN WALKER, Study Director
SARAH CAPOTE, Senior Program Assistant

Preface

The use of chemical dispersants as an oil spill countermeasure in the United States has long been controversial. In the late 1980s, the National Research Council was asked to conduct a study to "assess the state of knowledge and practice about the use of dispersants in responding to open-ocean spills." The resulting report, published in 1989, became an important summary of the effectiveness and possible impacts of dispersants and dispersed oil. In the early 1990s, there was a major initiative to get pre-approval for dispersant use in open waters, generally outside of 3 nautical miles and/or in water depths greater than 10 meters, and many regions have such pre-approval plans in place. Dispersants have not, however, been used frequently, with one of the limitations being the need to mobilize available dispersant and application equipment within the narrow (1–2 days) window of opportunity during which dispersants are most effective.

In the late 1990s, the U.S. Coast Guard began to review the regulatory planning requirements for dispersant use in vessel response plans, resulting in a proposed rulemaking that will require the availability of dispersants and equipment where dispersant use has been pre-approved. Anticipating that the ready availability of dispersants would lead to increased desire to use dispersants at all types of spills, the U.S. Coast Guard also began conducting workshops to assist planners in comparing the ecological consequences of response options, especially in nearshore or estuarine situations. During these workshops, it became clear that there were significant gaps in the knowledge needed to make sound decisions regarding the use of dispersant in areas that were nearshore, shallow, or with restricted flushing rates. In these areas, the simplifying assumptions that were used in the risk analysis for open-water setting were insufficient.

Realizing that there are limited funds to support oil spill research in general, and dispersant use in particular, the Minerals Management Service, National Oceanic and Atmospheric Administration, U.S. Coast Guard, and American Petroleum Institute requested that the National Research Council review and evaluate the existing and ongoing research, and make recommendations on the information needed to support risk-based decisionmaking.

A committee of twelve scientists and responders, representing a wide range of technical backgrounds, was appointed by the National Research Council to prepare the requested report. Taking to heart the emphasis on risk-based decisionmaking, the committee decided to frame its assessment and recommendations around the questions that planners and responders must answer when faced with the decision as to whether or not dispersants should be used at a given spill. This approach, I believe, has improved the value of the study by linking the recommended research to the needs of decisionmakers.

I wish to thank the committee members for their dedication and hard work during the preparation of the report. They conducted a fresh and thorough review of the existing and ongoing research that should make the report a significant contribution to understanding the current knowledge on the effectiveness and effects of dispersants. The Study Director, Dr. Dan Walker, did an outstanding job of keeping the committee focused on the statement of task and the importance of the decision-making framework approach to the report. I would like to personally thank him for his insight, technical knowledge, and professionalism. The committee members wish to especially thank the hard work of Ms. Sarah Capote who greatly helped the committee develop what I think is a high-quality final report. The sponsors are to be commended for their vision in providing funding for this study—a study that will likely influence both the direction of dispersant-related research and the actual use of dispersants as an oil spill countermeasure in the coming years.

Jacqueline Michel, *Chair*

Acknowledgments

This report was greatly enhanced by the participants of the three workshops held as part of this study. The committee would first like to acknowledge the efforts of those who gave presentations at meetings. These talks helped set the stage for fruitful discussions in the closed sessions that followed.

Don Aurand, Ecosystems Management & Associates
Mace Barron, Environmental Protection Agency
C.J. Beegle-Krause, National Oceanic and Atmospheric Administration (NOAA)
James Clark, ExxonMobil
Merv Fingas, Environment Canada
Deborah French-McCay, Applied Science Associates, Inc.
Jerry Galt, Genwest Sytems, Inc.
Charlie Henry, NOAA Scientific Support Coordinator for the Gulf of Mexico
Alun Lewis, Oil Spill Consultant
Carol-Ann Manen, NOAA
Joe Mullin, Minerals Management Service
Bob Pond, U.S. Coast Guard
Robin Rorick, American Petroleum Institute
Michael Singer, University of California-Davis
Al Venosa, Environmental Protection Agency
Glen Watabayashi, NOAA
Jim Weaver, Environmental Protection Agency

The committee is also grateful to a number of people who provided important discussion and/or material for this report: **Alan Allen** (Spiltec), **Randy Belore** (S.L. Ross), **James Clark** (ExxonMobil), **Per Daling** (SINTEF), **Don Davis** (Louisiana State University), **Dave DeVitis** (Ohmsett Site Manager), **Dave Evans** (Environmental Protection Agency), **Merv Fingas** (Environment Canada), **R. Lloyd Gamble** (Environment Canada), **Julien Guyomarch** (Cedre), **Kurt Hansen** (U.S. Coast Guard), **Charlie Henry** (NOAA), **Robin Jamail** (Texas General Land Office), **Jim Lane** (Minerals Management Service), **Ken Lee** (Fisheries and Oceans Canada, Bedford Institute of Oceanography), **Carol-Ann Manen** (NOAA), **Joseph Mullin** (Minerals Management Service), **Leslie Pearson** (Alaska Department of Environmental Conservation), **Robin Rorick** (American Petroleum Institute), and **Mike Sowby** (California Department of Fish and Game Office of Spill Prevention and Response).

This report has been reviewed in draft form by individuals chosen for their diverse perspectives and technical expertise, in accordance with procedures approved by the NRC's Report Review Committee. The purpose of this independent review is to provide candid and critical comments that will assist the institution in making its published report as sound as possible and to ensure that the report meets institutional standards for objectivity, evidence, and responsiveness to the study charge. The review comments and draft manuscript remain confidential to protect the integrity of the deliberative process. We wish to thank the following individuals for their participation in the review of this report:

Mace Barron, Environmental Protection Agency, Gulf Breeze, Florida
James Bonner, Texas A&M University, Corpus Christi
James Clark, ExxonMobil Research and Engineering Company, Fairfax, Virginia
Merv Fingas, Environment Canada, Ottawa, Ontario
Robert "Buzz" Martin, Texas General Land Office, Austin
Judy McDowell, Woods Hole Oceanographic Institution, Woods Hole, Massachusetts
Robert Paine, University of Washington, Seattle
Mark Reed, SINTEF, Trondheim, Norway
Susan Saupe, Cook Inlet Regional Citizens Advisory Council, Kenai, Alaska

Although the reviewers listed above have provided many constructive comments and suggestions, they were not asked to endorse the conclusions or recommendations nor did they see the final draft of the report before its release. The review of this report was overseen by **Mahlon "Chuck" Kennicutt**, Texas A&M University, College Station. Appointed by the Divison on Earth and Life Studies, he was responsible for making certain that an independent examination of this report was carried out in accordance with institutional procedures and that all review comments were carefully considered. Responsibility for the final content of this report rests entirely with the authoring committee and the institution.

Contents

Executive Summary

Approximately 3 million gallons (10,000 metric tons [tonnes]) of oil or refined petroleum product[1] are spilled into the waters of the United States every year (NRC, 2003). This amount represents the total input from hundreds of spills, many of which necessitate timely and effective response. When these oil spills occur in the United States, the primary response methods consist of the deployment of mechanical on-water containment and recovery systems, such as booms and skimmers.

Under the Oil Pollution Act of 1990 (OPA 90), the U.S. Coast Guard (USCG) passed rules for vessel and facility response plans that specified the minimum equipment and personnel capabilities for oil containment and recovery. This requirement has significantly expanded mechanical response capability above that which existed in 1989 at the time of Tanker Vessel (T/V) *Exxon Valdez* spill (the event that led to passage of OPA 90). Mechanical recovery, however, is not always sufficient because conditions at the spill are often outside of the effective operating conditions of the equipment. OPA 90 also called for national and regional response teams to develop guidelines to address the use of other on-water response strategies, specifically the use of chemical dispersants and *in-situ* burning.

Throughout the Unites States, many regional response teams have identified zones where dispersants and *in-situ* burning are "pre-approved" for use. This pre-approval means that the response and re-

[1]The terms oil, refined product, or petroleum hydrocarbon are used interchangeably in this report.

source agencies have determined that the Federal On-Scene Coordinator has the authority, as outlined under the pre-approval definitions, to decide to use dispersants without additional consultation. In general, these pre-approval zones are in waters beyond 3 nautical miles (nm; roughly 5 kilometers [km]) of the shoreline and in water depths greater than 30 feet (10 meters). Even with establishment of these pre-approval zones, dispersant use has been infrequent, in part reflecting the difficulty of mobilizing available equipment and dispersants within a narrow window of opportunity in which they can be effective. In areas where dispersants are not often considered, it takes more time to identify, contract, and mobilize the specialized resources needed for dispersant application.

To address the concerns regarding requisite equipment and personnel capabilities, the U.S. Coast Guard in 2002 proposed changes to the oil spill contingency planning regulations measuring the minimum capabilities for dispersant application in all pre-approved zones within acceptable time frames. With implementation of the regulations, dispersant application resources will become more readily available. The potential, therefore, for using dispersants in nearshore and shallow waters, when appropriate, will increase as well.

Oil spill dispersants do not actually reduce the total amount of oil entering the environment. Rather, they change the inherent chemical and physical properties of oil, thereby changing the oil's transport, fate, and potential effects. Small amounts of spilled oil naturally disperse into the water column, through the action of waves and other environmental processes. The objective of dispersant use is to enhance the amount of oil that physically mixes into the water column, reducing the potential that a surface slick will contaminate shoreline habitats or come into contact with birds, marine mammals, or other organisms that exist on the water surface or shoreline. Conversely, by promoting dispersion of oil into the water column, dispersants increase the potential exposure of water-column and benthic biota to spilled oil. Dispersant application thus represents a conscious decision to increase the hydrocarbon load (resulting from a spill) on one component of the ecosystem (e.g., the water column) while reducing the load on another (e.g., coastal wetland). Decisions to use dispersants, therefore, involve trade-offs between decreasing the risk to water surface and shoreline habitats while increasing the potential risk to organisms in the water column and on the seafloor. This trade-off reflects the complex interplay of many variables, including the type of oil spilled, the volume of the spill, sea state and weather, water depth, degree of turbulence (thus mixing and dilution of the oil), and relative abundance and life stages of resident organisms.

Each spill is a unique event that unfolds over a variety of time scales. Properties of petroleum hydrocarbons immediately start to change when spilled onto water. This natural "weathering" makes the oil more difficult to disperse through time; consequently, the window of opportunity for effective dispersant application is early, usually within hours to 1–2 days after a release under most conditions, though there are exceptions. The decision to apply dispersants is thus time sensitive and complex. Given the potential impacts that dispersed oil may have on water-column and seafloor biota and habitats, thoughtful analysis is required prior to the spill event so that decisionmakers understand the potential impacts with and without dispersant application. Thus, decisionmaking regarding the use of dispersants falls into two broad temporal categories: (1) before the event during spill contingency planning; and (2) shortly after the initial event, generally within the first 12 to 48 hours.

In recognition of the increased potential to use dispersants in a variety of settings, the Minerals Management Service (MMS), the National Oceanic and Atmospheric Administration (NOAA), the USCG, and the American Petroleum Institute (API) asked the National Academies to form a committee of experts to review the adequacy of existing information and ongoing research regarding the efficacy and effects of dispersants as an oil spill response technique in the United States.[2] Emphasis was placed on understanding the limitations imposed by the various methods used in these studies and on recommending steps that should be taken to better understand the efficacy of dispersant use and the effect of dispersed oil on freshwater, estuarine, and marine environments. Specifically, the committee's task was to:

- review and evaluate ongoing research and existing literature on dispersant use (including international studies) with emphasis on (a) factors controlling dispersant effectiveness (e.g., environmental conditions, dispersant application vehicles and strategies, and oil properties, particularly as the spilled oil weathers), (b) the short- and long-term fate of chemically or naturally dispersed oil, and (c) the toxicological effects of chemically and naturally dispersed oil;
- evaluate the adequacy of the existing information about dispersants to support risk-based decisionmaking on response options for a variety of spatially and temporally defined oil spills;
- recommend steps that should be taken to fill existing knowledge gaps, with emphasis to be placed on how laboratory and mesoscale ex-

[2]A similar request was put to the National Academies in the mid 1980s, leading to the publication of the 1989 NRC report *Using Oil Spill Dispersants on the Sea*. The current report is not truly an update of the 1989 report, as it selectively revisits some topics while including discussions on issues that have emerged since that time. Many readers may, therefore, find the assessments and summaries in *Using Oil Spill Dispersants on the Sea* of value.

periments could inform potential controlled field trials and what experimental methods are most appropriate for such tests.

OVERARCHING CHALLENGE TO EFFECTIVE DECISIONMAKING

In general, the information base used by decisionmakers dealing with spills in areas where the consequences of dispersant use are fairly straightforward has been adequate (for example, situations where rapid dilution has the potential to reduce the possible risk to sensitive habitat enough to allow the establishment of pre-approval zones). Many of the technical issues raised in this report, however, deal with settings where greater confidence is needed to make effective decisions regarding potential benefits or adverse impacts associated with dispersant use. In many instances where a dispersed plume may come into contact with sensitive water-column or benthic organisms and populations, the current understanding of key processes and mechanisms is inadequate to confidently support a decision to apply dispersants. Thus, such decisions regarding the potential use of dispersants in nearshore settings are creating a demand for additional information.

Research funds in the United States to support oil spill response options in general are extremely limited and declining (as discussed in Chapter 1, the total amount is less than $10 million annually). Consequently, despite the complex and numerous variables involved in risk-based decisionmaking regarding the potential use of dispersants, efforts to fill knowledge gaps must be thoroughly grounded in the recognition that no amount of research or environmental monitoring will eliminate uncertainty entirely. Failure to make a timely decision regarding dispersant application is in actuality a decision not to use dispersants, and in some instances may place some natural resources at an increased and unnecessary risk. Given the limited funding available to carry out needed research in this area, it is particularly important that research be carried out as efficiently as possible and that the research process focuses on efforts that result in sound, reproducible results that support decisionmaking. In many instances, efforts to reduce experimental complexity to ensure reproducibility or to secure cost savings have led to results that have very limited utility for making decisions in natural settings. **NOAA, the Environmental Protection Agency (EPA), the Department of the Interior (including MMS and U.S. Geological Survey), USCG, relevant state agencies, industry, and appropriate international partners should work together to establish an integrated research plan which focuses on collecting and disseminating peer-reviewed information about key aspects**

of dispersant use in a scientifically robust, but environmentally meaningful context (see Chapter 6 for more detail).

SETTING PRIORITIES IN DISPERSANT RESEARCH

Key components of an effective and integrated research effort should include efforts to further improve understanding of dispersant effectiveness and the potential impact of dispersed oil at meaningful scales to support decisionmaking in a broader array of spill scenarios, especially those scenarios where potential impacts on one portion of the ecosystem (e.g., water column) must be weighed against benefits associated with reducing potential impact on another (e.g., coastal wetland). In an effort to provide some prioritization, the following research recommendations are presented in order of significance. The most pressing or widely relevant issues are listed first, with less pressing or narrowly relevant issues raised later.

With the proposed USCG regulations requiring the availability of dispersants in pre-approval zones, the issue of availability will no longer be a limiting factor; thus the main questions to be addressed by responders in the pre-approval zones are: (1) Will mechanical recovery be effective and sufficient? (2) If not, is the oil dispersible? (3) If so, are the environmental conditions conducive to the successful application of dispersant and its effectiveness? and (4) If so, will the effective use of dispersants reduce the impacts of the spill to shoreline and water-surface resources without significantly increasing impacts to water-column and benthic resources? Better information is needed to determine the window of opportunity and percent effectiveness of dispersant application for different oil types and environmental conditions. **Relevant state and federal agencies, industry, and appropriate international partners should develop and implement a focused series of studies that will enable the technical support staff advising decisionmakers to better predict the effectiveness of dispersants for different oil types and environmental conditions based on climatological data supplemented with real-time *in-situ* observations.** (Detailed and specific recommendations are discussed at length in Chapters 3 and 4.)

Oil trajectory and fate models used by the technical support staff advising on-scene decisionmakers for dispersed oil behavior are not adequate in terms of: (1) their representation of the natural physical process involved, (2) verification of the codes, and (3) validation of the output from these models in an experimental setting or during an actual spill. Thus, their ability to predict the concentrations of dispersed oil and dissolved petroleum hydrocarbons of concern in the water column with sufficient accuracy to aid in real-time spill decisionmaking has yet to be fully determined. **Oil trajectory and fate models used by government agencies during spill response to predict the behavior of dispersed oil should be improved, verified, and then validated in an appropriately designed experimental setting or during actual spills.** Two general types of modeling efforts and products should be recognized: (1) output intended to support decisionmaking during preplanning efforts, and (2) output intended to support emergency response to provide "rough-cut" outputs in hours. (Detailed and specific recommendations are discussed at length in Chapters 4 and 5.)

The mechanisms of both acute and sublethal toxicity from exposure to dispersed oil are not sufficiently understood. Recent studies in the literature suggest that toxicity from physically and chemically dispersed oil appears to be primarily associated with the additive effects of various dissolved-phase polynuclear aromatic hydrocarbons (PAH) with additional contributions from heterocyclic (N, S, and O) containing polycyclic aromatic compounds. Additional toxicity may be coming from the particulate, or oil droplet, phase, but a particular concern stems from potential synergistic effects of exposure to dissolved components in combination with chemically dispersed oil droplets. **Relevant state and federal agencies, industry, and appropriate international partners should develop and fund a series of focused toxicity studies to determine the mechanisms of both acute and sublethal toxicity to key organisms from exposure to dispersed oil.** With a better understanding of the mechanisms of toxicity, toxicity tests can be refined to generate data on toxic levels and thresholds for use by decisionmakers. (Detailed and specific recommendations are discussed at length in Chapters 5 and 6.)

The factors controlling rates of the biological and physical processes that determine the ultimate fate of dispersed oil are poorly understood. Of particular concern is the fate of dispersed oil in areas with high suspended solids and areas of low flushing rates. There is insufficient information to determine how chemically dispersed oil interacts with suspended sediments, both short- and long-term, compared to naturally dispersed oil. There are many important, unanswered questions about how dispersed oil might be consumed by plankton and deposited on the seafloor with fecal matter or otherwise passed through the food chain. **Relevant state and federal agencies, industry, and appropriate international partners should develop and fund a focused series of studies to quantify the weathering rates and final fate of chemically dispersed oil droplets compared with undispersed oil.** (Detailed and specific recommendations are discussed at length in Chapters 3 and 4.)

There is insufficient understanding of the actual concentrations and temporal or spatial distributions and behavior of chemically dispersed oil from field settings (from either controlled experiments or actual spills). Data from field studies (both with and without dispersants) are needed to validate models and provide real-world data to improve knowledge of oil fate and effects. In the future, wave-tank or spill-of-opportunity studies should include efforts to measure total petroleum hydrocarbon (TPH) and PAH concentrations in both the dissolved phase and particulate/oil droplet phase for comparison to TPH and PAH thresholds measured in toxicity tests and predicted by computer models. **Relevant state and federal agencies and industry should develop and implement detailed plans (including pre-positioning of sufficient equipment and human resources) for rapid deployment of a well-designed monitoring effort for actual dispersant applications in the United States.** (Detailed and specific recommendations are discussed at length in Chapters 2, 3, and 4.)

To date, there have been no wave-tank or laboratory studies that can be used reliably to predict the performance of dispersants on water-in-oil emulsions (i.e., mousse) generated from the weathering of oil on the water surface. **Relevant state and federal agencies, industry, and appropriate international partners should initiate a detailed investigation of wave-tank studies that specifically address the chemical treatment of weathered oil emulsions.** (Detailed and specific recommendations are discussed at length in Chapters 3 and 4).

One of the most significant weaknesses in correlating laboratory-scale and meso-scale experiments with conditions in the open ocean results from a lack of understanding of the turbulence regime in all three systems. Likewise, one of the biggest uncertainties in computer modeling of oil spill behavior (with and without dispersant application) comes from obtaining horizontal and vertical diffusivities. **Relevant state and federal agencies, industry, and appropriate international partners should initiate a detailed investigation of upper sea-surface turbulence with particular emphasis on quantifying horizontal and vertical diffusivities and the rate of energy dissipation.** (Detailed and specific recommendations are discussed at length in Chapter 4.)

Finally, serious consideration should be given to determining the value and potential role of field testing. The body of work done to date has provided important, but still limited understanding of many aspects of the efficacy of dispersants in the field and the behavior and toxicity of dispersed oil. Developing a robust understanding of these key processes and mechanisms to support decisionmaking in nearshore environments will require taking dispersant research to the next level. This new work will require systematic analysis using rigorous experimental design and execution, making use of standard chemical and other measurement techniques carried out by trained, certified personnel. Many factors will need to be systematically varied in settings where accurate measurements can be taken. It is difficult to envision the proper role of field testing in a research area where investigators have yet to reach consensus on standard protocols for wave-tank testing. The greater complexities (and costs) of carrying out meaningful field experiments suggest that greater effort be placed, at least initially, on designing and implementing a thorough and well-coordinated bench-scale and wave-tank research program. Such work should lead to more robust information about many aspects of dispersed oil behavior and effects. When coupled with information gleaned through more vigorous monitoring of actual spills (regardless of whether dispersants are used effectively in response), this experimental work should provide far greater understanding than is currently available. Upon completion of the work recommended in this report, the value of further field-scale experiments may become obvious. If deemed valuable, such field-scale work would certainly be better and more effectively designed and executed than is currently possible. **Future field-scale work, if deemed necessary, should be based on the systematic and coordinated bench-scale and wave-tank testing recommended in this report.** (Detailed and specific recommendations are discussed in Chapters 3, 4 and 5.)

1

Introduction

Although significant steps have been taken over the last 15 years to reduce the size and frequency of oil spills, the sheer volume of petroleum consumed in this country and the complex production and distribution network required to meet the demand make spills of oil and other petroleum products inevitable (NRC, 2003). Oil spill contingency plans, therefore, specify appropriate response to spills whenever and wherever they occur. Spill response in the United States has traditionally focused primarily on physical containment and recovery approaches. For spills on water, these approaches emphasize controlling and recovering spilled oil or petroleum products through the deployment of mechanical equipment such as booms and skimmers.

The effectiveness of mechanical response techniques is variable and highly influenced by the size, nature, and location of the spill as well the environmental conditions under which the response is carried out. Essentially, mechanical response works satisfactorily under a finite subset of all possible spill scenarios. The spill response community has worked to expand the subset of spill scenarios where effective response can be mounted, through improving the quality, and to some degree, the quantity of mechanical equipment available to respond to a spill, and training and coordination of efforts. In addition, other non-mechanical techniques have been developed and tested. The two most commonly considered non-mechanical techniques include *in-situ* burning and the use of chemical dispersants.

In-situ burning refers to the controlled burning of oil close to where the spill occurred. For spills on open water, the oil must be collected and held by fire-resistant booms or trapped in ice to ensure that the oil has a minimum thickness to be ignited and sustain burning. The advantages of *in-situ* burning include rapid removal of oil and no need for oil recovery, transport, storage, and disposal. The major disadvantages of *in-situ* burning include the black smoke, difficulties of collecting and containing a large amount of the oil to burn, lower effectiveness as the oil weathers (spreads, emulsifies), and sensitivity to sea state and weather conditions that reduce the viability of all response options (Michel et al., 2004). Worldwide, there have been 43 known intentional *in-situ* burns of oil on water (Fingas, 1999b; Michel et al., 2004). Of these, only thirteen were actual spills (the rest were planned tests). Of these, four were in ice, two were attempts to burn the oil inside the holds of the ship (Torrey Canyon and New Carissa), and four were of uncontained slicks. In the United States, the only on-water *in-situ* burning at a spill was the 1989 test burn during the *Exxon Valdez* oil spill, which was the first time a fire-resistant boom was used at a spill (Michel et al., 2004).

Dispersants are chemical agents (surfactants, solvents, and other compounds) that reduce interfacial tension between oil and water in order to enhance the natural process of dispersion by generating larger numbers of small droplets of oil that are entrained into the water column by wave energy. The small dispersed oil droplets tend not to merge into larger droplets that quickly float back to the water surface and reform into surface slicks. Instead, the small droplets stay suspended in the water column, spreading in three dimensions instead of two and being distributed by turbulent diffusion.

The use of chemical dispersants, as well as *in-situ* burning, revolves around changing the fate of spilled material within the environment, as opposed to attempting recovery or removal of that material from the environment. They are therefore generally viewed in the United States as secondary options intended to support or supplement mechanical response, and requiring risk-based decisionmaking at the time of a spill.

Early efforts to disperse oil slicks on water and along shorelines used degreasing agents or detergents that contained highly toxic components and resulted in high mortality to rocky shore communities (Smith, 1968). Recent formulations are much less toxic such that the toxicity associated with dispersed oil droplets is essentially a function of the toxicity of the oil itself. As a consequence, U.S. policymakers have been exploring the potential for dispersant use for nearly two decades. In 1989, the National Research Council released *Using Oil Spill Dispersants on the Sea*. That report focused on the possible effects and effectiveness of using dispersants to combat spills in open waters. Highlighting a number of specific research efforts that should be pursued, one of the report recommendations was that "dispersants be considered as a potential first response option" to large spills in the open ocean. Because dispersant effectiveness diminishes as the spilled oil weathers, it was recommended that regulations and contingency planning make rapid response possible. It was recognized that the availability of both dispersant and the equipment needed to apply it greatly influenced the potential to use dispersants during the critical window of opportunity following a spill. Many countries, including France, South Africa, Canada, New Zealand, Norway, and the United Kingdom, have established standards regarding the use of chemical dispersants and adopted specific decision-making processes to evaluate the appropriateness and effectiveness of use under given situations and with specified types of oils. Approaches vary among countries, reflecting biophysical differences as well as differing cultural values regarding the appropriateness of using chemical dispersants to combat oil spills.

FOCUS OF CURRENT STUDY

Although the chemical processes by which dispersants work are generally well understood, their effectiveness is limited to varying degrees by the type of oil spilled and the environmental conditions at the time and location of a spill, as well as the timing and method of application. In general, the information base used by decisionmakers dealing with spills in areas where the consequences of dispersant use are fairly straightforward, has been adequate (for example, situations where rapid dilution has the potential to reduce the possible risk to sensitive habitat enough to allow the establishment of pre-approval zones). Many of the technical issues raised in this report, however, deal with settings where greater confidence is needed to make effective decisions regarding potential benefits or adverse impacts associated with dispersant use. In many instances where a dispersed plume may come into contact with sensitive water-column or benthic organisms or populations, the current understanding of key processes and mechanisms is inadequate to confidently support a decision to apply dispersants.

While laboratory experiments over the last decade or so have shed some light on how, when, and where dispersants can be effective, the use of non-standardized laboratory or mesocosm testing and monitoring techniques, lack of sufficiently coordinated effort, and misinterpretation of available information, have limited development of consensus about dispersant efficacy in some settings (e.g., freshwater, estuarine, coastal, and high-latitude environments). The lack of standardized procedures, when coupled with an insufficient number of well-designed tank or field-scale tests, has limited the value of this research for decisionmaking. In addition, there has been insufficient research into the fate of both chemically and naturally dispersed oil to evaluate concerns about its long-term impact. There are many unanswered questions about what happens to the oil droplets after they mix into the water column, such as the extent to which they will they bind to sediment or be ingested by organisms, how quickly they degrade, and what are the final degradation products.

Typically, the effects of oil spills have been very apparent on shorelines and coastal megafauna (e.g., birds and otters). The potential effects of dispersed oil on benthic flora and fauna (e.g., seagrasses, corals, and clams), fish populations, and the trophic relationships among these species, are less documented. Thus, while the use of dispersants is assumed to reduce the threat posed to some portions of the ecosystem (e.g., marine mammals and birds that frequent the air-water interface where oil slicks form), it is not clear in many instances how changing the fate of oil alters the potential threat to other portions of the ecosystem (e.g., fish and other fauna in the water column and on the seafloor) that are exposed to the dispersed oil plume. In deep open-water settings (deeper than 10 m or roughly 30 feet)[1] where there is rapid dilution of the dispersed oil, impacts to water-column and benthic resources are likely to be low, thus most of the pre-approval zones are defined in terms of distance offshore and minimum water depths.

While nearly every marine Area Contingency Plan (development of the plans was mandated by the Oil Pollution Act of 1990 [OPA 90]) includes flowchart-like decision trees and pre-approval for use of dispersants in offshore waters (e.g., >3 nautical miles [just over 5 km] and deeper than 10 m [about 30 ft]), there appears to be minimal risk assessment and decision-making procedures in place for spills in coastal and estuarine waters. Improving response to nearshore spills is particularly important as the majority of spills of all sizes in the United States occur within 3 nm of the shoreline (see Figure 1-1; NRC, 2003). Kucklick and Aurand (1997) conducted an assessment of the spills from 1973 to 1994 to determine the number of historic spills where dispersant use might have been appropriate, using the following criteria: >1,000 barrels (roughly 130 tonnes) of a dispersible oil, weather conditions, and distance from shore. Of the 69 crude oil spills meeting their criteria, only 10 percent were greater than 3 miles offshore, thus dispersant use in nearshore waters will be a common consideration. While dispersant use generally presents greater risks in shallower, nearshore settings, the likelihood that untreated nearshore

---

[1]Conversions reported in the text conserve the number of significant figures of the original reported value using rules consistent with the NRC report *Oil in the Sea III: Inputs, Fates and Effects* (NRC, 2003) and available on the following Massachusetts Institute of Technology website: http://web.mit.edu/10.001/Web/Course_Notes/Statistics_Notes/Significant_Figures.html. See Appendix D for additional information on definitions and unit conversions used through out this report.

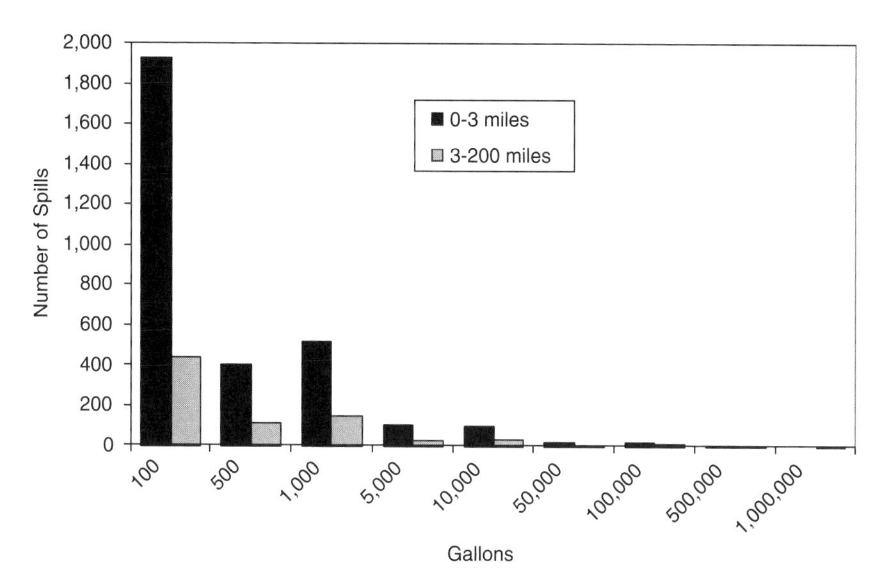

FIGURE 1-1 Frequency distribution of oil spills of various size classes (in gallons) in U.S. marine waters for 1990–1999. Note that for all size classes, spills more frequently occur in nearshore locations (less than 3 miles from shore) than offshore (3–200 miles from shore). These are the same data used to develop trends reported in *Oil in the Sea III* (NRC, 2003).
SOURCE: Data and figure provided by D. Etkin, Environmental Research Consulting.

spills may impact coastal resources is also much greater. Thus, the increased complexity of dispersant use decisions in nearshore settings is accompanied by a greater need to make the most appropriate overall decision.

SELECTING AMONG VARIOUS SPILL RESPONSE OPTIONS

Approximately 3 million gallons (roughly 10,000 metric tons [tonnes]) of oil or refined petroleum product are spilled into waters of the United States every year (NRC, 2003). These spills occur anywhere from nearshore to the open sea and range from small spills of refined products such as diesel fuel to thousands of gallons of crude oil. Once a spill occurs, the slick remains at the surface until it evaporates, disperses naturally into the water column, is recovered, strands along a shoreline, or breaks up into a field of tarballs.

When containment and recovery are not possible, practical, or sufficient, the application of dispersants may help to break up the oil slick prior to contact with sensitive habitats and resources. In general, the use of dispersants is recommended if: (a) an oil slick threatens a sensitive coastal area and mechanical recovery is not feasible, (b) there is sufficient wave energy to break up the surface slick and mix the oil droplets into the water column, (c) the oil is of a type know to be dispersible (i.e., the type and properties of the oil favor chemical dispersion), and (d) there is sufficient potential for rapid dilution of the dispersed oil, and (e) in the course of spraying, dispersants are not applied directly to birds and mammals.

Although these general rules provide the decisionmaker with some guidance in determining when to use dispersants (i.e., when they may be effective), there is still insufficient scientific information upon which to make decisions about likely benefits and consequences of dispersant use as an oil spill countermeasure. As previously stated, the fate and effect of chemically and naturally dispersed oil has not been well documented in field trials, although there have been several published intertidal studies in tropical (Tropical Oil Pollution Investigations in Coastal Systems [TROPICS] study in Panama; Ballou et al., 1987), temperate (Searsport, Maine; Gilfillan et al., 1986) and arctic (Baffin Island Oil Spill [BIOS] Project; Blackall and Sergy, 1981) regimes. Additionally, there is disagreement about how to interpret the results of laboratory, mesocosm, and the limited field tests to date because of the difficulty of simulating or capturing an adequate range of realistic exposure conditions. There remain basic issues that need to be resolved before dispersants are more fully accepted as a response tool in a wide variety of settings. For example, the effectiveness of dispersants is sensitive to certain environmental factors (e.g., wave energy, water temperature, salinity) and certain oil properties (e.g., viscosity, degree and type of emulsification), and it cannot be accurately predicted with sufficient consistency to support decisionmaking over a wide variety of settings. Further, the acute and chronic toxicity of dispersed oil under realistic conditions has not been adequately studied to support robust decisions involving the balancing of risks among various components of the ecosystem when sensitive species or habitat may be exposed.

OPA 90 required the development of Area Contingency Plans and specifically charged Area Committees to address the use of chemical countermeasures. Most area plans now include limited pre-approvals for dispersant use in offshore waters. Chemical dispersion could be considered as a viable method in supplementing mechanical response options in nearshore waters, but a lack of sufficient information regarding dispersant effectiveness and potential effects over the wide range of settings found in nearshore areas has precluded a similar broad policy change. In an effort to address this dilemma at an appropriate scale, Regional Response Teams (RRTs) are conducting workshops to assess the risks of using chemical countermeasures in shallow coastal waters from 0 to 3 nm (roughly 5 km) offshore. Additional robust scientific investigations should proceed at an accelerated rate so as to support these important decisionmaking efforts.

In recognition of the need to prioritize dispersant research, the Minerals Management Service, National Oceanic and Atmospheric Administration, U.S. Coast Guard, and American Petroleum Institute requested that the National Research Council's Ocean Studies Board undertake a study to explore existing and ongoing dispersant research and make recommendations for improving the knowledge base used to support dispersant decisionmaking in the United States (see Box 1-1). A similar request was put to the National Academies in the mid 1980s, leading to the publication of the 1989 NRC report *Using Oil Spill Dispersants on the Sea*. This current report is not truly an update of the 1989 report, as it focuses more

BOX 1-1
Statement of Task
Committee on Understanding Oil Spill Dispersants: Efficacy and Effects

This study will review and evaluate existing information and ongoing research regarding the efficacy and effects of dispersants as an oil spill response technique. Focus will be placed on understanding the limitations imposed by the various methods used in these studies and to recommend steps that should be taken to better understand the efficacy of dispersant use and the effect of dispersed oil on freshwater, estuarine, and marine environments. Specifically, the committee will:

- review and evaluate ongoing research and existing literature on dispersant use (including international studies) with emphasis on: a) factors controlling dispersant effectiveness (e.g., environmental conditions, dispersant application vehicles and strategies, and oil properties, particularly as the oil weathers), b) the short- and long-term fate of chemically or naturally dispersed oil, and c) the toxicological effects of chemically and naturally dispersed oil;
- evaluate the adequacy of the existing information about dispersants to support risk-based decisionmaking regarding response options for a variety of spatially and temporally defined oil spills;
- recommend steps that should be taken to fill existing knowledge gaps. Emphasis will be placed on how laboratory and mesoscale experiments could inform potential controlled field trials and what experimental methods are most appropriate for such tests.

specifically on information needs to support a decision-making process that was not in existence in the late 1980s. Thus the current report revisits some topics covered in the 1989 report, while including discussions on issues that have emerged since that time. Many readers may, therefore, find the assessments and summaries in *Using Oil Spill Dispersants on the Sea* of value.

STUDY APPROACH AND ORGANIZATION OF THIS REPORT

Despite the significant organizational and fiscal resources committed to responding to spills in the United States each year, fairly limited funding is available to support research geared to spill response or spill response decisionmaking. While determining specific funding levels for oil spill research nationwide is beyond the scope of this study, the trend described in the key programs in Box 1-2 suggest that the overall amount of

BOX 1-2
Funding for Oil Spill Research and Development

Making decisions based on "good science" requires that an adequate scientific foundation be available. In the case of dispersant decisionmaking, this scientific foundation has developed through research and development (R&D) funded through a variety of mechanisms and supported more or less independently by various federal and state programs and industry.

Federal Support

Title V.II of the Oil Pollution Act of 1990 (OPA 90) recognized the need for a comprehensive program of oil pollution research and technology development among the federal agencies, in cooperation and coordination with industry, universities, research institutions, state governments, and other nations. The legislation set up an Interagency Coordination Committee on Oil Pollution Research, (ICC; see Title VII of OPA 90, Executive Order 12777) comprised of members from the Departments of Commerce, Interior, Energy, Transportation, and Defense as well as the Environmental Protection Agency (EPA), the National Aeronautics and Space Administration, and the Federal Emergency Management Agency, to develop, implement, and coordinate such a plan. OPA 90 also authorized $19 million annually for the period from 1990 through 1995 for those R&D projects, in addition to those already underway under existing agency budgets, for this

funding available to carry out much of the work proposed in this study is very limited and may be decreasing. Given the greater need for scientific information to support decisionmaking in nearshore waters, it will be important to coordinate research efforts to the greatest degree possible. (see Box 1-2). The committee, therefore, recognized early on that the steps recommended to address any identified knowledge gaps would need to be prioritized in some manner. After some discussion during open sessions with federal and state resource trustees, representatives of industry, and sponsors of oil spill research, the committee determined that any recommendations for future work should be related to key decision points within the overall decision-making process used in spill contingency planning and during actual spill response. This grounding in the spill response decision process (discussed in Chapter 2) helps ensure that research recommendations put forward in subsequent chapters reflect most pressing information needs.

program. (Dispersant research competes with many other worthwhile R&D programs including fundamental research into other response technologies and oil spill effects.)

Federal funding for oil spill R&D has decreased with time and, for many agencies, is non-existent (C. Manen, National Oceanic and Atmospheric Administration, Silver Spring, Maryland, written communication, 2005). For example:

- U.S. Coast Guard Oil Spill research funding has remained level in the $3.5 million range. Emphasis is placed on oil spill prevention as well as oil spill response. Thus oil spill research is integrated into waterways management research, sensor systems, reduction in crew fatigue, etc., as well as oil spill clean up and response systems (K. Hansen, United States Coast Guard Research and Development Center, Groton, Connecticut, written communication, 2005).
- The EPA Oil Program has provided roughly $900,000 over the last few years to the EPA Office of Research and Development (ORD) for research under the Oil Program appropriation (a separate program element in EPA's budget). Roughly $800,000 of the total is for contract funding (D. Evans, Environmental Protection Agency, Washington, D.C., written communication, 2005).
- Starting in 2004, the National Oceanic and Atmospheric Administration's Office of Response and Restoration (OR&R) funds approximately $1.5

million in oil spill R&D projects a year. These funds come from a mix of congressional earmarks and in-house budgets (C. Manen, National Oceanic and Atmospheric Administration, Silver Spring, Maryland, written communication, 2005).

• Over the past five (5) years, the Minerals Management Service has received approximately $2.3 million dollars, annually, for oil spill response. Approximately $900,000 per year is spent on research, while approximately $1.4 million is spent on the operation and maintenance of OHMSETT (the National Oil Spill Response Test Facility located in Leonardo, New Jersey; J. Mullin, Minerals Management Service, Herndon, Virginia, written communication, 2005).

State Support

Several states (e.g., Texas, Louisiana, California, and Alaska) with particular interest in protecting natural resources from possible impacts from oil and gas development or transportation have invested funds in oil spill R&D. In several instances, these state programs rival the size of the federal programs mentioned above. For example, on March 28, 1991, the Oil Spill Prevention and Response Act (OSPRA) was adopted and signed into law by the Governor of Texas. One of the many innovative and new responsibilities mandated by OSPRA is the formation of a Research and Development component in the General Land Office (GLO) Oil Spill Prevention and Response division. Section 40.302 of OSPRA establishes the availability of $1.25 million dollars per fiscal year (R. Jamail, Texas General Land Office, Austin, written communication, 2005).

With passage of its Oil Spill Prevention and Response Act in 1993, Louisiana created the Louisiana Applied and Educational Oil Spill Research and Development Program (OSRADP). The program has an annual research budget of $500,000. Consequently, in the last 12 years the program has underwritten 91 projects from a highly diversified research agenda. (D. Davis, Louisiana State University, Baton Rouge, written communication, 2005).

Similarly, the Lempert-Keene-Seastrand Oil Spill Prevention and Response (OSPR) Act of 1991 established the oil spill response program in California and directed the Administrator of OSPR to develop a research program designed to examine the effects of oil and oil spill response technologies on the environment. Between 1993 and 1999 the program received $600,000 a year from OSPR and additional $50,000 to $200,000 a year from government and oil industry funding sources. The research program was temporarily discontinued between 2000 and 2002 and reinitiated in 2003 with an annual budget of approximately $300,000 (M. Sowby, California Department of Fish and Game Office of Spill Prevention and Response, Sacramento, written communication, 2005).

Alaska does not have a formally established oil spill research program;

however, it has invested considerable time and resources in updating the baseline knowledge regarding the use of dispersants. As a result of a legal judgment from the T/V *Exxon Valdez* oil spill, funds were appropriated for use by the state to enhance oil spill research and development. A total of $2,500,000 was made available to the Alaska Department of Environmental Conservation for projects under this program. Research projects carried out to date have focused on understanding the effectiveness of dispersants, fate and effect, and uncertainties associated with exposure tolerances of marine species to potentially acute, sublethal, and chronic toxicity levels from the dispersant and dispersed oil (L. Pearson, Alaska Department of Environmental Conservation, Juneau, written communication, 2005).

Industry Support (J. Clark, ExxonMobil Research and Engineering Company, Fairfax, Virginia, written communication, 2005; and R. Rorick, American Petroleum Institute, Washington, D.C., written communication, 2005)

Historically, the petroleum industry has supported research on dispersants as part of American Petroleum Institute (API) funded programs, R&D programs funded through spill response organizations such as the Marine Spill Response Corporation (MSRC), and through joint industry or joint industry/government projects funded to address specific issues. During 1970–1995, a broad diversity of oil spill R&D was organized and sponsored through API (Gould and Lindstedt-Siva, 1991; Aurand et al., 2001) with a budget of approximately $50 million, about one-third of which focused on dispersants. At the same time, individual petroleum companies, API, and MSRC also contributed funding for additional studies as part of large-scale, million dollar government/industry projects conducted in Europe. In addition, individual companies organized and/or conducted fundamental and applied research on dispersant use in oil spill response at a cost of several million dollars a year.

In subsequent years, API continued funding many of these dispersant projects, supporting them to completion over the next 5 years in the range of $200 to $400K per year. These studies included a variety of field, laboratory, and mesocosm tests (including supporting the construction of the test system now known as the Shoreline Environmental Research Facility at Texas A&M University, Corpus Christi) and targeted surveys and communication guides for spill responders and decisionmakers (Aurand et al., 2001).

In recent years, organized research programs at MSRC have ended and API support for research has been greatly reduced. As companies continue to evolve into organizations that manage and prioritize their technical information needs based on a global perspective, the focus of research projects has been on development and testing of basic principles and concepts that have broad applicability. It has become increasingly difficult to develop support for large research projects that may be driven by local or regional issues.

Chapters 3, 4, and 5 discuss lessons learned from an extensive review of the existing literature dealing with the effectiveness of dispersants, as well as the fate and effects of dispersed oil, with a major focus on studies completed since the release of the 1989 NRC report *Using Oil Spill Dispersants on the Sea*. Chapter 3 provides a detailed discussion of relevant petroleum properties and geochemical processes and the mode of action of various dispersants. In addition, it includes an in-depth discussion of the current understanding of dispersant effectiveness and provides specific recommendations for developing an adequate understanding of effectiveness to support more informed decisions regarding dispersant use in nearshore settings. Chapter 4 explores physio-chemical and biological processes that control the dispersion and fate of oil droplets and thus constrain the concentrations of various petroleum compounds in the water column. In addition, the role of modeling and monitoring to better support decisionmaking is explored, and specific recommendations to improve information needed to support decisionmaking are provided. Chapter 5 provides an in-depth analysis of toxicological studies that focus on dispersants or dispersed oil. By summarizing the salient points from existing reports and recommending specific additional, needed toxicological work, Chapter 5 provides guidance on efforts to better understand the effects of dispersed oil—a key component of effective decisions involving difficult trade-offs among sensitive species or habitats. Chapter 6 summarizes the key findings and recommendations of the previous chapters and organizes them into what is intended to be a coherent research plan to inform and coordinate to the degree possible research carried out or sponsored by federal and state entities, industry, and academia.

2

Making Decisions about Dispersant Use

A variety of perspectives exist about the value and potential of dispersing surface slicks of spilled oil or refined products. These perspectives reflect varying degrees of knowledge and opinions about dispersants and the fate and effects of dispersed oil in the environment. It is important to recognize, however, that avoiding a decision to apply dispersants due to lack of sufficient information or understanding may place some resources at risk that otherwise would be protected if dispersants were used effectively. Thus, the real key to effective decision-making regarding dispersant use is a fuller understanding of the implications of alternative outcomes in the decision-making process.

CURRENT FRAMEWORK FOR DISPERSANT APPROVAL AND USE IN THE UNITED STATES

Under OPA 90, the national response system is the federal government's mechanism for emergency response to discharges of oil into navigable waters of the United States. The system provides a framework for coordination among federal, state, and local responders and responsible parties. Structurally, the national response system is comprised of three organizational levels: National Response Team (NRT, co-chaired by the U.S. Coast Guard and the Environmental Protection Agency), Regional Response Teams (RRTs), and Area Committees. In addition to regional planning and response to federal incidents, the RRTs are vested with the authority over the use of chemical dispersants.

The U.S. Coast Guard is designated as the Federal On-Scene Coordinator (FOSC) responsible for ensuring a safe and effective response to all discharges of oil into the marine environment, Great Lakes, and major navigable rivers. The U.S. Coast Guard is also designated, along with the U.S. Environmental Protection Agency (EPA), as co-chairs for the RRT. At the time of an oil spill incident, a FOSC may authorize the use of dispersants on oil discharges upon concurrence of the federal co-chairs and the state representative to the RRT and in consultation with the federal natural resource trustee agencies, the U.S. Department of Commerce (DOC) and U.S. Department of the Interior (DOI). In an effort to compensate for the need to make a rapid decision regarding dispersant use early in the timeline of a spill, the NRT revised the National Contingency Plan to require both Area Committees and RRTs to address, as part of their planning activities, the desirability of using appropriate dispersants and the development of preauthorization plans (40 CFR 300.910). The status of pre-approval for dispersants in the United States, as of the publishing of this report, is presented in Appendix B and summarized in Figure 2-1. This information includes the status of dispersant-use approval zones; the conditions and zones where pre-approval exists (if applicable); and the status of monitoring and Section 7 consultation requirements. Section 7 of the Endangered Species Act requires consultation with the appropriate natural resource trustees prior to taking an action that may impact any federally listed species. Approval for use of dispersants, during both planning and emergency phases, falls into this category. Therefore, for purposes of dispersant use planning, any pre-approval agreement is subject to consultation with the trustee agencies prior to its implementation.

Pre-approval agreements are drafted at the local area and regional levels, either through the auspices of RRTs or through the Area Committee planning process; therefore some variations in terminology have developed in the agreements themselves or in the supporting literature. In this report the terms "case-by-case approval," "expedited approval," and "pre-approval" are used to describe the decision-making mechanism governing a given location, as defined below.

Case-by-Case Approval

(also referred to as incident-specific RRT approval)

The use of dispersants in each incident requires the FOSC to seek and gain approval from the RRT. The RRT reaches its approval through the concurrence of the U.S. Coast Guard and EPA co-chairs and affected state(s) and in consultation with DOI and DOC.

FIGURE 2-1 The status of pre-approval agreements, at the time of this writing, in the United States by Regional Response Team region.

[1]Agreements exist for Cook Inlet and Prince William Sound in Alaska.

[2]And/or borders with Oregon and Mexico within National Marine Sanctuary Boundaries.

[3]And/or borders with Oregon and Mexico outside of the National Marine Sanctuary boundary.

SOURCE: Modified from U.S. Coast Guard, http://www.uscg.mil/vrp/maps/dispmap.shtml.

Expedited Approval
(also referred to as quick approval)

The use of dispersants for each incident requires the FOSC to obtain the agreement of several key individuals (typically, the U.S. Coast Guard and EPA co-chairs, as well as state and federal trustee agencies). Expedited approval agreements usually limit the quantity and type of information the FOSC must provide in order to obtain concurrence, as well as the time agencies may take prior to approving or disapproving use. Expedited approvals are generally associated with a limited time in which a decision must be reached (typically less than 2 hours). Expedited approval may be limited to a particular geographic zone, distance from shore, depth of water, or season within a given area or region. RRT 6 has ratified an expedited approval for use of dispersants in the nearshore environments of Texas and Louisiana, defined as seaward of the shoreline but less than 3 miles (4.8 km) from shore. If concurrence is not given by all specified agencies then, by definition, the request for dispersant use does not meet the requirements of expedited approval. Any further consideration or request for the use of dispersants must be done as a case-by-case decision made by the RRT.

Pre-approval

(also referred to as pre-spill approval, pre-authorization, or pre-spill authorization)

The use of dispersants for each incident is at the discretion of the FOSC (in some cases, within the context of the Unified Command Structure) without further required approvals by other federal or state authorities. As pre-approval zones are generally limited to a particular geographic zone, distance from shoreline, water depth, or season, the FOSC must determine that a specific spill meets the criteria established for dispersant pre-approval. If any of these criteria are not met, the dispersant use falls outside the parameters of the pre-approval process and further consideration or request for approval must be sought as a case-by-case decision made by the RRT.

In order to ensure that dispersants are available for use on spills in preauthorized zones in a timely manner, the U.S. Coast Guard recently proposed mandatory capabilities to apply dispersants (where preauthorized) within 12 hours of the initial discovery of the release. While these rules are specifically directed to enhance spill response in preauthorized zones (generally 3 to 50 nautical miles [roughly 5 to 90 kilometers] offshore), they will have the secondary impact of making dispersants more widely available for use on spills in nearshore waters. As a consequence, greater attention is being given to the process needed to make rapid and informed decisions to use dispersants in nearshore settings. This process is complicated as dispersed oil is generally recognized as posing limited risks to open marine ecosystems, but the effects of dispersed oil on living marine resources in the water column or on or beneath the seafloor in nearshore ecosystems are less well understood.

THE DECISION-MAKING PROCESS

Every oil spill is a unique combination of events; therefore, decision-making should be flexible, rigorous, and timely in order to be effective. A decisionmaker should not only evaluate the response options available given the oil type, size and rate of release, and geographic location, but should also put these parameters into the larger environmental, social, and economic contextual needs of the overall society, as depicted in Figure 2-2. The interplay among stakeholders including responsible parties (e.g., tanker operators, oil and gas operators), elected officials, and local, state, and federal government representatives, is best seen as a perpetual state of dynamic tension. One of the most difficult challenges for an oil spill decisionmaker results from the fact that not all resources of public concern, be they environmental, economic, or historical, can possibly be protected either simultaneously or prior to the time that oil impact is likely to occur. Decisions regarding the use of dispersants can be particularly problematic, as they often involve trade-offs among natural resources whose protection falls under the jurisdiction of different government entities.

All response options, based on the rapidness of deployment and oil fate regimes, have consequences inherent in their selection. Decision-makers are forced, by the very nature of response to oil spills in the marine environment, to identify environmental and economic trade-off choices in real time, adequately assess the risk associated with each choice,

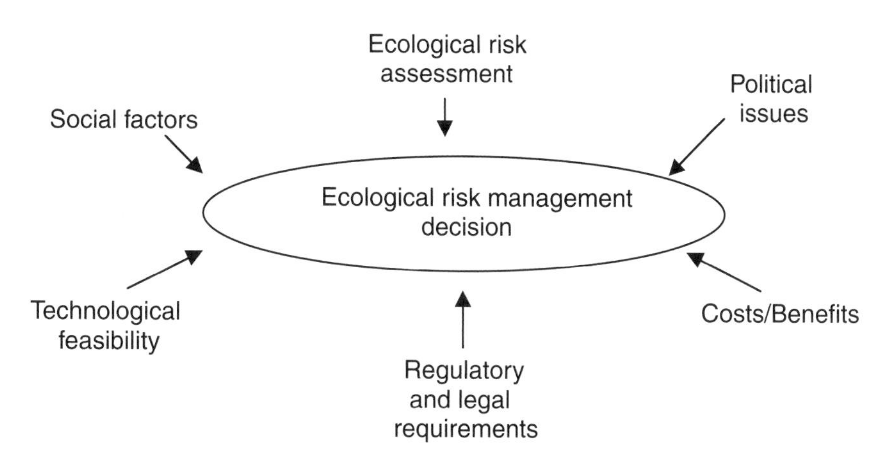

FIGURE 2-2 Relationship of ecological risk assessment to management decisions
SOURCE: Modified from Pittenger et al., 1998; courtesy of Alliance Communications Group.

and evaluate the potential magnitude of a negative unforeseen consequence of a given choice. Given the high public visibility of oil spills and oil spill response efforts, various stakeholders have become very interested in both the decision-making process and the information used in that process. Figure 2-3 is a generic, but typical decision flow chart for the

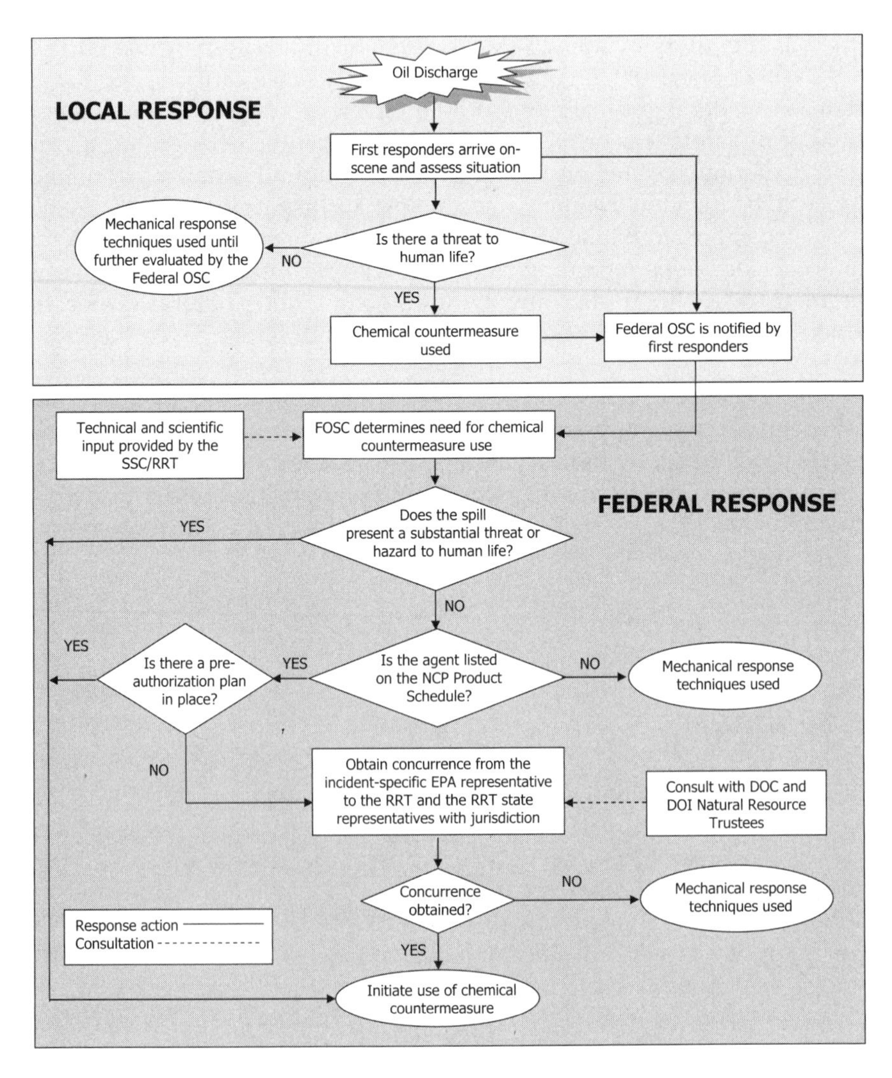

FIGURE 2-3 Example decision flow chart for using chemical countermeasures or dispersants showing federal and local responsibilities in the United States. SOURCE: National Response Team Response Committee, 2002.

United States showing specific decision-making points for evaluating any given response option, including chemical dispersants. Numerous decision processes are used around the world, including those used in New Zealand, Norway, France, Singapore, and the United Kingdom, and each is tailored specifically to address unique regional and regulatory considerations. In areas where weather and sea state often preclude the use of response options other than dispersants, or in island environments where deep water exists right up to the shoreline, the decision to use dispersants is even further expedited. The NRC (1989) report included a detailed comparison of four such flow charts in common use in the 1980s. However different these types of flow charts may appear at first glance, the actual thought process conducted by a decisionmaker required to evaluate the appropriateness of dispersant use is remarkably similar throughout the world. A decisionmaker should answer three basic questions before further considering the social or political implications of applying chemical dispersants:

(1) Will dispersants work? (i.e., predict chemical efficiency)
(2) Can the spill be treated effectively? (i.e., determine potential operational efficiency)
(3) What are the environmental trade-offs? (i.e., evaluate possible environmental consequences)

As depicted in Figure 2-4, many factors must be weighed and considerable information should be reviewed and evaluated in answering each of these questions. In many parts of the world, most of the decision points either are answered in advance or are answered in response to the limitations placed on response by the nature of the physical environment and subsequent logistical considerations. Figure 2-4 was developed to outline all the decision-making points that must be considered when evaluating the use of dispersants. Once the potential for damage caused by a particular oil spill has been established, the potential reduction in the amount of damage achievable by each of the response options (e.g., mechanical recovery, dispersants, *in-situ* burning, or do nothing) can be assessed. Such an assessment involves an evaluation of the expected effectiveness of each option within the constraining time limits and spill conditions. Because the purpose of any response is to minimize the damage caused by the oil spill, a quantitative set of criteria or measures of success needs to be defined so that decisionmakers can adequately compare response alternatives. Additionally, decisionmakers need real-time data to monitor and evaluate the effectiveness once a response option is undertaken, as well as a mechanism for determining when a response option is no longer effective or viable. A window of opportunity exists for any response decision,

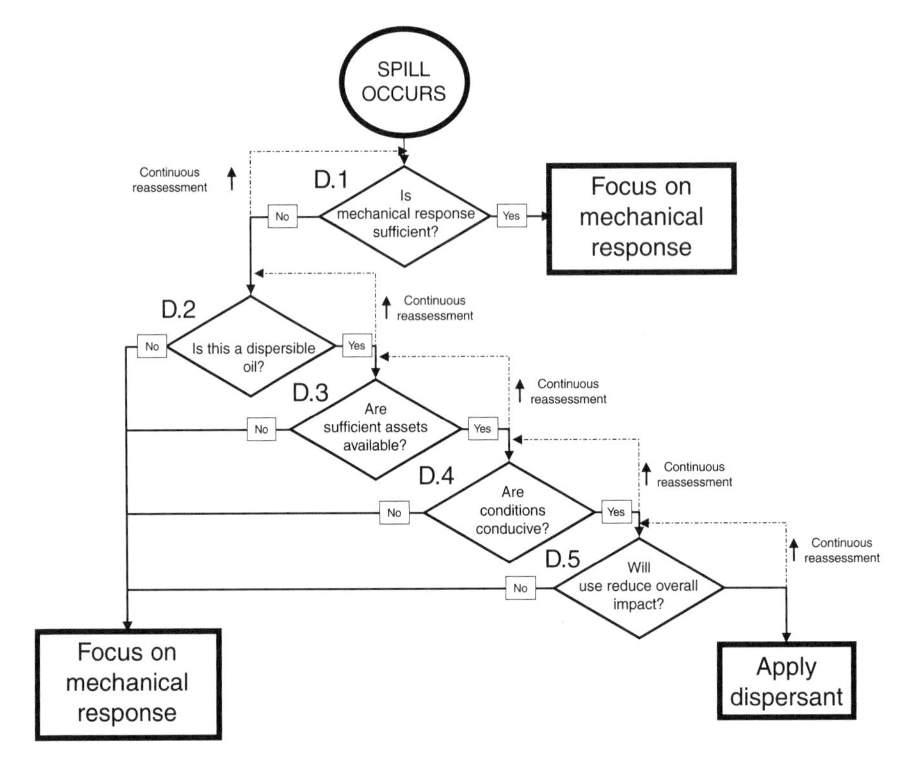

FIGURE 2-4 Idealized decision flow chart for evaluating the appropriateness of using chemical dispersants as a response option in the United States.

and any process, policy, or procedure that expedites reliable and accurate information to a decisionmaker will greatly improve not only the decision-making process but also a decisionmaker's comfort with making challenging trade-off choices. Responses to marine oil spills are conducted within the public arena and, as such, all response decisions can be and are reviewed and questioned by the general public, the media, governmental agencies, and the legal system. Within this social context of evaluation, a decisionmaker's choice should be seen as reasonable and prudent. Within the response context, decisions should also be viable and implementable within technically feasible constraints. Any actions that improve either the technical feasibility of a response option or the availability of timely and accurate information will enhance a decisionmaker's ability to make an appropriate and defendable response choice.

Predicting Chemical Effectiveness

As will be discussed at greater length in Chapter 3, the factors that limit the effectiveness of a given dispersant during a given spill are complex and the significance of each may change as time passes after the initial release. All crude oils and refined products have a unique and variable chemical composition and physical (rheological) properties that play a significant role in determining whether a specific dispersant will effectively disperse a surface slick under ideal conditions. Early in a spill response, decisionmakers should rapidly determine the nature of the fresh oil or product, how it will change over time, how effectively available dispersants are known to treat a specific oil or product under ideal circumstances, and how far from ideal circumstances does the particular spill deviate.

At present, real-time decisionmaking focuses on fairly conservative and simple tools to make the decision to attempt dispersant use, which are then verified by experimentation (e.g., dispersant is applied during a test flight, and the results are used to determine whether operations should continue). Decisions as to whether an oil is dispersible (Decision D.2, Figure 2-4) rely on databases on oil properties (e.g., density, American Petroleum Institute [API] gravity, viscosity, wax/asphaltene content, boiling point fractions), simple models that predict viscosity over time under forecast spill conditions, laboratory and field tests for a small set of oils, and (mostly) best professional judgment and experience of the response team. In fact, experienced responders are essential to the decisionmaking process. The more difficult assessment is predicting when various oil weathering processes, such as emulsification will render the oil no longer dispersible.

Determining Potential Operational Effectiveness

One of the first operational requirements for dispersing surface oil slicks with chemicals is that the dispersant must actually hit the target oil at the desired dosage. The ability to apply dispersants in a manner that satisfies this requirement is largely a reflection of environmental conditions and operational factors. The former can only be planned for; the latter can be addressed by making adequate preparations to have dispersant, appropriate equipment, and trained human resources available at the time of the spill. Dispersant and equipment availability have been, and continue to be, a key part of the decision-making process. The window of opportunity for effective dispersant application is often hours to a few days after a release; therefore, the logistics of getting resources to the spill site can be the driving factor in the decision to use dispersants. The longer the response time, the greater the spreading and chances of the oil stranding onshore, and the smaller the area of thicker dispersible oil. As the spreading of the oil occurs, there will be a corresponding increase in the weathering of the oil, governed by the oil type, sea conditions, and temperature. These processes will gradually reduce the effectiveness of chemical dispersants (as well as *in-situ* burning and mechanical response) to zero. The proposed changes in USCG regulations will, however, require the ability to apply dispersants within 12 hours after an oil release within 50 nm of shore. Getting dispersant resources to the spill site should, therefore, not be a limiting factor in the future. Logistical support for the operation should also be established. Fuel supplies, dispersant transfer equipment, and safety equipment will all need to be made available at the operating site.

The type of dispersant application platform used directly controls the operating distance offshore and amount of dispersant that can be applied in a day. Table 2-1 presents a summary of the capabilities, advantages, and disadvantages of different platforms that might be used to apply dispersants in the United States. The number of sorties per day and thus the amount of dispersants sprayed per day is a function of the operating distance; thus a C-130 can apply up to 67 tonnes (roughly 20,000 gallons) of dispersant if the target is 50 km (roughly 30 nm) from the airport, but only 55 tonnes (roughly 15,000 gallons) if the distance is 185 km (roughly 100 nm). The spill response management team needs to quickly work up a dispersant-use plan based on the volume of oil to be treated, available platforms, and other logistical factors.

Dosage control is another key operational factor. The planning goal is a dispersant:oil application rate of 1:20, though ratios of 1:40 or even 1:60 could be achievable with some dispersants and some oil types. Conversely dispersant:oil ratios of as high as 1:10 have been required with some of the more emulsified and viscous heavy oils. Assuming a uniform slick that is 0.1 mm thick (light brown or black as seen from aircraft; Figure 2-5), the dispersant application rate would be 5 gallons/acre (roughly 45 liters/hectare) for a 1:20 ratio of dispersant:oil. Realistically, slick thickness varies considerably, and most of the volume is in the thicker portions. The most efficient application strategy is to target the thicker portions of a slick, which will need higher application rates (1 mm thick oil would require 50 gallons/acre [roughly 450 liters/hectare]) and multiple passes to achieve these higher rates. Under ideal conditions, a spotter in a separate aircraft identifies the thicker portions of the slick and directs the spraying platform to these areas. There may be a case for reducing the application rate and making repeat applications of dispersant until dispersion is observed. This would be preferable to possibly overdosing an area and/or consuming available dispersant stocks prematurely.

TABLE 2-1 Characteristics of Dispersant Application Platforms for Example Operating Distances[a]

Platform	Operating Distance (nautical mi.)	Sorties per Day	Dispersant Sprayed/Day (gallons)
C-130 with ADDS-pack	30	4	20,000
	100	3	15,000
Advantages for C-130 ADDS-pack:	*Disadvantages for C-130 ADDS-pack:*		
High payload (5,000 gal)	Limited availability		
High speed (200 kt)	Needs longer runways		
Trained/dedicated crew	Start-up time is 24 hours		
DC-3/DC-4	30	5	10,000
	100	4	8,000
Advantages for DC-3 and DC-4:	*Disadvantages for DC-3 and DC-4:*		
Large payload (2,000 gal)	Limited availability		
High speed (200 kt)	Needs longer runways		
Air Tractor (AT-802)	30	7	5,500
	100	5	4,000
Advantages for Air Tractor (AT-802):	*Disadvantages for Air Tractor (AT-802):*		
Readily available	Small payload (800 gal)		
High speed (200 kt)			
Pilots with spraying experience			
Helicopter	10	21	5,000
	30	11	2,600
Advantages for helicopter:	*Disadvantages for helicopter:*		
Readily available	Smallest payload (150 gal)		
Good speed (90 kt)	Limited range		
Equipment easily adaptable			
High maneuverability			
High accuracy			
Vessels	10	2	2,000
	30	1	1,000
Advantages for vessels:	*Disadvantages for vessels:*		
Readily available	Slow speed		
Easily adapted	Low daily application rate		
High payload			
Maneuverability			

[a]The number of sorties per day and thus the amount of dispersants sprayed per day is a function of the operating distance. See text for further explanation.

SOURCE: Modified from Trudel, 2002.

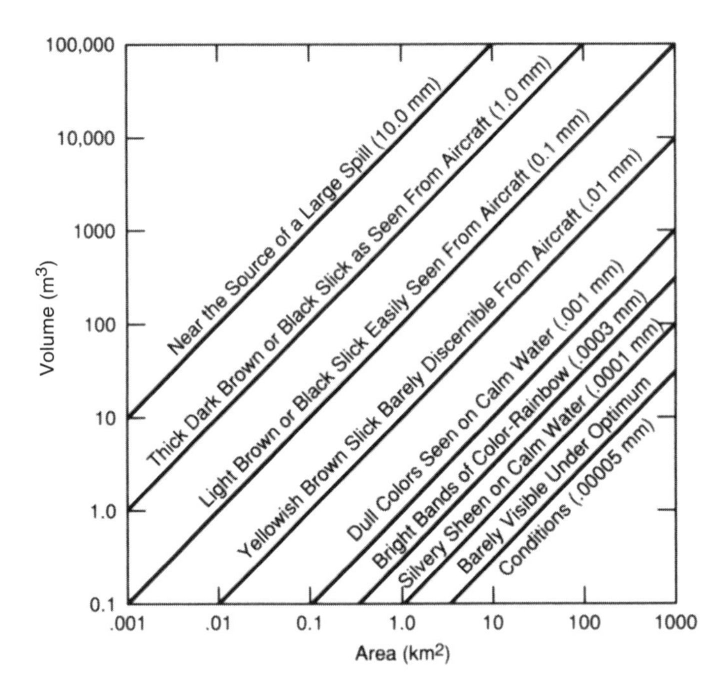

FIGURE 2-5 Chart showing the area of coverage and the relationships among slick thickness, color, and volume of oil.
SOURCE: Exxon, 1992; courtesy of Exxon-Mobil Corporation.

Table 2-2 presents a summary of the operational factors that affect the effectiveness of dispersant applications. Dispersant droplet size is important, because it will affect the overall performance and accuracy of the system. Tests have shown that an optimum droplet size of 600–1,000 microns is required for spray systems (NRC, 1989). Too small a droplet size will lead to an aerosol effect that will cause the dispersant to drift off target. Too large a droplet will result in the droplet passing through the oil layer into the sea rendering it ineffective. Dispersant droplets are also subject to evaporative loss of solvents during their descent to the sea surface, with an average drop out time of 5–7 seconds from an altitude of 15 m.

Assuming adequate resources are available (Decision D.3; Figure 2-4), the next question in the decision-making process (Decision D.4) is whether sea state condition and weather allow for dispersant use. Waves provide most of the mixing energy needed to break surface oil into droplets and mix them into the water column. Dispersion, both naturally and chemically enhanced, increases with wave energy, which is driven by wind speed. Wind speed should be at least 5 m/s to generate waves for

TABLE 2-2 Operational Factors That Influence the Effectiveness of Dispersants on Spilled Oil

Logistics	Dosage Control	Droplet Size and Spray Drift	Monitoring
Platform availability and capacities affect operating distance, transit speed, swath width, sorties per day, pump rate, and the total amount of dispersant that can be applied daily	Goal is 1:20 dispersant:oil ratio; target thick portions of the slick (>1 mm) with 50 gal/acre; uniform spraying over/under doses thin/thick slicks; use spotter to direct spraying of thicker portions	Drop size too small causes wind drift away from slick; drop size too big causes poor slick coverage, drop penetration through slick, and herding	Provides rapid feedback on whether or not the application is being conducted as planned and if the dispersant is effective; supports go/no go decision to continue dispersant applications

good dispersion. Even waves 15–20 cm in height can provide sufficient mixing energy. At wind speeds greater than about 25 knots (roughly 45 kilometers per hour), the dispersant droplets will not hit the oil. It is not known how long a dispersant applied under calm conditions would remain in the oil and still be effective during later periods of increased wind speed and wave energy. For small spills, mixing energy can be added by driving boats through the treated slick or applying water spray, such as from a fire hose or spray system, after the dispersant has been successfully applied to the slick.

Evaluating Possible Ecological Consequences

To adequately evaluate the use of dispersants on marine oil spills, environmental managers and decisionmakers need to assess the ecological risk and consequences associated with any given decision. Once it is determined that a specific oil spill is conducive to dispersant use (i.e., the oil is dispersible; appropriate dispersants, equipment, personnel are available; and weather/environment conditions are favorable), a decisionmaker should evaluate the potential environmental consequences for dispersant use (e.g., how such use will adversely impact some habitats and biological resources while reducing or preventing impacts to others). The need for such a comparative analysis of risk and benefits of dispersant use has been raised by several researchers (e.g., Lindstedt-Siva, 1987; Walker and Henne, 1991; Wiechert et al., 1991). Within the United States, Lindstedt-Siva (1991) proposed implementing a national goal for spill response: to minimize the ecological impacts of a spill. The implications of such a goal on planning and response, using the 1985 *Arco Anchorage* spill in Port Angeles, Washington, as an example, also were investigated. In that case, integration of response options to protect sensitive habitats—rather than to optimize cleanup—proved to be effective and acceptable to the regulatory community. However, none of these researchers developed a specific methodology for optimizing all potential response options in an integrated program. A solution is to integrate a simplified ecological risk assessment approach into the pre-spill planning process. Once an appropriate risk assessment is available, it can be used to support environmentally sound, integrated response plans and provide quantitative criteria for decisionmaking.

The EPA proposed a framework that groups the activities involved in ecological risk assessment into three phases: problem formulation, analysis, and risk characterization (EPA, 1992). A risk evaluation occurs whenever a decisionmaker needs to approve or disapprove an action. Belluck (1993) defined three classes of ecological risk assessment (scientific, regulatory, and planning) that lie along a continuum from most to least quantitative. Cost (and usually time) increases with the level of scientific detail incorporated; therefore, the desire to improve the analysis should always be weighed against the cost of the additional information. An ecological risk assessment follows a defined methodology that:

- uses quantitative data to define effects whenever possible;
- defines uncertainty;
- incorporates this information into conceptual or mathematical models of the affected system; and
- interprets information against clear, consistent endpoints that are related to the protection of resources.

Lewis and Aurand (1997) proposed a methodology for the application of risk assessment protocols to planning for dispersant use. In the case of oil spill planning, the goal is different from ecological risk assessments of proposed projects since the spilled oil cannot be prevented from entering the environment—the goal is to minimize adverse effects. Environmental planners and decisionmakers should evaluate scenarios for the expected range of incidents and focus on providing information tailored to meet the circumstances of a particular spill.

Modified ecological risk analyses with the goal of evaluating and quantifying a "net environmental benefit" have been undertaken throughout the United States (Pond et al., 2000; Addassi et al., in press) as well as other parts of the world (IPIECA, 2000), for use both as a part of oil spill planning and during actual spill response.

Ecological Risk Assessment Applications for Oil Spill Response in the United States

This section summarizes a process of a cooperative ecological risk assessment (ERA[1]) currently utilized in many regions of the United States to evaluate the ecological trade-offs associated with the use of each of five potential oil spill response options: natural recovery, on-water mechanical cleanup, shoreline cleanup, dispersant use, and on-water *in-situ* burning. The desired outcome of the evaluation is identification of the optimum mix of response options in reducing injury to specific environments. The evaluations are usually conducted during a series of workshops where technical experts, resource managers, and stakeholders come together to develop relative ecological risk evaluations for response options. Much of the work completed during this process is later incorporated in the dispersant-use planning process.

Community Participation

Two critical elements of an ERA are that the process must involve the active participation of both response operations personnel (risk managers) and response impact assessors (risk assessors) and be conducted to achieve consensus (Kraly et al., 2001). In addition, other groups such as local governments, concerned private citizens, and the press must have access to and an understanding of the process. This broad involvement by the informed public is essential if decisions and resultant actions are to withstand scrutiny. Public trust requires that the public have confidence in the decision-making process as well as the information used to support decisionmaking.

Phases of the ERA Process

The consensus ERA generally involves a step-by-step process to help participants logically order information and, in so doing, enable participants to collect all relevant data, identify conflicts and data gaps, and

---

[1]Providing recommendations for conducting risk assessments in the United States is beyond the committee's charge. The following discussions of the process, or sample figures or tables from various workshops carried out using this process, are included for information purposes.

determine the optimum course of action based on consideration of trade-offs in resolving those conflicts. The process is conducted in three phases: problem definition (formulation), analysis, and risk characterization (Figure 2-6).

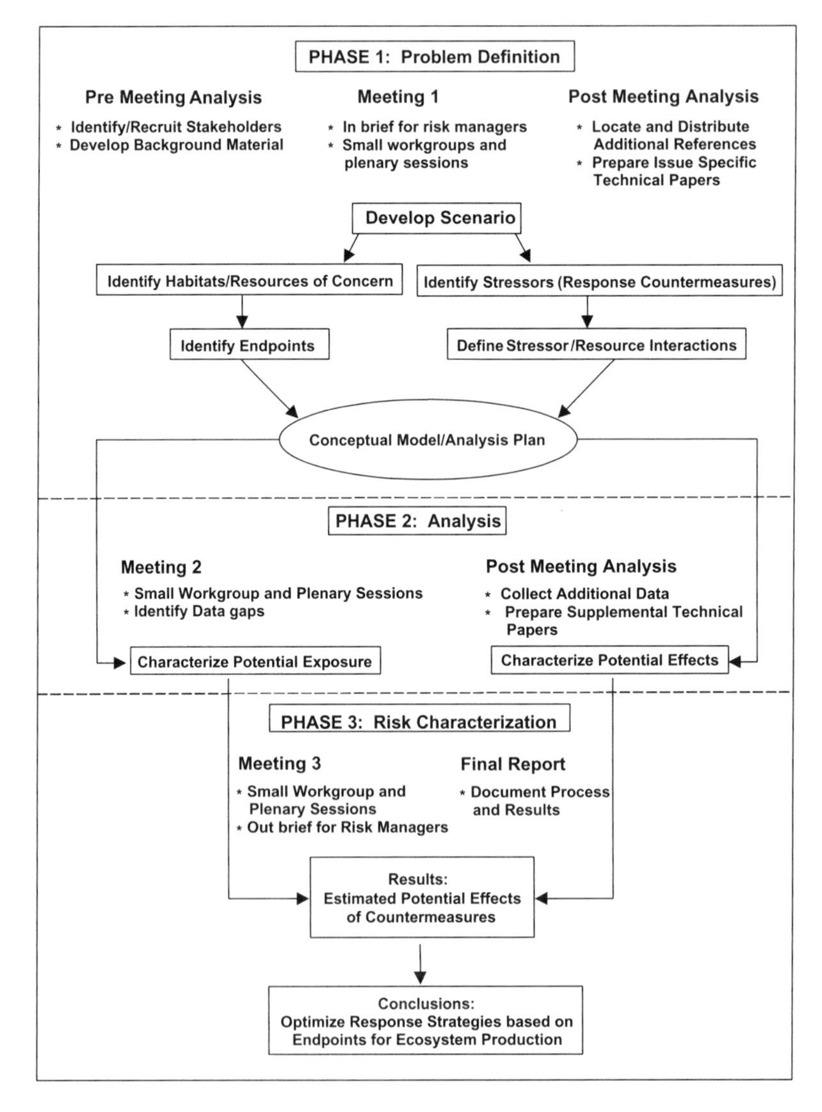

FIGURE 2-6 Diagram showing interaction among the three phases of a "representative" ERA.
SOURCE: Aurand et al., 2000.

Phase 1: Problem Definition An ERA is intended to analyze the potential environmental impacts of an oil spill and evaluate how response options can influence the nature and magnitude of those impacts. Therefore, selection of a specific scenario is critical to the risk assessment process because the scenario establishes the spatial and temporal parameters of the risk analysis. Scenario parameters include spill location, oil type, spill size, weather, seasonality, and established assessment objects.

Identify Habitats and Resources of Concern Once a scenario is established, the next step usually employs trajectory modeling to identify potentially impacted segments of the environment and the quantities of oil that may impact those segments. Additionally, because trajectory models show oil movement over time, they will also drive the determination of which response options might be appropriate in mitigating the spill. To ease evaluation, oil budgets are often developed for each response option.

The next step is identification of potentially affected natural resources. Typically, trustee agency representatives and environmental advocates with responsibility for specific resources and habitats examine the impacted areas to identify each habitat and resource category. The degree of specificity in habitat identification is dependent upon the concerns of the risk assessors and may focus on representative resources, endangered resources, or keystone resources in a particular habitat.

Identify Stressor (Response Options) The term stress can be defined as the "proximate cause of an adverse effect on an organism or system" (Suter, 1993). Although the primary stressor may be the oil itself, unique environmental stressors result from human intervention through on-water mechanical recovery, shoreline cleanup, dispersant application, *in-situ* burning, or any other response options. Typically five potential stressors are chosen for an ERA analysis: natural recovery, on-water mechanical recovery, shoreline cleanup, chemical dispersion, and on-water *in-situ* burning.

Identify Stressor/Resource Interaction While every response option is a source of potential ecological stress, the mechanisms that cause this stress are not always of the same type or magnitude. Exposure pathways that link stressors to resources are termed hazards (Kraly et al., 2001) and include air pollution, aquatic toxicity, physical trauma (mechanical impact from foot and vehicular traffic), oiling or smothering, thermal (heat exposure from *in-situ* burning), waste, and indirect (a secondary effect such as ingestion of contaminated food). Each stressor can be evaluated through the use of a conceptual model to show the hazards posed by that stressor on the environment and the pathways of exposure to those stressors. Fig-

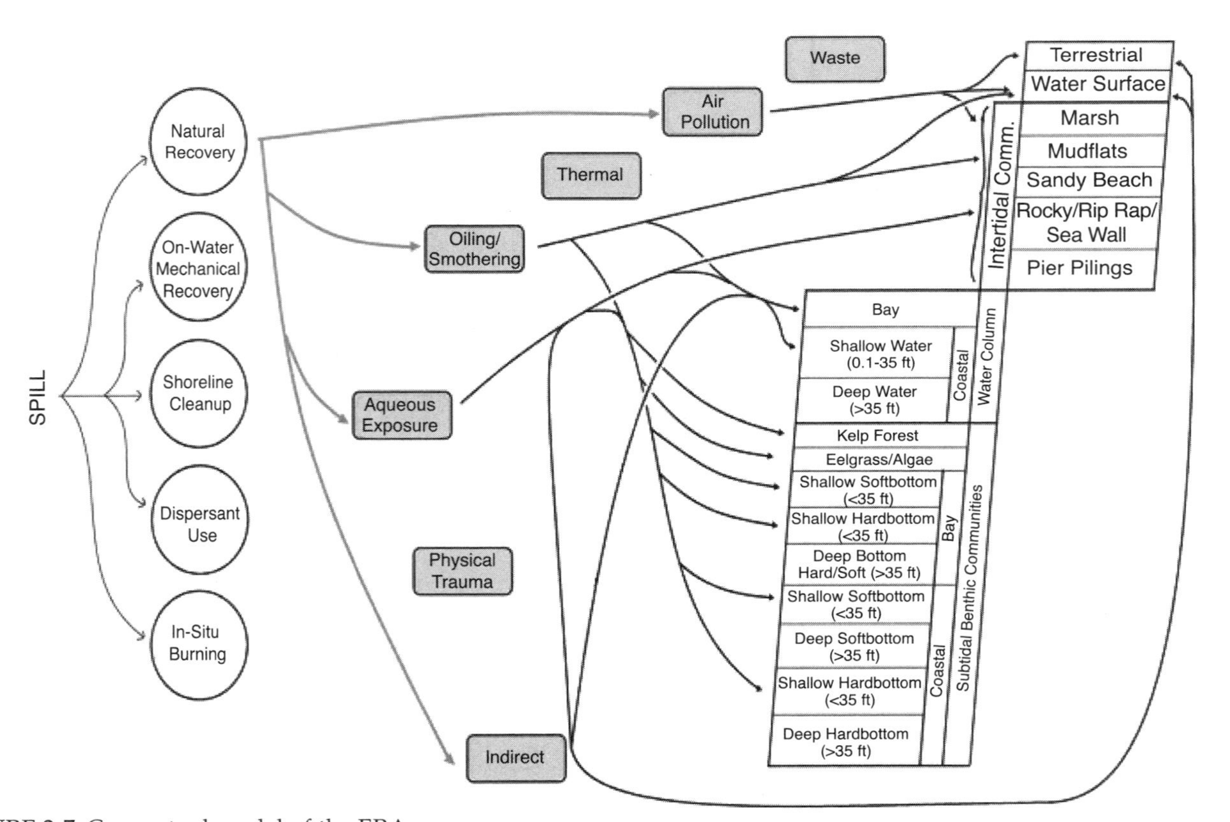

FIGURE 2-7 Conceptual model of the ERA process.
SOURCE: Pond et al., 2000.

ure 2-7 shows such a conceptual model, demonstrating the multi-layered connections between the many steps that must be completed to finish the ecological risk assessment. Figure 2-8 provides an example of a stressor-matrix developed for a surface microlayer habitat in a 500 barrel (roughly 21,000 gallons, 71.4 tonnes) spill scenario. Throughout the ERA process, any adverse impacts resulting from a response option are always compared against natural recovery. The stressors identified in the matrix in Figure 2-8 are those in addition to the natural recovery option and the hazards identified are for all habitats, not just the surface microlayer.

Habitats: / Subhabitats:	Surface (microlayer)					
Resources:	Algae	Birds	Fish	Mammals	Microlayer associated plankton	Reptiles/amphibians
Stressors:						
Natural recovery	2, 4	1, 4, 7	4, 7	1, 4	2, 4	1, 4, 7
On-water recovery	NA	NA	NA	NA	NA	NA
Shoreline cleanup	2, 4	4, 7	4, 7	4	2, 4	4, 7
Oil + dispersant	2, 4	4, 7	4, 7	4	2, 4	4, 7
In-situ burning	5	1, 5	5	1, 5	5	1, 5
Shoreline bioremediation	2	7	2, 7	7	2, 7	7

Hazards:

1. Air pollution
2. Aquatic toxicity
3. Physical trauma
4. Oiling/smothering
5. Thermal (heat exposure from *in-situ* burning)
6. Waste
7. Indirect (food web, etc.)

NA: Resource and stressor do not come in contact with each other.

FIGURE 2-8 Sample stressor-hazard matrix for the surface microlayer habitat used in the Texas ERA.
SOURCE: Kraly et al., 2001; courtesy of the American Petroleum Institute.

Phase 2: Analysis In the analysis phase, the degree of exposure for each response option on each segment of the environment is first examined, followed by a comparative analysis of the individual response option impacts. This analysis is accomplished by construction of a matrix with potential stressors listed on the vertical axis and habitats and resources listed across the horizontal axis. The objective is to score the potential severity of impact posed by each stressor on each resource and habitat.

Several supporting pieces of information are necessary to facilitate completion of this scoring matrix (Figure 2-8):

- The trajectory model provides an indicator of which habitats will be impacted and to what degree by various stressors.
- Scientific literature provides for estimates of potential acute and chronic impacts of oil and different response methods (e.g., on-water mechanical recovery, dispersants, *in-situ* burning, shoreline cleanup) on individual resources and habitats.
- Assessor discussion allows development of estimates regarding the potential effects of dispersed oil in the water column. In some regions, specific tables, like the one below, were used to provide guidelines in assessing dispersed oil toxicity (Table 2-3).
- A risk square provides a method of scoring and evaluating relative resource concern.

TABLE 2-3 Workshop Consensus on Exposure Concentration Thresholds of Concern for Dispersed Oil in the Water Column in the Texas Ecological Risk Assessment (ERA)

Level of Exposure	Level of Concern	Sensitive Life Stages	Adult Fish	Adult Crustacean/ Invertebrates
0–3 hours	Low	1	10	5
	Medium-low	1–5	10–50	5–10
	Medium-low	5–10	50–100	10–50
	High	10	100	50
24 hours	Low	0.5	0.5	0.5
	High	5	10	5
96 hours	High	0.5	0.5	0.5

NOTE: All numbers are in parts per million (ppm). Values are intended to indicate threshold levels of concern for resources. For example, if adult fish are exposed to a dispersed oil plume of 100 ppm for 3 hours, concern should be high. If they are exposed to a 10 ppm plume for 3 hours, concern should be low because there is little or no potential for acute effects.

SOURCE: Modified from Kraly et al., 2001.

Each axis of the square can be used to describe risk. Figure 2-9 provides a simplified risk square, with the x-axis representing rates of "recovery" and ranges from reversible to irreversible while the y-axis evaluates "magnitude" and ranges from severe to trivial. In its simplest form, a risk matrix is divided into four cells. Each cell is assigned an alphanumeric value to represent relative impacts. Thus a "1A" represents an irreversible and severe effect, while a "2B" represents a reversible and trivial effect. Most regions develop more expansive risk matrices to increase the level of sensitivity of evaluation of the two primary parameters—severity of exposure versus length of recovery for a specific resource. Severity of exposure includes level of effect, ranging from community level effects at the high level to the loss of a few individuals at the low level. Recovery includes both time and function expressed as lost services. Figure 2-10 is an example of such an expanded matrix.

The actual analysis involves assigning scores from the risk square to each sub-habitat block of the risk matrix. Often workshop participants are divided into three groups. Working separately, each group scores impacts of each stressor on the environment. Group scores for each stressor are then scored in plenary sessions and combined into a single matrix, reflecting consensus of all participants. When the groups have significantly different conclusions, this comparison helps make sure that areas of confusion or limited data are identified and addressed.

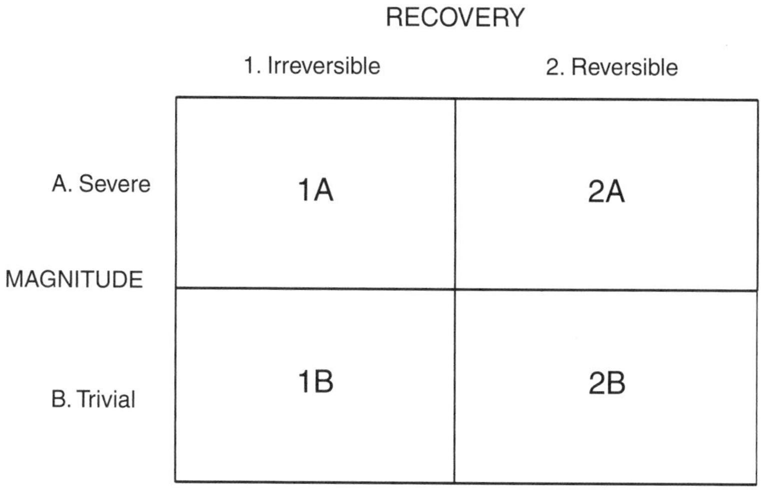

MAGNITUDE \ RECOVERY	1. Irreversible	2. Reversible
A. Severe	1A	2A
B. Trivial	1B	2B

FIGURE 2-9 Basic ecological risk matrix design.
SOURCE: Kraly et al., 2001; courtesy of the American Petroleum Institute.

PERCENT OF RESOURCES AFFECTED	Recovery (years)	>6 yrs (1)	3-6 yrs (2)	1-3 yrs (3)	<1 yr (4)
	> 50 (A)	1A	2A	3A	4A
	30-50 (B)	1B	2B	3B	4B
	10-30 (C)	1C	2C	3C	4C
	<10 (D)	1D	2D	3D	4D

FIGURE 2-10 Expanded matrix for a hypothetical spill.
SOURCE: Kraly et al., 2001; courtesy of the American Petroleum Institute.

Phase 3: Risk Characterizations The final phase of the ERA involves interpreting the data and analysis results. In Phase 2, resources/habitat impacts are scored on a stressor-by-stressor basis working horizontally across the matrix. The result is a snapshot of each stressor in isolation and provides no insight regarding the relative merits of any one stressor compared to any others. In Phase 3, the participants begin to examine the matrix vertically, comparing relative impacts of each stressor on a given segment of the environment, allowing determination of which response option or combination of options should provide optimum protection of the environment as a whole.

The first step in risk characterization revisits the risk square to determine whether individual scores represent a high, medium, or low threat to the environment. For convenience in reading the final characterization matrix, the high, medium, and low determinations are represented by different colors, as shown in Figure 2-11. Clear cells represent a "minimal" level of concern; light gray cells represent a "moderate" level of concern; and dark gray cells represent a "high" level of concern.

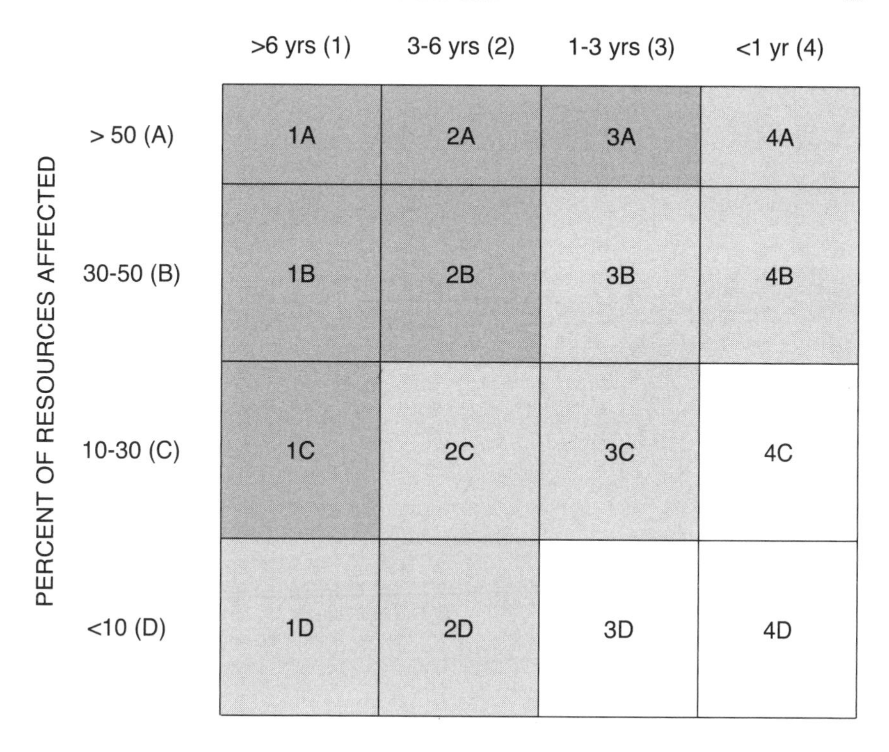

FIGURE 2-11 Final definition of levels of concern. Note: clear cells represent a "minimal" level of concern; light gray cells represent a "moderate" level of concern; and dark gray cells represent a "high" level of concern.
SOURCE: Kraly et al., 2001; courtesy of the American Petroleum Institute.

Once the high, medium, and low threat scores are determined, the scores in the completed risk matrix are then colored. This provides a visual display for assessors to use in reaching consensus on optimizing response. Figure 2-12 is a sample scenario summary score sheet from an ERA conducted in Texas (Kraly et al., 2001).

The ERA workshops for dispersant decisionmaking have been useful tools to bring stakeholders together to discuss the trade-offs of dispersant use in specific settings. Lessons learned from past workshops include: the difficulty of dealing with uncertainty; limited ability to use toxicological data to quantify impacts; the lack of quantitative data on the benefits of reduced shoreline oiling; the constraints posed by utilizing a specific model scenario and limited ability to extrapolate results to other scenarios; the sensitivity of the process to strong opinions by a few participants,

Response Options	Intertidal									Subtidal														
	Marsh/Tidal Flat			Sand/Gravel Beach			Rip rap/Manmade			Shallow < 3 feet			Open Bay 3-10 feet			Channel > 10 feet			Reef (not intertidal)			Submersed Aquatic Vegetation		
Natural Recovery	2C	3C	3C	3C	4D	3C	4D	4D	4D	4D	3C	4D	4D	4D	NA	4D	4D	NA	4D	4C	NA	4D	NA	NA
On-water Recovery	3C	3D	4C	3D	4D	3C	4D	4D	4D	4D	3C	4D	4D	4D		4D	4D		4D	4C		4D	NA	NA
Shoreline Cleanup	2C	3B	4C	3D	4D	3C	4D	4D	4C	4D	3C	4D	4D	4D	NA	4D	4D	NA	4D	4D	NA	4D	4D	NA
Oil + Dispersants	4D	4D	4D	4D	4D	4C	4D	4D	NA	4D	4D	4C	4D	4D	4C	4D	4D	4C	4D	4D	4C	4D	4D	NA
In-situ burning	3C	3C	4C	3D	4D	3C	4D	4D	4D	4D	3C	4D	4D	4D	4D	4D	4D	4D	4D	4C	4D	4D	NA	NA

FIGURE 2-12 Sample form summary score sheet for a hypothetical 500 barrel (roughly 21,000 gallons, 71.4 tonnes) spill used during an ERA for Texas. Note: clear cells represent a "minimal" level of concern; light gray cells are intermediate between "minimal" and "moderate," and dark gray cells represent a "moderate" level of concern.
SOURCE: Modified from Kraly et al., 2001; courtesy of the American Petroleum Institute.

often those with either the time or the money to participate; and the importance of participation by all stakeholders.

Real-Time Decisionmaking

Carrying out some type of risk assessment prior to a spill, such as the one discussed above, allows stakeholders to explore a finite set of scenarios, raise general and specific questions about dispersant use, understand the various concerns held by both the public and specific decisionmakers, and gain valuable experience working together to reach a consensus decision. An ERA may support development of guidelines or policies about where or when dispersants may be used (such as the designation of pre-approval zones), but because actual spill conditions will likely deviate in some way from the finite set of scenarios used, it cannot, in all instances preclude real-time decisionmaking. However, the awareness and understanding that a specific group of decisionmakers share by participating in a risk assessment process greatly facilitate real-time decisionmaking.

IDENTIFYING INFORMATION NEEDED TO SUPPORT EFFECTIVE DECISIONMAKING

As depicted in Figure 2-4, the availability of different pieces of information (about the environment, spilled oil or refined product, and response assets available) plays a role in the overall decision to apply dispersants. By understanding why this information is needed, how it is currently provided, and how well it meets the needs of the decisionmaker, one can gain a greater understanding of the current limitations in the decision-making process and how these limitations are addressed. If, as discussed earlier, the greater availability of dispersants, equipment, and personnel needed to respond to spills in pre-approved areas leads to greater consideration of their use in spills closer to shore, it is quite likely that current or readily available spill-specific information may prove to be inadequate. In an effort to set the stage for the subsequent chapters that examine current understanding of various technical aspects of spill research, each of the major decision points depicted in Figure 2-4 will be reviewed.

D.1 Will Mechanical Response Be Sufficient?

Current federal regulations specify that mechanical response is the primary option to be considered in response to a spill in U.S. waters. Thus, the U.S. Coast Guard has established minimum capabilities required to respond to spills in U.S. waters. Although mechanical response techniques have the advantage of removing spilled oil from the environment, their ability to do so is somewhat limited. Under ideal conditions, some portion of the spill cannot be recovered and under adverse environmental conditions (e.g., high sea state), the effectiveness of mechanical response can be very low. Often timing is a critical consideration; hence, oil spill trajectory analyses (generally provided by the National Oceanic and Atmospheric Administration's [NOAA] Office of Response and Restoration) are used to identify where the surface slick will move and how fast. Maps (Environmental Sensitivity Index maps provided by NOAA's Office of Response and Restoration) showing the distribution of sensitive habitats or species are used to document the resources at risk. When mechanical response is unlikely to sufficiently reduce impacts from a spill, the use of alternative response techniques (e.g., dispersant application) is considered.

Some information needed to support analysis of the potential effectiveness of mechanical response is readily available. For example, when a spill occurs beyond the operational limits of vessels or in conditions that exceed safe or effective operation of those vessels, mechanical response is not feasible. Because conditions change through both time and space, a forecast of conditions is also required. In the majority of instances, however, there is an adequate understanding of the future location and conditions along the surface slick's projected trajectory to allow decisionmakers to make reasonable inferences about the effectiveness of mechanical response in the hours following a spill.

D.2 Is the Spilled Oil or Refined Product Known to Be Dispersible?

Several aspects of a given crude oil or refined product may make it difficult to disperse under even ideal conditions. As will be discussed more fully in Chapter 3, the chemical composition of crude oil or refined product dictates a number of rheological properties (e.g., viscosity, pour point) that determine whether a specific dispersant will be under ideal conditions. Much of the work to date to understand the effectiveness of dispersants has involved laboratory tests designed to measure the effectiveness of a specific dispersant formulation in dispersing crude oils or petroleum compounds. Thus, some general but informal guidance has been developed that may help a decisionmaker (or those charged with providing technical assistance) reject dispersant application as a response to spills of certain types of crude oil or refined products (e.g., heavy oils). Again, as will be discussed at some length in Chapters 3 and 4, these rheological properties change through time as the spilled material weathers, requiring decisionmakers to constantly monitor the character of the surface oil and continuously reassess the decision to apply, or continue to apply, dispersant.

Product testing required by EPA for dispersant products provides an indication of how well a specific formulation will disperse one of two specific oils (e.g., Prudhoe Bay or South Louisiana Crude) under laboratory conditions. Thus, for the majority of spills in U.S. waters, once the nature of the spilled oil or refined product is accurately known, some reasonable conclusions can be drawn regarding its dispersibility. Under ideal conditions, some uncertainty remains regarding how effective a given dispersant formulation may be in some environmental conditions (e.g., cold temperatures; see discussion D.4 below).

D.3 Are Sufficient Chemical Response Assets (i.e., Dispersant, Equipment, and Trained Personnel) Available to Treat the Spill?

The size and location of a spill will dictate the platforms needed (e.g., aircraft, boats) to effectively treat the spill. There are various tools, such as NOAA's Dispersant Mission Planner, to help define requirements. The proposed U.S. Coast Guard rulemaking would set minimum capabilities for spill responders to be able to treat spills in U.S. waters with chemical dispersants in pre-authorized zones. Once in place, these rules would increase the likelihood that sufficient physical assets would be available to treat a spill, though there will always be a need for trained personnel. With such a narrow window of opportunity, there is little time to make adjustments, particularly in nearshore settings with lots of restrictions.

The proposed U.S. Coast Guard rulemaking would establish mandatory capabilities to apply dispersants in preauthorized zones within 12 hours of the initial discovery of the discharge within 50 nm of shore. As discussed previously, while these rules are specifically directed to enhance spill response in preauthorized zones (generally 3 to 50 m [roughly 5 to 92 km] offshore), they will have the secondary impact of making dispersants more widely available for use on spills in nearshore waters. Thus, once these rules are in place, there should exist a capability to treat the vast majority of spills in U.S. waters. If and when dispersant application capabilities are required, it will be necessary to implement methods and procedures to ensure the readiness of response equipment and supplies for dispersant use, similar to the requirements for mechanical response equipment. In the Notice to Public Rulemaking, the U.S. Coast Guard recommended the development of American Society for Testing and Materials (ASTM) standards for testing of dispersant application equipment. ASTM Standard Guides have been prepared for design of boom and nozzle systems (ASTM, 1992), calibration of boom and nozzle systems (ASTM, 1993), and maintenance, storage, and use these systems during spill response

(ASTM, 1996); however, standard guides should be developed for ensuring that dispersant stockpiles meet minimum efficacy standards.

D.4 Are the Environmental Conditions Conducive to the Successful Application of Dispersant and Its Effectiveness?

Water temperature, wind velocity, wave height, and other environmental factors play key roles in determining whether dispersant can be applied safely and effectively. Just as these environmental factors define a safe and effective operational window for mechanical response techniques, they also define an operational window for dispersant application. Generally, these operational windows are often dissimilar and sensitive to different environmental parameters. For example, booming and skimming (standard mechanical response techniques) work well in calm conditions and weak currents, whereas dispersants require some minimum wave energy to disperse the surface slick and entrain individual oil droplets. There are guidelines for minimum/maximum conditions for wind speed, sea state, and temperature, and conditions often change during the actual application (Fingas and Ka'aihue, 2004a). Thus, decision-makers should continuously monitor the character of the surface slick and on-site conditions and frequently reassess the decision to apply, or continue to apply, dispersant.

Existing capabilities to characterize and predict evolving environmental conditions beyond sea state and weather are limited. Unlike surface slicks that are affected primarily by surface winds, the nature and trajectory of subsurface dispersed oil plumes are more sensitive to currents. Even wave height, a critical component for predicting dispersant effectiveness, may be difficult to predict more than a few hours in advance.

D.5 Will the Effective Use of Dispersants Reduce the Impacts of the Spill to Shoreline and Water Surface Resources without Significantly Increasing Impacts to Water-Column and Benthic Resources?

As discussed throughout Chapters 3, 4, and 5, there are still many uncertainties about the fate of dispersed oil droplets and the many different factors and processes that control that fate in different biophysical settings. Understanding the relative risk posed to various portions of the ecosystem at a spill, however, requires an adequate understanding of the physical and toxicological effects that dispersed oil may have on many different components of that ecosystem. In open, offshore waters, physical mixing processes tend to rapidly dilute a plume of dispersed oil droplets, reducing the potential for significant impacts on organisms in the water column or associated with the seafloor. The effective use of dispersants, therefore, reduces the threat posed by a surface slick to organisms on the surface or, eventually, nearer to shore by altering the fate of that oil. As a consequence, a more limited and less robust set of information is needed to support the decision to use dispersants in such offshore conditions.

Use of dispersant in treating nearshore spills, however, raises many questions that are difficult to answer with the current understanding of the dispersed oil fate and effects. As pointed out in the previous discussion of environmental risk assessment, decisions regarding the use of dispersants in the nearshore settings often involve trade-offs and, therefore, call for more diverse and robust information (e.g., toxicological and population-level information about a particular species). As a consequence, questions about the fate and possible effect of dispersed oil or refined products make up a significant portion of the discussion in Chapters 4 and 5. Environmental monitoring of the operations usually focuses on preventing the direct application of dispersants onto wildlife or sensitive habitats. Additional monitoring is used in post-dispersant evaluations and model validation.

The models most commonly used to support real-time decision-making were designed to predict the trajectory of a surface slick, not a three-dimensional dispersed plume. Such models, which are in active use in the North Sea (Reed et al., 1999) and under development in the United States, are particularly sensitive to the quality of information about the subsurface current structure. In addition, current information is insufficient to evaluate dissolved components (e.g., toxic compounds) or concentrations of dispersed droplets for their impacts on nearshore environments. Ironically, as the effectiveness of dispersant increases, so does the potential threat to organisms exposed to the dispersed plume, due to the increased concentration of dissolved compounds and dispersed droplets in the water column. In open deep water, it may be reasonable to assume rapid dilution of the plume would take place. It is a generally held view, however, that such dilution should not be expected in shallower waters; hence a general avoidance of the use of dispersants in shallower waters exists. In addition, the current catalog of maps indicating the location and type of species or habitat that may be at risk from surface slicks is more adequate for areas along the shoreline. Information about the relative abundance of species in the water column or on the seafloor is inherently more difficult to obtain and tends to vary over shorter time scales. Greater capabilities to predict the trajectory of subsurface plumes of dispersed oil and the distribution of water-column and benthic species are needed, especially in shallower water where the impact of a dispersed oil plume may be more significant.

Overall

As the ability to apply dispersant to a variety of nearshore spills increases, the pressure to consider dispersant use in these waters will likely also increase. Consequently, the need for adequate and timely information to support decisions about dispersant use will become even greater. The remaining chapters examine the existing and needed capabilities to understand and predict the impacts of dispersed oil and recommend steps that should be taken to expand capabilities where needed.

3

Dispersant-Oil Interactions and Effectiveness Testing

Dispersants are mixtures of solvents, surfactants, and other additives that are applied to oil slicks to reduce the oil-water interfacial tension (NRC, 1989; Clayton et al., 1993). *Interfacial tension* is the free energy change that is associated with a change in the contact area at the interface between two immiscible phases (e.g., solid-liquid, liquid-liquid, liquid-gas). The term *surface tension* is also used to describe this phenomenon. Although these two terms are often used interchangeably, interfacial tension is considered to be the more general term, which can be applied to describe the free energy at the interface between any two phases, whereas surface tension applies specifically to those cases in which one of the phases is a gas (Lyklema, 2000). Reduction of the interfacial tension between oil and water by addition of a dispersant promotes the formation of a larger number of small oil droplets when surface waves entrain oil into the water column. These small submerged oil droplets are then subject to transport by subsurface currents and other natural removal processes, such as dissolution, volatilization from the water surface, biodegradation, and sedimentation resulting from interactions with suspended particulate material (SPM).

For the purpose of this and subsequent discussions, it is important to define two terms that are used interchangeably in the dispersant literature: entrainment and dispersion. In this report, entrainment is specifically the transport of oil from a surface slick into the water column by wind and waves, while dispersion includes both entrainment and subsurface transport (mixing and advection) by turbulent forces. It should also be mentioned that in the hydrodynamics literature the term dispersion (sometimes shear dispersion) refers to a specific mixing process resulting from the combination of shear in the mean velocity coupled with turbulent mixing (or other transport mechanism) in the direction of the shear. This process will be discussed in Chapter 4 and will be denoted as hydrodynamic dispersion to avoid confusion.

The following sections address dispersant chemistry, the physical and chemical interactions of dispersants with oil slicks and droplets, oil chemistry and weathering behavior and how they affect the window of opportunity for effective dispersant applications, and the importance of turbulence for introducing the energy necessary to entrain oil droplets into the water column as well as their subsequent transport by dispersive and advective processes. Next is a discussion of effectiveness testing and related issues, including laboratory systems, wave-tank tests, field studies, and studies involving spills of opportunity. Several of these topics are only considered briefly because there are a number of excellent reviews that consider the mechanisms of dispersant action and laboratory and field testing of dispersant performance (e.g., Meeks, 1981; Rewick et al., 1981; Mackay et al., 1984; Nichols and Parker, 1985; NRC, 1989; Clayton et al., 1993; Trudel, 1998; Etkin, 1999). Topics for which there are still major uncertainties or where data gaps exist are considered in greater detail, along with explicit findings and recommendations for areas requiring additional research.

COMMERCIAL DISPERSANT PRODUCTS AVAILABLE FOR USE IN U.S. WATERS

A typical commercial dispersant is a mixture of three types of chemicals: solvents, additives, and most importantly, surface-active agents (i.e., surfactants). Solvents are added primarily to promote the dissolution of surfactants and additives into a homogeneous dispersant mixture. In addition to keeping the surfactants in solution, these solvents reduce the product's viscosity and affect the dispersant's solubility in oil. Also, solvents determine to what extent the dispersant may be premixed with water for some spraying applications. Because aqueous-based solvent systems freeze in spray nozzles at ambient temperatures below 0° C (roughly 32° F) their usefulness is often limited in arctic or subarctic environments. Additives may be present for a number of purposes, such as improving the dissolution of the surfactants into an oil slick and increasing the long-term stability of the dispersant formulation.

Surfactants are compounds containing both oil-compatible (i.e., lipophilic or hydrophobic) and water-compatible (i.e., hydrophilic) groups. Because of this amphiphatic nature (i.e., opposing solubility tendencies), the surfactant molecules will reside at the oil-water interface as shown in

Figure 3-1. The surfactant reduces the oil-water interfacial tension by orienting with the hydrophilic groups interacting with the water phase and the hydrophobic groups interacting with the oil. Reduction of the oil-water interfacial tension facilitates the formation of a large number of small oil droplets that can be entrained into the water column.

Commercial formulations of modern chemical dispersants are usually comprised of two or more surfactant molecules that have differing solubilities in both water and oil. One parameter that has been used to characterize these different solubilities is the hydrophile-lipophile balance (HLB). The HLB ranges from 0 (no hydrophilic group) to 20 (no hydrophobic group), and the specific value characterizes the tendency of the surfactant to preferentially dissolve in either the oil phase (low HLB) or the aqueous phase (high HLB). The dominant group of the surfactant molecule will tend to orient in the outer phase to form a droplet of either oil or water (Porter, 1991). Therefore, a predominantly lipophilic surfactant (with a HLB below 7) will favor water-in-oil emulsions (mousse) where oil forms the continuous phase with discrete water droplets entrained within it (Porter, 1991). Natural components that promote the for-

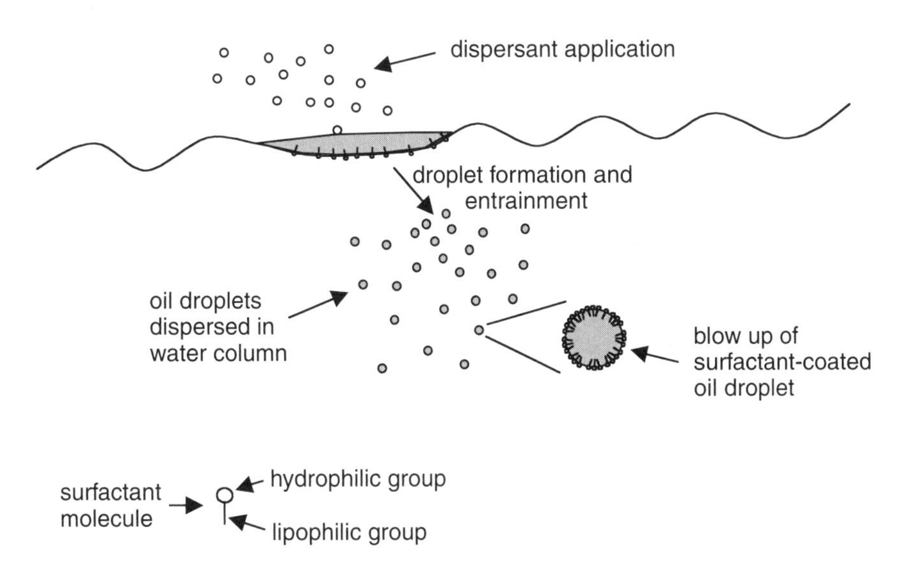

FIGURE 3-1 Mechanism of chemical dispersion: surfactant accumulates at oil-water interface, facilitating formation of small oil droplets that become entrained in the water column. Blow-up of oil droplet shows orientation of surfactant at the droplet surface with the hydrophilic group projecting into the water phase and the lipophilic group projecting into the oil phase.

mation of mousse (e.g., the resin and asphaltene fractions of crude oil) are generally lipophilic. In contrast, a predominantly hydrophilic surfactant (with an HLB greater than 7) will favor oil-in-water dispersions (i.e., entrained oil droplets in a water body) (Porter, 1991). The blend of surfactants in commercial dispersant formulations tend to be hydrophilic and the current formulations usually consist of surfactant mixtures with an overall HLB in the range of 9 to 11 (Clayton et al., 1993).

An example of the orientation of surfactant molecules at the oil-water interface is presented in Figure 3-2. Compound A is sorbitan monooleate (HLB = 4.3; predominantly lipophilic). Compound B is similar to A but has been ethoxylated with molecules of ethylene oxide to make it more hydrophilic (HLB = 15). The dispersant formulation shown in Figure 3-2 contains more compound B than A. Such a balance will promote formation of stable oil-in-water dispersions (entrained oil droplets in the water column) because the dominant hydrophilic group of the surfactant mixture favors the formation of oil droplets in water. The use of two or more surfactants with differing HLB values, but an overall average HLB in the range of 9-11, allows for closer physical interactions (i.e., packing) of the surfactant molecules at the oil-water interface compared to a single surfactant with an HLB value in this range (Porter, 1991). This produces a stronger interfacial surfactant film. Although ionic surfactants can inhibit coalescence of small droplets into larger droplets that would resurface more quickly by providing an electrostatic repulsion barrier (Porter, 1991), recent measurements suggest that this barrier is too small to significantly affect the collision efficiency (i.e., the fraction of collisions that result in coalescence), at least for dispersants (e.g., Corexit 9500) that consist mainly of nonionic surfactants, even when the dispersant-to-oil ratio (1:10) is relatively high (Sterling et al., 2004c).

Exact compositions for commercial dispersant formulations are proprietary, but their generic chemical characteristics are broadly known (e.g., Wells et al., 1985; Brochu, et al., 1986; NRC, 1989; Fingas et al., 1990; Singer et al., 1991, 1996; George-Ares and Clark, 2000). In general, a limited number of surfactant agents are currently used. Current dispersant formulations consist of mixtures of one or more surfactants, which may be either nonionic or anionic. Cationic (positively charged) surfactants are not used in current formulations (Clayton et al., 1993) because they are usually quaternary ammonium salts that are inherently toxic to many organisms.

The Corexit products are by far the most prevalent of all dispersants held in industry stockpiles within the United States, making up as much as 95 percent is some instances (J. Clark, ExxonMobil Research and Engineering Company, Fairfax, Virginia, written communication, 2005). Corexit 9527 was developed in the 1980s; it was supplemented in the 1990s

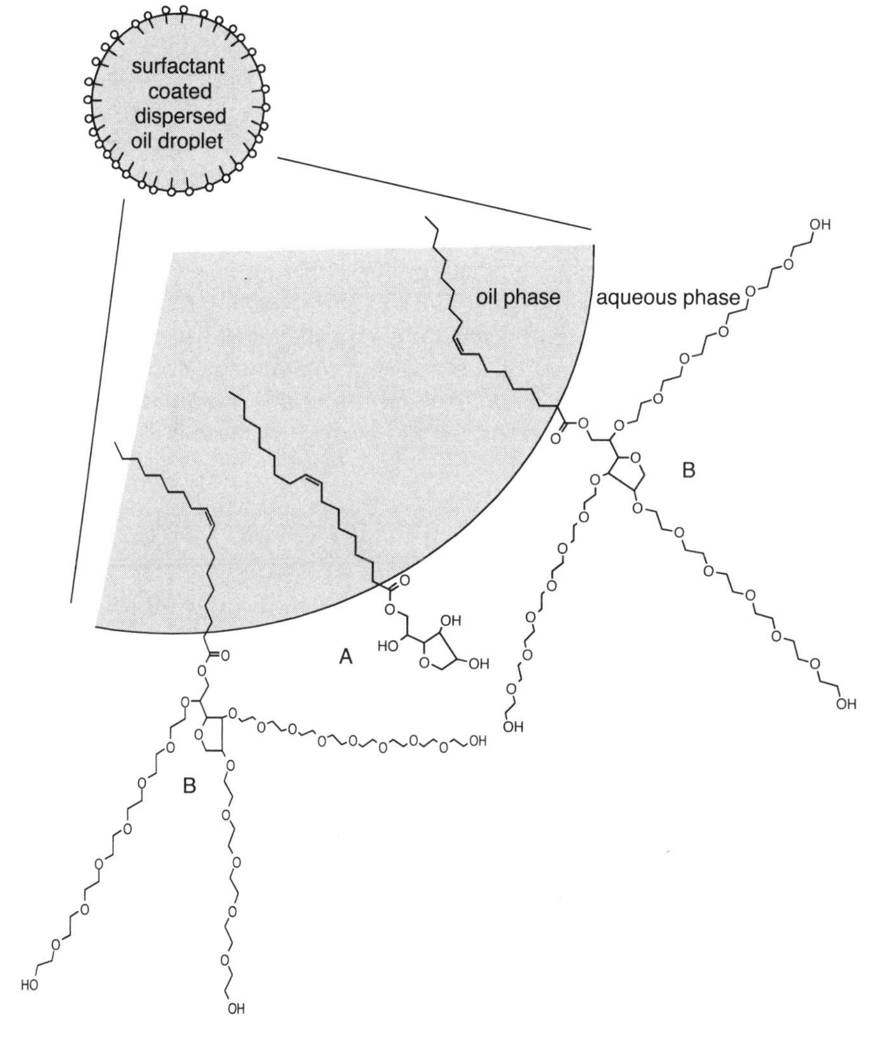

FIGURE 3-2 Orientation of surfactants at oil-water interface in dispersed oil droplets. Surfactant A is sorbitan monooleate (a.k.a., Span 80; HLB ≈ 4.3); surfactant B is ethoxylated (E_{20}) sorbitan monooleate (a.k.a., Tween 80; HLB ≈ 15).

by the introduction of Corexit 9500, which includes the same surfactants incorporated into a different solvent (George-Ares and Clark, 2000). Both products contain a mixture of nonionic (48 percent) and anionic (35 percent) surfactants. The major nonionic surfactants include ethoxylated sorbitan mono- and trioleates and sorbitan monooleate; the major ionic surfactant is sodium dioctyl sulfosuccinate (Singer et al., 1991). Neither Corexit product contains polyethoxylated alkyl phenols (J. Clark, Exxon-Mobil Research and Engineering Company, Fairfax, Virginia, written communication, 2004). A different solvent was used in Corexit 9500 for two reasons. First, prolonged exposure to Corexit 9527 caused adverse health effects in some responders. These effects were attributed to its glycol ether solvent (2-butoxyethanol). Therefore, the solvent was replaced by a mixture of food-grade aliphatic hydrocarbons (Norpar 13; n-alkanes ranging from nonane to hexadecane) in Corexit 9500 (Varadaraj et al., 1995). The second reason for changing the solvent in the reformulated dispersant was to extend the window of opportunity for dispersant use. This window of opportunity is limited by the effects of weathering on the chemical and physical properties of the spilled oil, especially the increase in oil viscosity. Corexit 9500 has been shown to be slightly more effective with high-viscosity oils than Corexit 9527.

THE PHYSICAL CHEMISTRY OF DISPERSANT-OIL INTERACTIONS AND THE ENERGY REQUIREMENTS FOR EFFECTIVE OIL-DROPLET ENTRAINMENT AND DISPERSION

The objective of an oil-spill dispersant application is to lower the oil/water interfacial tension to enhance entrainment of small oil droplets into the water column at lower energy inputs. Entrainment of small oil droplets into the water column (by either physical or chemical means) increases the oil-water interfacial area, which as shown in Eq. (3-1), requires energy:

$$W_K = \gamma_{o/w} A_{o/w} \qquad (3\text{-}1)$$

where W_K is the mixing energy (ergs or g-cm^2-s^{-2}; 1 erg equals 10^{-7} joule (kg-m^2-s^{-2})), $\gamma_{o/w}$ is the oil-water interfacial tension (dynes-cm^{-1}, where 1 dyne equals 1 g-cm-s^{-2}; equivalent to ergs-cm^{-2}), and $A_{o/w}$ is the oil-water interfacial area (cm^2). Therefore, reduction of the oil-water interfacial tension allows creation of a larger amount of interfacial area for the same level of energy input. Note that Eq. (3-1) provides an estimate of the *minimum* energy input that is required to disperse oil as droplets in the water column. Additional energy, which is proportional to viscosity, will be required to form droplets by stretching a continuous oil layer to the point at which it breaks.

The seven requirements for a chemical dispersant to enhance the formation of oil droplets (NRC, 1989) are:

(1) The dispersant must hit the target oil at the desired dosage.

(2) The surfactant molecules in the dispersant must have time to penetrate and mix into the oil.

(3) The surfactant molecules must orient at the oil-water interface with the hydrophilic groups in the water phase and the lipophilic groups in the oil phase.

(4) The oil-water interfacial tension must decrease due to the presence of the surfactant molecules at the oil-water interface, thereby weakening the cohesive strength of the oil film.

(5) Sufficient mixing energy must be applied at the oil-water interface (by wind and/or wave action) to allow generation of smaller oil droplets (with a concomitant increase in interfacial surface area).

(6) The droplets must be dispersed throughout the water column by a combination of diffusive and advective processes to minimize droplet-droplet collisions and coalescence to form larger droplets (which can resurface in the absence of continued turbulence).

(7) After entrainment, the droplets must be diluted to nontoxic concentrations and remain suspended in the water column long enough for the majority of the oil to be biodegraded.

Turbulent energy is the environmental parameter most responsible for generating and transporting dispersed oil droplets in the ocean. Delvigne and Sweeney (1988) studied natural dispersion and argue that the smallest scales of turbulence, with the greatest shear, are responsible for initial droplet formation, while the larger eddy scales are responsible for the subsequent vertical transport (described in more detail in Chapter 4—Transport and Fate). Conversely, Li and Garrett (1998) argue that natural dispersion is generated mainly by dynamic pressures associated with larger eddy scales, resulting in the creation of relatively large droplets (i.e., order of 100 μm diameter) that resurface relatively quickly. They suggest that reduction of the oil-water interfacial tension by chemical dispersants allows the mechanism of turbulent shear to govern droplet formation, which leads to smaller droplets (i.e., order of 10 μm diameter), which is more consistent with the diameters observed for "permanently dispersed" droplets. Unfortunately, the droplet-size distributions of chemically dispersed oil have only rarely been compared directly to those produced when untreated oil was dispersed under identical conditions (see Box 3-1). In the few cases where direct comparisons were made, however, the volume mean diameter was reduced by 30–40 percent by dispersants (Jasper et al., 1978; Lunel, 1995b). Figure 3-3, which was reconstructed from data presented by Lunel (1995b), shows the effect of a chemical dispersant (premixed Dasic Slickgone NS) on the droplet-size distribution produced when Forties crude oil was dispersed at sea: the number of small droplets (<50 mm) increased by about 5- to 30-fold,

BOX 3-1
Droplet-Size Distributions: What Are They and Why Are They Important?

When oil is entrained in the water column due to input of turbulent energy, droplets of various sizes are produced, regardless of whether the process is enhanced by addition of dispersants. Droplet-size distributions describe the relative abundance of droplets of various sizes, which may range from <1 μm to >100 μm in diameter. These distributions can be based on either droplet number or volume, although the volume distribution may be most informative, because the relationship between droplet volume and oil mass is constant regardless of droplet size (i.e., the proportionality constant is the density), whereas the relationship between droplet number and oil mass is not. The most common metrics for characterizing the central tendency of droplet-size distributions are the mean and median diameter, which will be approximately the same if the droplet sizes are normally distributed. The number mean diameter (NMD) is a simple average of droplet diameters, whereas the volume mean diameter (VMD) is the diameter of a droplet with the average volume (i.e., the mean of the volume distribution):

$$\mathrm{NMD} = \frac{\sum n_i D_i}{\sum n_i} \qquad \mathrm{VMD} = \left(\frac{\sum n_i D_i^3}{\sum n_i} \right)^{1/3} \tag{3-2}$$

where n_i is the number of droplets with diameter D_i. The VMD is larger than the NMD. Number and volume median diameters (also commonly referred to as NMD and VMD) are those droplet diameters that divide the number and volume distributions in half (i.e., 50 percent of the oil volume is present as droplets smaller than the volume median diameter).

whereas the number of large droplets (>50 mm) produced from dispersant-treated and untreated oil were similar. Note that although there were relatively few very large droplets produced from either treatment, these represented a significant fraction of the oil mass in both treatments, because the volume of oil in each droplet is proportional to the diameter cubed. Therefore, the volume distribution is extremely sensitive to uncertainty in the number of large droplets. This uncertainty can be seen in the reconstructed volume distribution shown in Figure 3-3. It is not clear whether the differences in characteristic droplet size are statistically significant, but if real, they would result in a 50–65 percent decrease in droplet rise velocity. Therefore, this phenomenon is potentially important and should be investigated further.

Droplet-size distributions result from the interaction of two processes: (1) droplet formation due to turbulent shear and (2) size fractionation due to differential rise velocities (Lunel, 1995b). Although the mechanism of droplet formation has not been proven, the initial size distribution of chemically dispersed oil droplets is thought to be related to the scale of the smallest eddies (i.e., microscale turbulence; Delvigne and Sweeney, 1988; Lunel, 1995b; Li and Garrett, 1998), but the distribution will be shifted toward smaller droplets following a period of quiescence due to resurfacing of larger droplets (Daling et al., 1990; Lunel, 1995b). Lunel (1995b) has suggested that dispersant effectiveness tests should be conducted in laboratory-scale systems and wave tanks that generate microscale turbulence similar to that which prevails in surface seawater, because such similarity suggests that the droplet-formation mechanisms will also be similar. Therefore, effectiveness testing should include measurement of droplet-size distributions, preferably in the presence of turbulent mixing energy, so that the observed size distribution will not be affected by size fractionation. Although droplet-size distributions have been measured in some lab-scale effectiveness-testing systems (Byford et al., 1984; Daling et al., 1990a; Lunel, 1995b; Fingas et al., 1995d), the effects of energy dissipation rate, oil and dispersant characteristics, and dispersant treatment should be more thoroughly investigated, because the existing database is not sufficient to support general conclusions regarding how (or whether) these factors affect the droplet-formation mechanisms and kinetics. Even fewer data are available regarding droplet-size distributions formed during dispersant effectiveness tests in wave tanks (Lunel, 1995b). Since one argument for increased use of these systems is their presumed ability to simulate sea surface conditions, it would be prudent to test this hypothesis by measuring droplet-size distributions and comparing them to those measured at sea.

More effort has been focused on studying the relationship between droplet size and dispersant effectiveness, but conflicting results have been obtained. For example, one study demonstrated an inverse relationship between dispersant effectiveness and the volume median droplet diameter (Byford et al., 1984), whereas others observed no correlation between effectiveness and characteristic droplet size (Daling et al., 1990a; Fingas et al., 1995d; Lunel, 1995b). Although the relationship between effectiveness and droplet-size distribution is uncertain, the droplet-size distributions clearly vary among different experimental systems: volume mean diameters of about 3 mm were observed in a system that was mixed by a six-blade vaned-disk turbine (Jasper et al., 1978), whereas significantly larger diameters (volume median diameters of 20 to 45 μm) were observed in

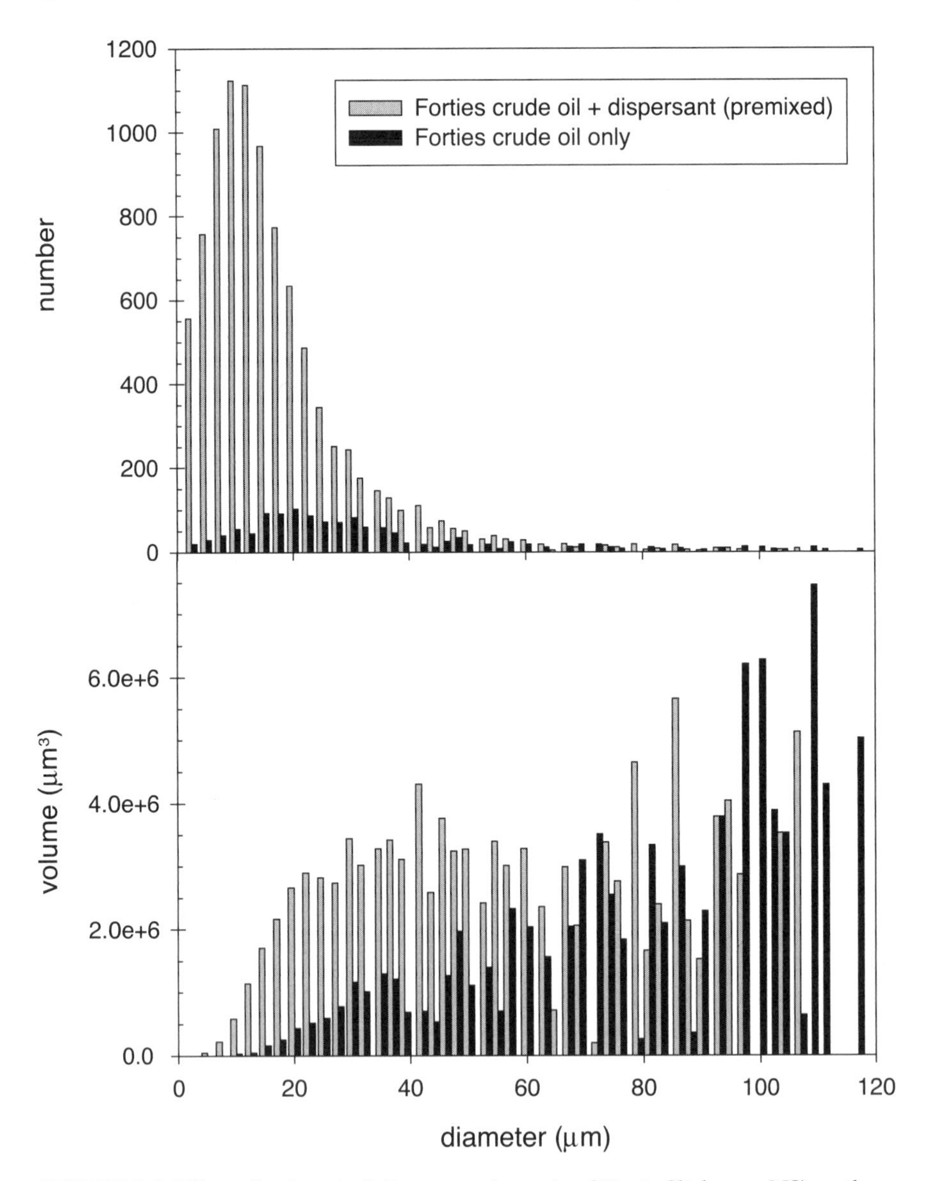

FIGURE 3-3 Effect of a chemical dispersant (premixed Dasic Slickgone NS) on the droplet-size distribution produced when crude oil (Forties) was dispersed at sea. SOURCE: Reconstructed from data presented by Lunel (1995b).

experimental apparatuses that are more commonly used in dispersant testing (e.g., the Warren Springs Laboratory, Mackay-Nadeau-Steelman, and swirling flask tests) (Daling et al., 1990a; Fingas et al., 1995d). The strong dependence of droplet-size distributions on the characteristics of the experimental system are consistent with the hypothesis that they reflect microscale turbulence (Delvigne and Sweeney, 1988; Lunel, 1995b; Li and Garrett, 1998), and Lunel (1995b) suggested that laboratory-scale or wave-tank effectiveness tests should be evaluated based on their ability to produce size distributions similar to those observed at sea.

In the ocean, turbulent energy is provided mainly by the wind, either by its direct action in shearing the water surface, or through the generation of surface waves. Above a critical wind speed, waves break, creating local areas of intense mixing. Internal waves, bottom shear stress caused by tidal or wind-driven currents interacting with a fixed bottom, and river inflows may also provide turbulent energy. Because of the variety of energy sources and mechanisms for oil droplet generation, it is unlikely that any single parameter can completely characterize the mixing energy responsible for oil dispersion. This is particularly true when including consideration of bench-scale lab tests (see below) in which mixing is produced by other mechanical means such as stirring, swirling, or tumbling. Nonetheless, the parameter that is most likely to be correlated with effective entrainment and dispersion is energy dissipation rate.

Turbulent energy enters a water body at large length scales and is transferred to smaller scales by the process of vortex stretching until it is dissipated by viscosity into thermal energy at the smallest scales. At equilibrium, the rate of energy input equals the rate of energy transferred at each scale, and hence the rate of energy dissipation (Tennekes and Lumley, 1972). Energy dissipation rates can be expressed in units of energy loss per volume per time, e ($J\text{-}m^{-3}\text{-}s^{-1}$) where J is joules ($kg\text{-}m^{2}\text{-}s^{-2}$). So, the volumetric energy dissipation rate, e, can also be expressed as $kg\text{-}m^{-1}\text{-}s^{-3}$. The energy dissipation rate can also be expressed as energy loss per unit *mass* per time, denoted by ε ($J\text{-}kg^{-1}\text{-}s^{-1}$ or $m^{2}\text{-}s^{-3}$). The latter is numerically smaller than e by a factor of the water density (about 10^3 $kg\text{-}m^{-3}$). Table 3-1, adapted from Delvigne and Sweeney (1988), gives approximate ranges of e and ε for a variety of water bodies.

In-situ values of the dissipation rate can be determined from highly resolved velocity measurements. Doron et al. (2001) describe several methods involving either evaluation of fine-scale velocity gradients or finding a fit to the spectrum of turbulent kinetic energy

$$E(k) \sim \varepsilon^{2/3} k^{-5/3} \tag{2}$$

where $E(k)$ is the turbulent kinetic energy density as a function of wave

TABLE 3-1 Energy Dissipation Rates for Different Water Bodies

Water Body	e (J-m^{-3}-s^{-1})	e (m^2-s^{-3})
Deep sea	10^{-4} to 10^{-2}	10^{-7} to 10^{-5}
Estuary	10^{-1} to 1	10^{-4} to 10^{-3}
Surface layer	1 to 10	10^{-3} to 10^{-2}
Breaking waves	10^3 to 10^4	1 to 10

SOURCE: Modified from Delvigne and Sweeney, 1988.

number, k. The turbulent kinetic energy (i.e., the integral of $E(k)$ over k), expressed per unit mass (units of J-kg^{-1} or m^2-s^{-2}), equals

$$1/2\sum_i \overline{u_i^2}$$

where u_i are the turbulent velocity fluctuations in up to $i = 3$ coordinate directions (Tennekes and Lumley, 1972).

Turbulent velocities themselves have been measured using a variety of techniques, some more appropriate to the lab and others more appropriate to the field. Point measurements can be made using airfoils, acoustic time-of-travel current meters, drag-sphere devices based on the instantaneous acceleration of a small sphere, hot-wire anemometers, and acoustic and laser Doppler velocimeters (Osborn, 1974; Agrawal et al., 1992; Terray et al., 1996; Doron et al., 2001). A one-dimensional velocity field can be determined using an acoustic Doppler current profiler (Veron and Melville, 2001), or by attaching probes to a vertical profiler, glider, or moving vessel. A two-dimensional velocity field can be obtained simultaneously using particle image velocimetry where a laser is used to illuminate a plane, and velocities are determined by correlating the displacement of natural particles observed in successive images captured with a charge-coupled device camera (Doron et al., 2001; Bertuccioli et al., 1999). In a laboratory flask, column, or tank, the rate of energy dissipation can also be determined *indirectly* by the rate of energy input by assuming that all input energy turns into turbulence. For example, in their "grid column," Delvigne and Sweeney (1988) determined ε by measuring the hydraulic resistance of their oscillating grid, while in their wave flumes, they determined ε by measuring the decline in wave energy as a function of distance along their tank. To the extent that ε uniquely determines oil dispersion, designing a laboratory experiment with values of ε equal to those expected in the field allows one to directly apply observations of dispersion effectiveness in the laboratory to predict dispersion effectiveness in the field. Unfortunately, this approach has not been typically utilized in laboratory and flume studies to date.

FACTORS THAT AFFECT THE OIL/DISPERSANT INTERACTION—THE WINDOW OF OPPORTUNITY AS CONTROLLED BY OIL CHEMISTRY AND WEATHERING STATE

When crude oil or refined petroleum products are released at sea, they are immediately subject to a wide variety of weathering processes that affect the resulting oil's chemical composition and physical (rheological) properties. These properties, including the chemical components responsible for stabilizing water-in-oil emulsions, are described more fully in Chapter 4. With regard to interactions with dispersants, the two most important weathering factors include evaporation and the formation of stable water-in-oil emulsions, because they both affect the spilled oil's *in-situ* viscosity on the water surface. Not surprisingly both of these processes are influenced by temperature (evaporation occurs more rapidly at higher temperatures, while emulsification can occur more rapidly at lower temperatures). Figure 3-4 summarizes the changes in bulk physical properties and water content in weathered Prudhoe Bay crude oil measured in experiments conducted in three 2,800-liter outdoor flow-through wave tanks over a 13 month period at Kasitsna Bay (lower Cook Inlet), Alaska (Payne et al., 1984, 1991a). The residence time of water flowing through the tanks was 4 hours, and the water temperature ranged from about 2° C (roughly 35° F) in the winter to 14° C (roughly 57° F) in the summer. Note the rapid change in properties after as little as 1–2 days of weathering under subarctic conditions. Although the initial oil-water ratio in these experiments was relatively high (1:175) and surface spreading of the oil was limited by the walls of the tank, the changes in oil chemistry and rheological properties that occurred in this oil over time were remarkably similar to those that were observed in the Alaska North Slope crude oil released from the T/V *Exxon Valdez* oil spill in Prince William Sound, Alaska (Payne et al., 1991a).

Viscosity is typically reported in dynamic units of centipoise (cP; 0.01 dyne-s-cm^{-2} or 0.01 g-cm^{-1}-s^{-1}). It may also be reported in kinematic units of centistokes (cSt; 0.01 cm^{2} s^{-1}) by dividing the dynamic viscosity by the fluid density. Because the density of oil is usually between 0.8 and 1.0 g-cm^{-3}, viscosities reported as cP and cSt are numerically similar, but the kinematic viscosity may be up to 25 percent larger than the dynamic viscosity. To provide perspective on the viscosity of weathered oil, Table 3-2 presents data for the water-in-oil emulsions from the wave-tank

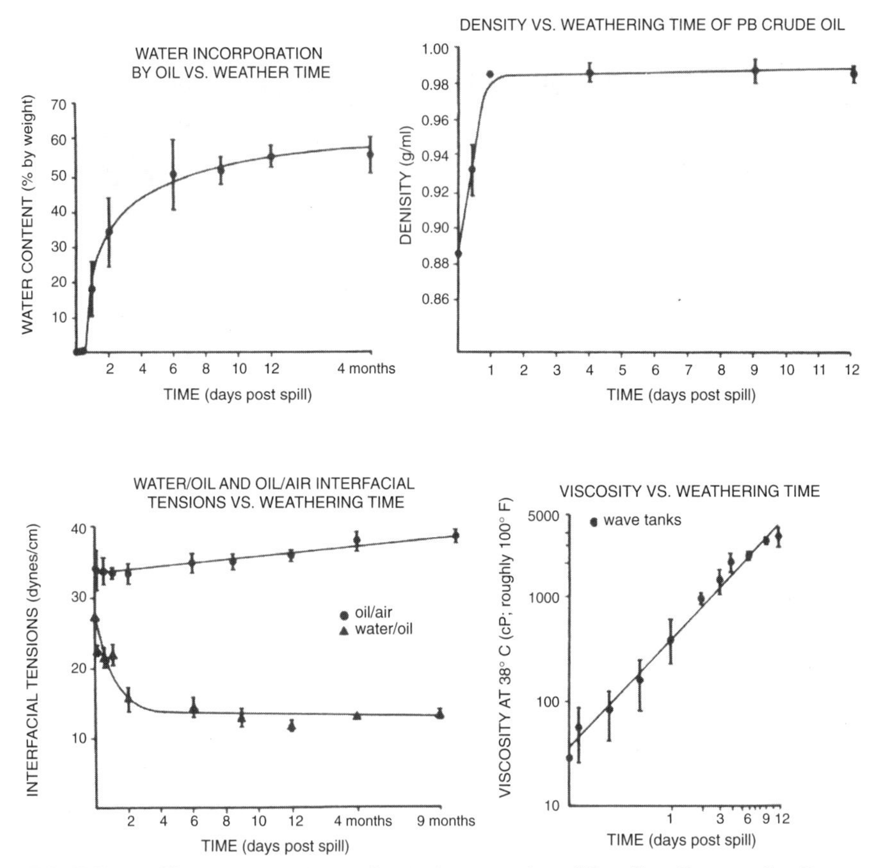

FIGURE 3-4 Changes in various physical properties of Prudhoe Bay crude oil as a function of weathering time. The values given are means from the three replicate summer wave tank experiments ± one standard deviation.
SOURCE: Payne et al., 1991a.

studies along with examples of the viscosities for several common food and household items.

During most of the 1980s, oils or emulsions with viscosities greater than 2,000 cP were considered to be difficult or impossible to chemically disperse (NRC, 1989). More recent studies (Fiocco et al., 1999; Guyomarch et al., 1999a) have shown that a number of intermediate fuel oils and weathered water-in-oil emulsions with viscosities approaching 20,000 cP can at least be partially dispersed in laboratory and field trials with multiple applications of newer hydrocarbon-solvent-based dispersants and demulsifiers (e.g., Corexit 9500, Inipol IP 90, Slickgone NS, Alcopol,

TABLE 3-2 Example Viscosities of Foods and Other Liquids.

Product	Temperature (°C)	Viscosity (cP)
Water	20	1.0
Ethyl alcohol	20	1.2
Olive oil	40	36
Fresh Prudhoe Bay crude oil (PBCO)	14	68
Olive oil	20	84
Olive oil	10	138
Castor oil	20	986
48-hr weathered PBCO water-in-oil emulsion	14	1,080
72-hr weathered PBCO water-in-oil emulsion	14	2,350
Pancake syrup	20	2,500
144-hr weathered PBCO water-in-oil emulsion	14	5,400
Honey	20	10,000
Chocolate syrup	20	25,000
Ketchup	20	50,000
Peanut butter	20	250,000

SOURCE: Data from CRC (1967), Transtronics (2000), and outdoor subarctic wave-tank experiments described by Payne et al. (1984, 1991a).

Demoussifier, Gamabreak, and Demulsip). As a result, these researchers have concluded that there is no hard and fast rule for the upper viscosity limit for dispersibility of water-in-oil emulsions.

If the pour point of the oil or refined product is above the ambient temperature encountered during a spill, the oil will not flow (it behaves as a semi-solid plastic-like material) and cannot be chemically dispersed. It has also been noted that certain highly paraffinic (waxy) crude oils can form a surface film due to evaporation of light ends (Berger and Mackay, 1994) and that photooxidation can lead to the formation of tar and gum residues (Payne and Phillips, 1985a,b; NRC, 1985, 2003), and it has been suggested that such surface layers may inhibit dispersant penetration into those oils.

Extensive research has been undertaken on the numerous factors responsible for the formation of stable water-in-oil emulsions with different oils (Bridie et al., 1980a,b; Zagorski and Mackay, 1982; Payne and Phillips, 1985b; Mackay, 1987; Bobra, 1990, 1991; Fingas and Fieldhouse, 1994, 2003, 2004a,b; Fingas et al., 1995a,b, 1996b, 1998, 1999, 2000a,b; Walker et al., 1993a,b, 1995), and the major findings from much of this research are briefly summarized in Chapter 4. The current consensus among researchers is that the type and stability of the emulsions is controlled by the properties of the starting oil, especially the asphaltene and resin content and initial oil viscosity (Fingas and Fieldhouse, 2003). Notwithstanding these advances, most of the existing knowledge on whether or not a particular oil will emulsify under given environmental conditions is empirical, and Fingas and Fieldhouse (2003) compiled a comprehensive data set that was used to develop a model of emulsification rate and stability (Fingas and Fieldhouse, 2004a, b). Although the predictions of this model are reasonably accurate, it is not always possible to predict whether a particular oil will emulsify under specified environmental conditions in the field and what the final water content will be. Often at the time of a spill, the critical compositional data (percent saturates, asphaltenes, resins, etc.) for the oil are not immediately available, and as a rough approximation the >343° C (roughly >650° F) boiling point fraction has been used as a surrogate in predicting whether mousse formation is likely (NRC, 1989). If that fraction is greater than 40 percent, the oil may emulsify and be difficult to disperse.

Empirical models of oil dispersibility with Corexit 9500—as measured in the swirling-flask laboratory test—were also recently developed (Fingas et al., 2003b). These models, which range in complexity from two (viscosity and density) to fourteen parameters, were developed by determining the effects of twenty-nine physical and chemical properties on oil dispersibility. Viscosity was found to be the most important physical property in determining dispersibility, but various aspects of chemical composition (e.g., the concentrations of n-dodecane, n-hexacosane, and naphthalenes) were more highly correlated. The most effective models were used to predict the dispersibility of 295 oils in the Environment Canada oil properties catalog (Environment Canada, 2005). Although these correlations may be useful for predicting and ranking the dispersibility of a large number of oils, the authors caution that the laboratory tests (upon which the correlations are based) may not provide a direct representation of what can be obtained in the field where different salinity and energy regimes are likely to be encountered.

Based on the above considerations and from practical experience, it is evident that response actions using dispersants should be initiated as soon as possible, and every effort should be made to apply the dispersants before significant oil weathering has occurred (usually within 24–72 hours in temperate conditions and possibly within 12–24 hours during the winter and under arctic conditions) to improve the probability of success. It should be noted that increased viscosity and water content in an emulsion also affect the ability to treat spilled oil by other response methods. For example, increased viscosity makes oil harder to pump, and increased water content increases the volume of material that must be handled and stored. For heavier oils, water contents above 20–30 percent make *in-situ* combustion essentially impossible (Twardus, 1980; Fingas and Punt, 2000).

Weather Considerations and the Window of Opportunity

Another important factor to be considered in evaluating the window of opportunity for effective dispersant applications is the energy regime at the time of dispersant application. As discussed in the previous section, a certain minimum energy (i.e., wind speed of 5 m/s; Allen, 1988; Fingas and Ka'aihue, 2004a), is required to break up the oil slick into small droplets, but applications under higher energy conditions can be plagued by other factors, such as:

(1) dispersant drift in the wind (missing the target as discussed in Chapter 2),
(2) possibly washing the dispersant off the slick before it penetrates into the oil phase, and
(3) the fact that the benefits of dispersant application begin to diminish compared to natural dispersion at wind speeds of 12–14 m/s (Allen, 1988; Fingas and Ka'aihue, 2004a).

Likewise, when dispersants are applied under low-energy conditions (little or no wind and/or reduced sea states), there may be a time lag between dispersant application and a subsequent increase in sea state (energy regime) to enhance dispersion. This delay also can lead to the potential for leaching of the dispersant from the oil phase before there is sufficient energy to promote droplet dispersion. It is believed that this problem might be avoided with some of the newer hydrocarbon-solvent-based dispersant formulations (or by additional adjustments to the HLB), but no studies on leaching of surfactants from the oil phase have been conducted at realistic oil-to-water ratios and under different energy regimes to test this hypothesis. In particular, the effects of surfactant leaching on the effectiveness of initial oil dispersion and the potential for dispersed oil droplet coalescence should be understood better. Recent laboratory studies have shown that surfactants do not appear to inhibit droplet coalescence, but the behavior of dispersed droplets and the concomitant leaching of surfactant under conditions of high dilution have not been studied. This is important because it will eventually affect dispersed oil behavior and the potential for re-surfacing in the field.

HISTORY OF DISPERSANT USE IN THE UNITED STATES

At the time of this writing, dispersant use in the United States had been limited to spills in Alaska (i.e., the T/V *Exxon Valdez* spill in 1989) and a series of smaller spills in the Gulf of Mexico (spanning 1999 to 2004). Understanding the circumstances and results of these actions provides some insight into the consequences of dispersant use, and thus is summarized below.

T/V *Exxon Valdez* Oil Spill (EVOS), Prince William Sound, Alaska (1989)

Spilled Oil Type/Volume/Conditions

An estimated 38,000 tonnes (roughly 250,000 bbls) of Alaska North Slope crude oil were released from T/V *Exxon Valdez* when it grounded on Bligh Reef in northeast Prince William Sound, Alaska, on March 23, 1989. Alaska North Slope crude has an API gravity of 29.8, a relatively high asphaltene content, and tends to form stable emulsions. Weather conditions were calm and clear.

Physical and Biological Setting

Prince William Sound includes many narrow fiords with deep, cold (<5° C [roughly 41° F]) seawater of low salinity and modest circulation. Rocky outcroppings and gravel beaches are common. There are an extensive local fisheries for both finfish and shellfish, as well as robust sportfishing and tourist industries throughout the Sound.

Dispersant Application

Two weeks prior to the spill, the Alaska Regional Response Team had adopted the first pre-approval zones for dispersant use in the United States. The spill occurred in Zone 1, where the state and federal coordinators could approve dispersant use on their own authority. Stockpiles of Corexit 9527 were available locally in Valdez, Anchorage, and Kenai. Both helicopters and large military C-130s were available within the state. However, there were no large capacity application packages (e.g., ADDS pack) in Alaska, and only a single helicopter bucket spray system was stored in Kenai (Alaska Oil Spill Commission, 1990)

Twelve hours into the spill, the helicopter bucket system arrived in Valdez and was immediately loaded with Corexit 9527 and used on the evening of March 24, and again on the morning of March 25. A third attempt on the morning of March 26 failed due to applicator malfunction. A fourth and final helicopter application occurred late in the afternoon of March 26. The first large-scale dispersant application occurred on the morning of March 27, 80 hours into the spill. In total, 5,500 gallons (roughly 20,800 liters) of Corexit 9527 were applied by C-130 (Alaska Department of Environmental Conservation [ADEC], 1993).

Monitoring Results

The U.S. Coast Guard and State of Alaska agreed that, on the first two days of helicopter applications, calm conditions did not supply sufficient mixing energy to achieve any noticeable effects. On the evening of the third day, visibility was poor and visual monitoring of the final helicopter application was inconclusive. Nevertheless, with the weather picking up, the decision was made to allow full-scale application in Zone 1 with a one-mile exclusion zone around the grounded tanker. Unfortunately, both T/V *Exxon Valdez* and the lightering tanker *Baton Rouge* were heavily sprayed during the next application, forcing a suspension of this extremely vital and difficult operation in order to decontaminate both personnel and equipment. No other effects of this dispersant application were observed. The State of Alaska, citing Exxon's inability to "accurately and effectively target the dispersant," declined to allow further dispersant application outside of Zone 1 (Alaska Department of Environmental Conservation, 1989). In any event, a large storm arrived with 40–70 knot (roughly 74–129 kilometers per hour) winds. The window for dispersant use was closed.

In its final report on the T/V *Exxon Valdez* oil spill, ADEC felt it necessary to state, "There was never a case in which loaded dispersant planes were held on the ground because the government couldn't or wouldn't make a decision" (Alaska Department of Environmental Conservation, 1993, p. 58).

Gulf of Mexico (1999 to 2004)

Between 1999 and 2004, dispersants were used seven times to combat oil spills in the Gulf of Mexico. In six of these cases, dispersants were used under the existing pre-approval plan for oil spills greater than 3 nautical miles offshore and in waters of greater than 10 m depth. Four of these dispersant cases are summarized below.

High Island Pipeline Spill (January 1998)

Approximately 360 tonnes (roughly 2,500 bbls) of South Louisiana crude (API gravity 38.2) were treated with Corexit 9527 using DC3 and DC4 aircraft. The application was very successful, based on aerial observations, SLAR measurements that showed decreased slick size, and SMART monitoring using field fluorometers that showed increased dispersed oil concentrations under the treated slick (Gugg et al., 1999).

BP-Chevron Pipeline Spill (October 1998)

Between 530 and 1,070 tonnes (roughly 3,700–7,500 bbls) of South Louisiana crude (API gravity 28.6) were released during a routine pipeline transfer operation at an offshore oil platform. Approximately 12,000 L of Corexit 9500 and 6,650 L of Corexit 9527 were applied to two of the three oil slicks over a period of two days using DC3 and DC4 aircraft. Visual observations suggested that the dispersant application was successful, but no confirmatory water-column data were obtained due to malfunction of the *in-situ* fluorometer that was deployed with the on-water monitoring team. Chemical analysis of water samples collected from the area of one of the treated slicks on the second day of dispersant operations showed only low concentrations of dispersed oil in the water column. British Petroleum estimated that approximately 160 tonnes (15 to 31 percent) of oil were chemically dispersed based on an assumed 80 percent effectiveness on the first day and 60 percent effectiveness on the second day, but these values were not independently confirmed by NOAA or the U.S. Coast Guard. ADIOS modeling predicted that about 33 percent of the oil was removed by evaporation. Only about 3 bbls of oil were recovered by mechanical response. This dispersant operation was considered to be successful due in part to the quick and aggressive chemical-treatment response and the good dispersibility of the oil (C. Henry, National Oceanic and Atmospheric Administration, New Orleans, Louisiana, written communication, 2004).

M/V Blue Master *Spill (August 1999)*

Approximately 17 tonnes (roughly 100 bbls) of IFO 180 (specific gravity of 0.988) were released from the M/V *Blue Master* following a collision with a fishing vessel 55 km south of Galveston, Texas. With light winds and calm seas, the oil was concentrated in a current-generated convergence zone. Within 12 hours after the spill and just before dark, 2,660 L of Corexit 9500 were applied (ratio of 1:6). Next-day observers reported a marked reduction in heavy concentrations of oil. It was considered a "cautious success" because only 0.25 tonne (roughly 1.8 bbls) of tarballs stranded onshore two weeks later (Kaser et al., 2001). Water-column oil concentrations were not measured to confirm that dispersion occurred.

Poseidon Pipeline Spill (January 2000)

Approximately 290 tonnes (roughly 2,000 bbls) of S. Louisiana crude (API gravity 31.5) were released from a 24 inch (roughly 60 cm) pipeline 65 miles (roughly 110 km) south of Houma, Louisiana that was caught and dragged by a large anchor. Due to 1–2 m seas, mechanical recovery was determined to be ineffective. Within 7 hours after the release, 11,400 L of Corexit 9527 were applied by DC3 and DC4 aircraft, resulting in an estimated 75 percent effectiveness, based on visual observations and fluorometry measurements. The next day, another 3,800 L were applied to the remaining patches of dispersible oil. There was no visual observation of a dispersed oil plume, but fluorometry did detect increased oil concentrations in the water under the treated oil. By the end of the second day of the release, it was determined that the remaining oil slicks were not dispersible. The applications were considered to be highly successful (Stoermer et al., 2001).

In summary, dispersants have been used successfully on oil spills in the Gulf of Mexico on several occasions in the past seven years. Because of the close proximity of dispersant application resources, responders were able to mobilize dispersant operations relatively quickly, which may have contributed to the overall success. Effectiveness, however, was evaluated primarily by visual observation, and not all operations included confirmation by measurement of dispersed oil in the water column. Therefore, the reliability of effectiveness estimates is unknown.

EFFECTIVENESS TESTING AND EFFECTIVENESS ISSUES

The overall effectiveness of oil dispersion has three components: (1) operational effectiveness, which describes the encounter probability of the dispersant application and the ability of the dispersant to become incorporated into the floating oil, (2) chemical effectiveness, which is measured by the fraction of treated surface oil that becomes stably entrained as small droplets in the water column, and (3) hydrodynamic effectiveness, which describes the transport of the chemically dispersed oil plume and its dilution by turbulent diffusion through horizontal and vertical mixing processes. The main focus of this section is a review of the experimental methods that have been used to investigate the chemical effectiveness of oil dispersants, but because the effectiveness that would be realized during spill-response operations at sea is determined by the interaction of all three components, those aspects of operational and hydrodynamic effectiveness that can be studied in effectiveness tests are identified and discussed where appropriate.

Operational effectiveness is determined by site-specific parameters, such as the patchy distribution of oil on the water surface, the ability to accurately target and hit the thicker parts of oil slicks with the dispersant spray, and the size distribution and impact velocity of dispersant droplets that hit the floating oil (as discussed in Chapter 2). It is difficult to simulate important characteristics of dispersant application in laboratory-scale experimental systems due to their relatively small size. Some large wave tanks can investigate many, but not all aspects of operational effectiveness. Operational effectiveness can be tested best in studies conducted at sea, provided the scale of the experiment is sufficient. Monitoring of operational effectiveness is the primary objective during real spill applications.

Hydrodynamic effectiveness is discussed primarily in Chapter 4 because it is governed by the transport of the dispersed oil plume. Hydrodynamic effectiveness cannot be tested in laboratory-scale systems or wave tanks, because significant dilution can only occur due to externally imposed flow through the system, not due to eddies of varying scales (e.g., turbulent diffusion and hydrodynamic dispersion). In principle, full-scale field studies can test hydrodynamic effectiveness, but appropriate measurements can be difficult and this is not always done.

Chemical effectiveness has been investigated in the laboratory, in wave tanks, and at sea. In many of these studies, effectiveness was defined based on chemical effectiveness, which was quantified as the mass fraction of oil that was measured in samples collected from the water column or the mass fraction that was not recovered from the water surface as floating oil. This definition has resulted in some confusion when attempting to compare studies conducted using different experimental systems, because these effectiveness metrics are operationally defined and measure different things in different systems. For example, some experimental designs include oil droplets that are large enough to resurface relatively quickly in the dispersed-oil concentration (e.g., those that measure water-column oil concentrations during periods of intense mixing), whereas others do not (e.g., those that include a settling period before measurement of dispersed-oil concentrations). Similarly, oil that is not recovered on the water surface may have been transferred to any of several compartments, of which the water column is only one. The droplet-size distribution of dispersed oil is a particularly important factor for chemical and hydrodynamic effectiveness, because it will determine whether the entrained oil will remain in the water column or float back to the surface under low energy conditions, which are unlikely to be the same during spill-response operations and effectiveness tests, regardless of the scale of the test. Future studies should include measurement of droplet-size distribution or some related metric to facilitate comparison among treatments. Lunel (1995b) has suggested that effectiveness tests should produce droplet-size distributions similar to those observed at sea, because this indicates similarity in the droplet-formation mechanisms.

Objectives of Effectiveness Testing

Dispersant effectiveness testing is performed using experimental systems that encompass a wide range of physical scales, from small (hundreds of milliliters) bench-scale systems to large (thousands of cubic meters) wave tanks, to open-ocean testing. All experimental systems used to evaluate dispersant effectiveness suffer from significant limitations; thus, it is important to clearly identify the objectives of the investigation before selecting an experimental system and designing an effectiveness study. Investigations of dispersant effectiveness are conducted for several common reasons: product screening; comparison of commercially available products for specific applications; fundamental investigations into the mechanisms that control dispersion of floating oil; and prediction of dispersant effectiveness under spill-response conditions. These objectives are quite different, and the experimental designs should reflect the differing requirements for data quality and application.

Effectiveness tests can be grouped into four broad categories: bench-scale tests; wave-tank tests; planned field studies; and spills of opportunity. Bench-scale tests often involve relatively common equipment, such as flasks and separatory funnels that are adapted or modified for the specific purpose of testing dispersants. They are also called laboratory-scale tests. Most wave tanks or hydraulic flumes are relatively small, but at least one, the Oil and Hazardous Materials Simulated Environmental Test Tank (OHMSETT), is very large. Although both are considered in the same category, the advantages and disadvantages of large vs. small wave tanks for dispersant effectiveness tests can be significant. Planned field studies and spills of opportunity also have many similarities, but the advantages and disadvantages are sufficiently different that they are considered separately. In general, as the physical scale of an effectiveness test increases, the cost and realism (i.e., the degree to which the test includes all three components of effectiveness) increase, but the degree to which the factors that affect dispersion effectiveness can be controlled and the ability to quantitatively measure effectiveness decrease. As a result of these competing trends, especially between realism and control, effectiveness tests at different scales are appropriate for achieving different objectives, and experimenters should be careful to match the objectives with the appropriate experimental scale.

Screening of dispersant products is often conducted for regulatory purposes. In the United States, dispersant products must be on the National Contingency Plan (NCP) Product Schedule to be considered for use as a response alternative for oil spills in U.S. marine and coastal waters (EPA, 2003). Inclusion on the NCP Product Schedule is contingent on demonstration that the candidate dispersant is capable of dispersing at least

45 percent of South Louisiana crude oil and Prudhoe Bay crude oil in the laboratory-scale swirling flask test. Although the specific method used in this procedure is likely to change in the near future, the objective of this test remains the simple demonstration of a prescribed degree of chemical effectiveness as measured by the concentration of oil in water samples collected from the bottom of the flask after a specified settling period to allow larger droplets to return to the water surface. The outcome is a pass-fail decision: if the product achieves the prescribed degree of dispersion, it may be included on the NCP Product Schedule (assuming it meets all other required criteria, such as successful toxicity testing); if it does not, the dispersant will not be included on the NCP Product Schedule, and it cannot be used in the United States as an oil spill countermeasure.

A related objective of effectiveness testing is comparison of available dispersants for specific applications, such as their ability to disperse specific crude oils or refined products under the environmental conditions that are known to prevail in certain regions (Blondina et al., 1997; Moles et al., 2002; White et al., 2002; Stevens and Roberts, 2003). These tests often attempt to compare the performance of specific oil-dispersant combinations under defined or standardized testing conditions. The results of these studies are intended to provide guidance for spill responders and regulators regarding selection of appropriate response actions or products. Due to the very large number of potential oil-dispersant combinations, the wide range of environmental conditions that may need to be considered, and the difficulty of extrapolating performance data beyond specifically tested conditions, these tests should be relatively simple. As a result, these comparisons are often conducted in bench-scale systems, but more limited testing has also been conducted in wave tanks and at sea.

Effectiveness tests may also be used in fundamental investigations of the mechanisms that control natural or chemically enhanced dispersion of oil into water (Belk et al., 1989; Fingas et al., 1991; Blondina et al., 1999; Canevari et al., 2001; Chandrasekar et al., 2003). Factors that have been investigated include dispersant-to-oil ratio (DOR), salinity, dispersant characteristics (e.g., hydrophilic-lipophilic balance, surfactant chemical structure, solvent characteristics), mixing energy, and the physical-chemical characteristics of the oil. Again, due to the wide range of conditions that may be of interest, the requirement for appropriate control treatments, and the need to rigorously control experimental conditions to facilitate testing of specific mechanisms, bench-scale systems are often used for these studies.

Ultimately, the objective of most effectiveness tests is to provide insight into the potential effectiveness of dispersants under actual spill-response conditions. Although most spill responders agree that quantitative prediction of dispersant performance is extremely difficult—if not impossible—based on current understanding of the factors that control it, the decision-making process during oil spill response involves implicit assumptions regarding expected effectiveness. For example, most oil transport and fate models that include an option for simulating dispersant application (e.g., French-McCay and Payne, 2001; Lehr et al., 2002; Simecek-Beatty et al., 2002; French-McCay, 2004) use effectiveness estimates as model inputs. These estimates are often based on experience and professional judgment rather than extrapolation from effectiveness tests and, as such, are not predictive. A major goal of chemical dispersant research should be development of quantitative tools for predicting dispersant performance (i.e., mathematical models) that can systematically incorporate many different types of information and the best current scientific understanding regarding droplet-formation and transport mechanisms. Ideally, dispersant effectiveness would be an output of a mathematical model, and the inputs would be factors such as oil characteristics, weather conditions, and other operational factors (e.g., dispersant type, effective DOR). Although multiple-regression models that relate oil dispersibility in a lab-scale effectiveness test to chemical composition have been proposed (Fingas et al., 2003b), these are completely empirical and cannot predict performance in the field—due at least in part to the inability to scale performance predictions from laboratory conditions to the field—and are not, therefore, useful for this purpose. Regardless of whether the predictions are quantitative (i.e., based on a mathematical model) or qualitative (i.e., based on the judgment of experienced professionals), effectiveness tests may provide the needed input parameters. In order to be useful, however, the effectiveness tests should be properly designed and the results should be interpreted with appreciation of their strengths and limitations.

Design of Effectiveness Tests

Effectiveness tests, regardless of the specific objectives or configuration of the experimental system, should explicitly consider how the experimental design will affect the results. Factors that are known to affect the extent of oil dispersion should be carefully controlled or characterized to the extent that is possible given the configuration of the experimental system. Examples of such factors include but may not be limited to the following: physical and chemical characteristics of the oil; physical characteristics of the surface slick; oil-water and dispersant-oil ratios; salinity and temperature; physical and chemical characteristics of the dispersant; method used to apply the dispersant to the oil; energy provided to disperse the oil; and the method used to measure effectiveness. In addition, the experimental design should include a clear description of the data analysis procedures that will be used, especially those used to estimate the random error term in the response variables, which is required in order to compare treatments. When possible, experimental designs should include independent replication of treatments and appropriate controls. Positive as well as negative controls (as discussed below) should be included whenever possible. Although these principles can be applied at all scales at which dispersant effectiveness can be tested, time and financial constraints will limit the degree to which they can be implemented as the scale of the test system increases. Such practical limitations, however, make clear definition of objectives and careful experimental design more —not less—important with increasing scale.

Among the factors that affect dispersion efficiency, the physical characteristics (e.g., pour point, viscosity, density) and chemical composition (especially aliphatic, aromatic, and asphaltic hydrocarbon concentrations) of the oil have received considerable attention. These characteristics are important because they can vary greatly among oils from different sources and change relatively quickly as oil weathers following a spill. Viscosity, which is roughly correlated with API gravity and density (Speight, 1991), has long been recognized to be an important parameter controlling the efficiency of oil dispersion (Daling, 1988), but viscosity alone is an insufficient predictor of dispersion efficiency (Fingas et al., 1991; Canevari et al., 2001). As a result, the chemical composition of the oil has also been considered, with various investigators identifying either positive or negative correlations between chemical effectiveness and the aliphatic, aromatic, polar, and asphaltene fractions of oil (Fingas et al., 1991, 2003b; Blondina et al., 1999; Canevari et al., 2001). Unfortunately, the nature of the relationships between composition and dispersion effectiveness is not well understood, and many of the results are contradictory. So, additional well-planned investigations are needed. It seems likely that some of the confusion may be due to unrecognized or unquantified differences among the experimental systems, such as the energy input or the characteristics of the oil droplets that are measured as dispersed. Therefore, future experiments should measure energy dissipation rates and the droplet-size distributions of dispersed oil. The dynamic changes that can occur in physical properties and chemical composition of oil during weathering make empirical investigation of these relationships particularly complex.

An important interaction likely exists between the physical characteristics of the oil and the method of dispersant addition. The dispersant must penetrate into the oil phase to effect dispersion, and certain physical characteristics (e.g., high viscosity) of the oil can prevent this from occurring efficiently (Canevari, 1984). Some investigators have suggested that evaporative weathering of waxy crude oils can lead to the formation of a viscous "skin" (Berger and Mackay, 1994) that may provide additional resistance to dispersant penetration. Few dispersant effectiveness tests use realistic weathering or dispersant application methodologies, and premixing the dispersant with oil is not uncommon. Many bench-scale tests add dispersant to floating oil, but the drop size is typically much larger (>1,600 μm diameter) than would be expected from a typical spray system (350–500 μm; NRC, 1989). For example, several studies involved addition of dispersant to floating oil in volumes ranging from 2 to 10 μL (Blondina et al., 1997, 1999; Venosa et al., 2001; Sorial et al., 2004a), which correspond to droplet diameters ranging from about 1,600 to 2,700 μm. Droplet velocity at impact with the oil is another important aspect of dispersant application that is not adequately simulated in existing bench-scale effectiveness tests. In general, this aspect of dispersant effectiveness, which would be considered operational, is not adequately characterized or controlled in most existing effectiveness tests at any scale. Wave tanks provide the most appropriate system for investigating the relationship between dispersant penetration and oil characteristics, because these systems are large enough to use realistic dispersant application systems (e.g., spray booms with typical nozzles) and can be controlled well enough to characterize the fraction of dispersant droplets that come into contact with floating oil. Therefore, the effects of oil characteristics (e.g., chemical composition, rheological properties, extent and mechanism of weathering) on the ability of dispersants to interact effectively with the oil should be investigated in future wave-tank studies and should be considered when interpreting the results of field-scale effectiveness tests.

The DOR and oil-to-water ratio (OWR), both typically measured on a volume-to-volume basis, are critical factors affecting dispersion effectiveness. Several investigators have shown a direct relationship between DOR and dispersion efficiency (Fingas et al., 1991; White et al., 2002); a DOR of 1:25 is commonly used, but this value can vary by a factor of two or more in either direction in some studies. The OWR of experimental systems for testing the chemical effectiveness of different dispersants can vary over a much larger range, with the values of lab-scale systems reportedly ranging from 1:1 to 1:120,000 (Fingas et al., 1989). The OWR affects the efficiency of oil dispersion in a variety of ways, some of which can have opposing effects. For example, anionic and nonionic surfactants with a high HLB will tend to partition into the aqueous phase where they cannot effectively promote formation of small oil droplets. The extent of partitioning will be determined in part by the OWR: when the OWR is high, more of the surfactant will be associated with the oil phase where it can facilitate droplet formation. Alternatively, high OWR could reduce the observed dispersion efficiency by increasing the rate of droplet coalescence, which is proportional to the number concentration of oil droplets (NRC, 1989). Droplet coalescence will produce larger oil droplets that can resurface more quickly and reduce the mass of oil entrained in the aqueous phase.

One of the most important factors in dispersant effectiveness testing is energy dissipation rate (e.g., mixing energy). Energy is required to create new oil-water interfacial area, which occurs when an oil slick breaks up into dispersed oil droplets. Successful oil dispersion will increase the oil-water interfacial area by a factor of ten or more, and sufficient energy should be provided to form the new oil-water interfacial area. Increased mixing energy, therefore, should result in the formation of smaller droplets (i.e., larger oil-water interfacial area). Because smaller droplets will have less tendency to resurface, higher mixing energy should result in more efficient and more stable dispersion. Energy dissipation rate is a parameter that varies widely among experimental systems, and differences among the results obtained with various systems are often attributed to differences in this parameter. Despite its importance, the energy dissipation rate is not measured in most dispersant effectiveness tests, and the relationship between mixing energy and effectiveness is only rarely investigated (Kaku et al., 2002; Fingas, 2004b). When it is, however, dispersion effectiveness is found to be directly proportional to mixing energy, but the proportionality varies among oil-dispersant combinations (Fingas et al., 1996a; Sorial et al., 2001; Chandrasekar et al., 2003). Predicting dispersant effectiveness for spill response based on bench-scale or wave-tank studies is hampered by our lack of understanding of the effect of mixing energy on oil dispersion for specific oil-dispersant combinations and the relationship between energy dissipation rates that prevail in common experimental systems and typical values at sea (1 to 10 $J\text{-}m^{-3}\text{-}s^{-1}$ in open-ocean surface) (Delvigne and Sweeney, 1988).

An important aspect of any experimental design is identification and measurement of the endpoint. For dispersant effectiveness testing, the endpoint is often defined to be the percent of added oil that is dispersed into the water column. For larger-scale systems, such as wave tanks and field studies, the water-column sample collection protocols can affect the observed effectiveness because the distribution of dispersed oil droplets is likely to be heterogeneous (Brown et al., 1987; Brown and Goodman, 1988; Lewis and Aurand, 1997). The concentration that is measured will depend on the location at which the sample is collected, and multiple samples will be required to characterize the distribution and estimate the total mass of dispersed oil. The mass of floating (non-dispersed) oil remaining on the surface is sometimes measured in wave tanks (Brown et al., 1987; Brown and Goodman, 1988; Louchouarn et al., 2000; Belore, 2003; Bonner et al., 2003) and field studies (Lewis et al., 1995a,b, 1998a), but many errors can be reflected in these measurements, including incomplete recovery of floating oil, unquantified losses due to evaporation, dissolution, or sorption to surfaces in the experimental system, and uncertainty in the distribution of floating oil (e.g., the size of the slick and variations in slick thickness with position; Fingas and Ka'aihue, 2004c).

In laboratory-scale tests, chemical effectiveness measured as percentage of oil dispersed into the water column is very sensitive to the settling time that precedes collection of samples, regardless of which method is used to measure the dispersed oil concentration (Fingas et al., 1989; Daling et al., 1990b; Venosa et al., 1999). This sensitivity is largely due to resurfacing of large oil droplets. Experimental methods that measure the dispersed oil concentration while mixing is occurring (e.g., the Institute Francais du Petrole and Mackay-Nadeau-Steelman tests) tend to result in greater "effectiveness" than those that involve a discrete settling period (e.g., the Labofina, swirling flask, and baffled flask tests). Coalescence of oil droplets, which is promoted by high oil-to-water ratios, can further decrease the measured effectiveness for tests that involve a settling period. Like mixing energy, settling periods vary among effectiveness tests, ranging from zero to about ten minutes. Because mixing energy affects the droplet-size distribution, which will affect the fraction of dispersed oil that resurfaces during the settling period, interpretation of dispersion effectiveness is difficult when the only endpoint is percentage of oil dispersed into the water column. As a result, more generally useful information would be obtained if effectiveness tests measured droplet-size distribution in addition to the mass fraction of oil dispersed into the water column or remaining on the water surface.

An objective of dispersant effectiveness testing at all levels is to determine whether addition of a chemical dispersant to a floating oil slick will increase the amount of oil that is transferred into the water column as small droplets relative to the amount that would be transferred from an untreated oil slick or from a slick treated with a different dispersant. This implies that a comparison should be performed to achieve the objectives of the experiment. For example, if one wishes to determine whether a particular dispersant is effective on a particular oil, the extent of dispersion that occurs for the oil-dispersant combination under specified conditions of temperature, salinity, and mixing energy should be compared to the extent of dispersion that occurs when the oil is exposed to the same conditions in the absence of dispersant. Such a comparison should involve estimation of the uncertainty in the amounts of dispersed oil measured in the presence and absence of the dispersant. The statistical significance of the effect of the dispersant is determined by estimating the probability that the difference in the amount of dispersed oil observed in the presence and absence of dispersant could be due to chance (i.e., the probability that a similar difference would be observed if the experiment were conducted without application of dispersant to either treatment).

The most reliable method for estimating the uncertainty in a measurement is to repeat it several times under identical but independent conditions. Independence of replicate measurements requires, at a minimum, that they be performed in separate experimental units (Hurlbert, 1984; Ruxton and Colegrave, 2003). In addition, some experimental designs, especially those involving large physical scales (e.g., field studies, large wave tanks), may require replication over time (see Box 3-2). Because the experimental conditions (e.g., weather) may vary from day to day, the replicates for different treatments should be interspersed in time to preclude the possibility that factors other than the treatment(s) under investigation will result in endpoint differences that are correlated with the treatment.

When an experimental design requires tests to be conducted over a prolonged period of time or by different analysts, precautions should be taken to ensure that results are comparable. That is, a mechanism should

BOX 3-2
Basic Principles of Experimental Design

Dispersant effectiveness is often quantified by measuring the amount of oil that is transferred to the water column or remains on the surface (or both) following application of a dispersant and mixing energy. All measurements are subject to some error, thus, the measured effectiveness is an estimate of the true effectiveness. The quality of this estimate is determined by its *accuracy* and *precision*. Accuracy is a measure of the agreement between the estimate and the true value, whereas precision provides an estimate of the reproducibility of replicate measurements. Since the true effectiveness is unknown, the accuracy cannot be independently evaluated, but the precision is used to identify a range that is likely to contain the true value. This range of values, sometimes called a confidence interval, is often used to compare one estimate of dispersant effectiveness to another (e.g., the extent of dispersion observed for dispersant-treated oil might be compared to the extent observed for an untreated control or to a threshold value specified by a regulatory agency). Statistical analysis is used to determine the probability that the two values that are being compared both estimate the same true effectiveness and appear to be different only due to the effects of random errors.

Two types of errors can cause measured estimates of dispersant effectiveness to be different from the true values: systematic errors and random errors. Systematic errors affect all measurements in the same direction, and therefore, bias the estimate. Evaporation of volatile compounds and incom-

exist to identify errors caused by differences in procedures or reagents that are not related to the treatment that is under investigation. One such mechanism is the use of *controls*. Controls are treatments (i.e., tests) that are performed periodically throughout a study for the purpose of quality control. Positive controls usually involve treatment with a reagent or procedure that produces a well-known and predictable result. A negative control usually involves measurement of the background response variable in experimental systems that are either untreated or treated with a mixture containing the inert ingredients (e.g., solvent) but lacking the active ingredients (e.g., surfactants). For oil dispersant effectiveness tests, a positive control might involve treatment of a standard easily dispersible oil with a standard dispersant under standard conditions. A negative control might involve subjection of the same standard oil to the physical conditions that would be applied in the dispersant test but without application of a chemical dispersant. Positive and negative controls are often

plete recovery of floating oil are two examples of systematic errors that can introduce a positive bias in estimates of dispersant effectiveness when the mass of oil remaining on the surface after treatment is used as the measure of effectiveness (i.e., the measured effectiveness will be greater than the true effectiveness because processes other than dispersion can reduce the mass of recovered oil). Random errors, which can be introduced by uncontrolled (or uncontrollable) variations in experimental conditions or measurement technique, will reduce the likelihood that two independent measurements of dispersant effectiveness will produce the same result even when they are made under nominally identical conditions. For example, small variations in the energy input or dispersant-to-oil ratio may cause the measured extent of dispersion to be different in replicate effectiveness tests. If a sufficiently large number of independent replicate measurements are made, however, positive errors will be offset by negative errors, and the mean (or another appropriate measure of the central tendency of the distribution) will be approximately equal to the true effectiveness. The more replicate measurements that are made, the closer the mean of those replicates is likely to be to the true effectiveness. Statistics can be used to quantify and correct for the effects of random errors, but systematic errors can only be mitigated by proper experimental design (including using appropriate experimental systems, sample collection procedures, and measurement techniques) and careful experimental technique. Proper experimental design should include provisions that eliminate systematic errors, minimize the size of random errors by controlling known sources of variation, and quantify the magnitude of unknown or uncontrollable random errors.

compared to expected values, which may be determined from experience with the experimental system, and if the results are outside of a predetermined range, the tests should be repeated. Use of a standard oil in controls during dispersant effectiveness testing requires that the characteristics of the oil remain constant over time. As such, the oil should be stored under conditions that prevent evaporation, photooxidation, and other changes in the physical and chemical properties that can affect its dispersibility.

In the following sections, all of the four categories of tests for dispersant effectiveness are discussed, in terms of their roles and objectives, the types of systems and methods used, and advantages and disadvantages.

Bench-Scale Tests

Role of Bench-Scale Testing in Evaluating Dispersant Performance

Due to their relative simplicity, bench-scale tests are widely used to evaluate the performance of dispersants and the physical and chemical mechanisms of oil dispersion. Bench-scale testing has been used to screen dispersants for inclusion on both state and federal product lists (Blondina et al., 1997; Venosa et al., 1999; Sorial et al., 2001; Venosa et al., 2002), compare the relative effectiveness of specific dispersant-oil combinations (Fingas et al., 1991; Moles et al., 2002; Venosa et al., 2002; Stevens and Roberts, 2003), and investigate the effects of environmental conditions or oil composition on dispersion effectiveness (Belk et al., 1989; Fingas et al., 1991, 1996a; Blondina et al., 1999; Canevari et al., 2001; Moles et al., 2002; White et al., 2002; Chandrasekar et al., 2003). A critical review and comparison of bench-scale dispersant effectiveness tests was presented by Clayton and others (1993). In many cases, the ultimate goal of these studies was to provide guidance to spill responders regarding which dispersants are likely to work on which types of oil under what range of conditions. Although it is generally recognized that these results cannot be used to quantitatively predict dispersant effectiveness in the field, their use for the objectives described above implies a belief that they provide reliable relative rankings. Although the results of several bench-scale effectiveness tests may be weakly correlated, this assumed relationship has not been thoroughly investigated or subjected to rigorous peer review (Fingas et al., 1994; Fiocco et al., 1999).

Descriptions of Common Bench-Scale Testing Systems

Although bench-scale effectiveness testing is used to achieve a set of common objectives, several different types of experimental systems have been used. The most common of these include the Warren Springs rotating flask or Labofina test, the Exxon dispersant effectiveness test (EXDET), the Mackay-Nadeau-Steelman (MNS) apparatus, the swirling flask test, and the baffled flask test.

The MNS apparatus (Figure 3-5), which is the largest of these four test systems, contains about 6 L of seawater in a 29 cm (ID) by 29 cm (depth) glass vessel (Mackay et al., 1978). Mixing is provided by tangential airflow (velocity usually between 6 and 20 m/s) over the water surface, which creates a circular flow pattern and surface waves between 2 and 4 cm in height. Oil is added to the water surface in the center of the vessel inside a 9-cm diameter aluminum containment ring. Dispersant is added to the oil inside the ring and allowed to soak for one minute before starting the airflow. The containment ring is removed as soon as airflow is started, and the oil spreads over the water surface while dispersion occurs. After mixing for ten minutes, but without stopping the airflow, a 500-mL water sample is collected from a sample port located 3 cm from the bottom of the vessel and 2 cm from the wall. Dispersed oil is extracted from the water sample into methylene chloride and the concentration is determined by measuring the absorbance at 580 nm. The circular motion of the water tends to minimize losses of oil to the vessel walls, but Fingas et al. (1994) found that, in the absence of dispersant, 13 to 19 percent of a light Bunker C adhered to the walls. Another effect of the circular motion of the water is that the radial distribution of dispersed oil is not uniform (Mackay et al., 1978). Because the sample is collected from a discrete location, the measured concentration may not be representative of the volume-averaged dispersed oil concentration. Although the MNS apparatus provides mixing in a way that is similar to what would be expected at the ocean surface (i.e., wind- and wave-driven shear at the air-water interface), the observed dispersion efficiency is very sensitive to small differences in the air-flow rate and angle of entry (Mackay et al., 1978; NRC, 1989), and the reproducibility, as measured by the performance of two apparatuses operating side-by-side in the same lab, was reported to be unacceptable (Fingas et al., 1989; Fingas et al., 1994). Note that the basis for this conclusion was that the standard error of dispersant effectiveness in replicate MNS apparatuses was 9 percent versus 3 percent in replicate swirling flask tests (described below) (Fingas et al., 1989) with maximum errors of 40 percent and 20 percent, respectively. Others, however, have raised similar objections to the swirling flask test, finding an average coefficient of variance (CV) of 22 percent with a maximum CV of about 160 percent (Sorial et al., 2004a,b). So, the reproducibility of many of these tests may be highly operator dependent.

The Warren Springs Laboratory (WSL; Labofina) test involves mixing of 5 mL of oil with 250 mL of synthetic seawater in a conical separatory

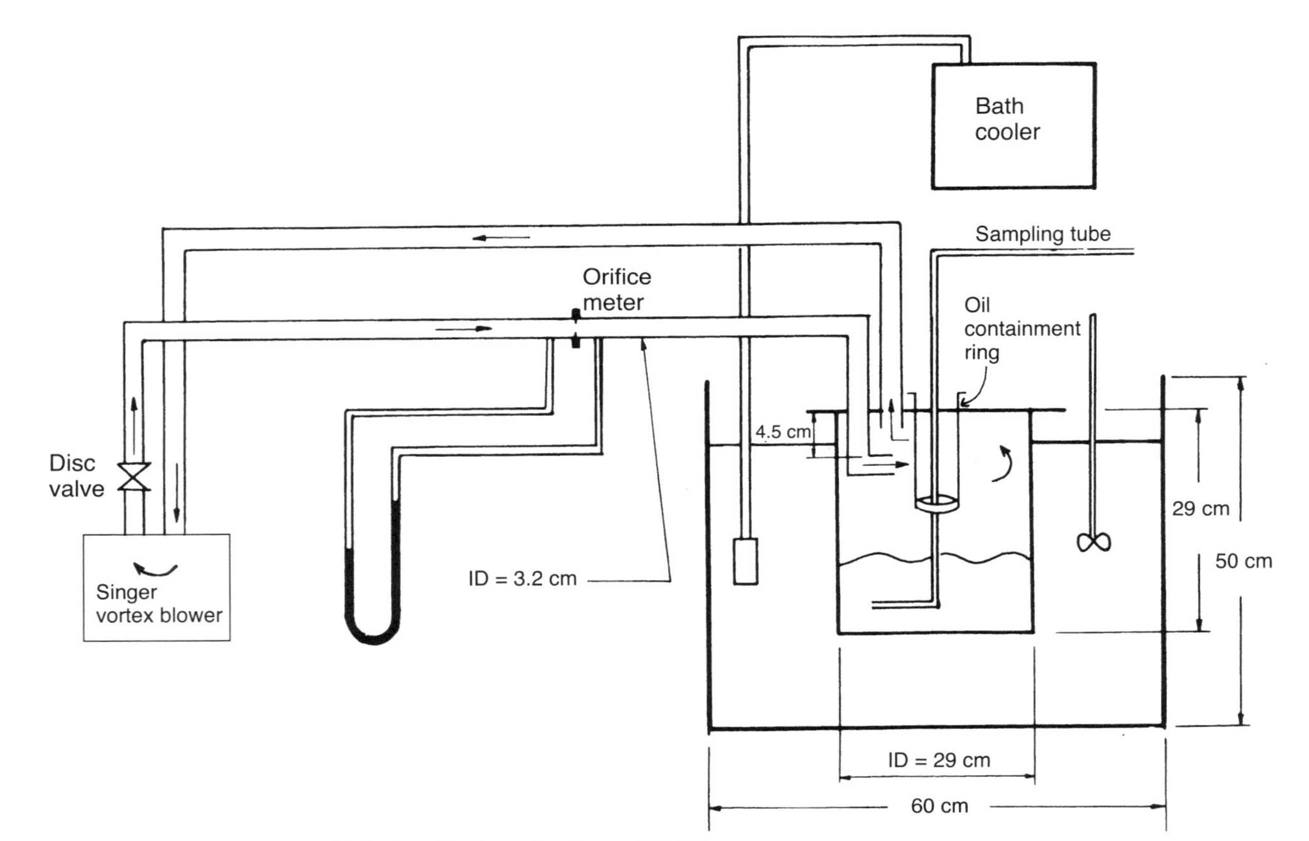

FIGURE 3-5 Schematic diagram of Mackay-Nadeau-Steelman (MNS) apparatus.
SOURCE: Mackay et al., 1978; courtesy of the American Society for Testing and Materials.

funnel (Figure 3-6). Dispersant is added to the oil surface dropwise, and the funnel is rotated end-over-end at 34 ± 2 rpm for five minutes. After mixing stops, the oil-water dispersion is allowed to stand for one minute to allow large oil droplets to rise, and a 50-mL sample is removed through the stopcock at the bottom. The oil in the sample is extracted into methylene chloride and the oil concentration is determined by measuring the absorbance at 580 nm. The WSL test is simple and reproducible, but observed performance was sensitive to the geometry of the separatory funnel. In addition, the standard test involves a very high oil-water ratio (1:50), which, as described above, tends to favor droplet coalescence. Alternatively the high oil-water ratio could result in unrealistically high aqueous-phase dispersant concentrations, which could increase the efficiency of dispersion by increasing the equilibrium concentration of the dispersant in the oil phase (NRC, 1989). Of these competing effects, droplet coalescence probably dominates in most cases, because Fingas et al. (1989) found that dispersion efficiency generally decreased with increasing oil-water ratio in the swirling flask test when the oil-water ratio was greater than 1:1000.

The Exxon dispersant effectiveness test (EXDET) is similar to the WSL test in that it is conducted in 250-mL separatory funnels, which are available in many laboratories. In the EXDET procedure, however, mixing energy is provided by a wrist-action shaker, which is also available in many

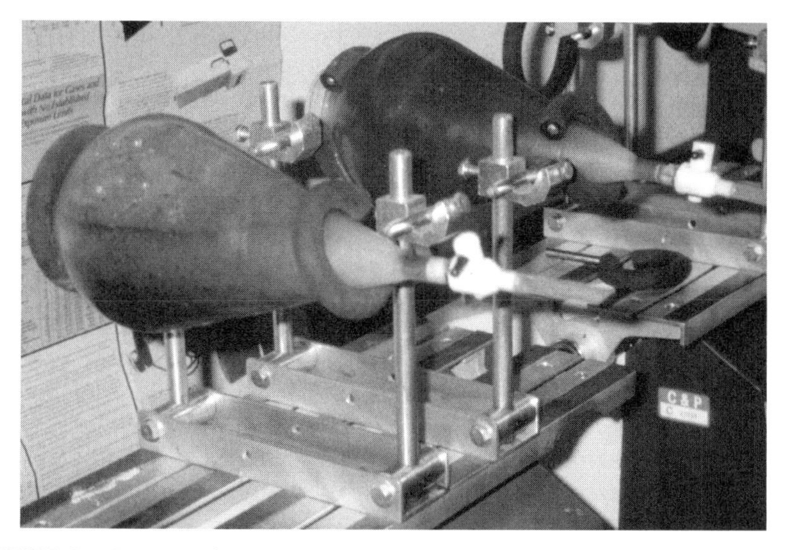

FIGURE 3-6 Picture of Warren Springs Laboratory (WSL; a.k.a., Labofina) dispersant effectiveness testing apparatus.
SOURCE: M. Fingas, Environment Canada.

laboratories (Becker et al., 1993; Clayton et al., 1993). So, EXDET has an advantage over many other dispersant effectiveness tests in that it does not require specialized equipment. Another advantage of the EXDET method is that it implicitly incorporates a mass balance. The procedure involves addition of a known volume of oil (e.g., 1 mL of oil premixed with dispersant at the desired DOR) to 250 mL of water in the separatory funnel, followed by shaking for 15 minutes on the wrist-action shaker. At the end of this mixing interval, but while shaking continues, a small absorbent pad is added to each funnel to collect the undispersed oil, and shaking continues for 5 minutes longer. Finally, the water is drained from each funnel and the dispersed oil is extracted with an appropriate solvent (e.g., methylene chloride, chloroform). The oil remaining in each 250-mL funnel, including that collected by the sorbent pad, is also extracted, and the concentrations of oil in both fractions (dispersed and undispersed oil) is measured by colorimetry (Becker et al., 1993; Clayton et al., 1993). The fraction dispersed is calculated by taking the ratio of the oil recovered in the aqueous phase to the sum of the oil recovered in both fractions. As long as the absorbances of the dispersed and undispersed oil fractions are within the linear range of the instrument, there is no need to calculate the exact mass of oil in each fraction, so calibrations curves are unnecessary. Of course, this procedure assumes that the oil collected by the absorbent pad is recovered completely and that no undispersed oil is transferred with the aqueous fraction, and any deviations from these assumptions will tend to cause the measured effectiveness to be higher than the actual effectiveness.

The swirling flask test (Figure 3-7) was developed to provide a simple method for screening dispersants (ASTM, 2000) and was adopted by the Environmental Protection Agency (EPA) for testing products for inclusion on the NCP Product Schedule (EPA, 2003; Sorial et al., 2004a,b). This test involves addition of 0.1 mL of oil to 120 mL of synthetic seawater in a modified 125-mL Erlenmeyer flask. Dispersant may be either premixed with oil (Sorial et al., 2001) or added to oil floating on the water surface (Blondina et al., 1997; Venosa et al., 1999). The flasks are mixed by swirling at 150 rpm on a gyratory shaker table, then allowed to settle for 10 minutes before a sample of the aqueous phase is collected by pouring through a glass spout that extends from the bottom of the flask upward to the neck. A recent variant of the swirling flask test, that involves collection of samples by draining water through a stopcock installed in the bottom of the Erlenmeyer flask, was developed to avoid reintroduction of oil droplets into the water phase when the flasks are tilted to pour from the spout (Blondina et al., 1997; Sorial et al., 2004a,b). This modification significantly improved the reproducibility of dispersion effectiveness measurements and reduced the extent of dispersion that was observed for five

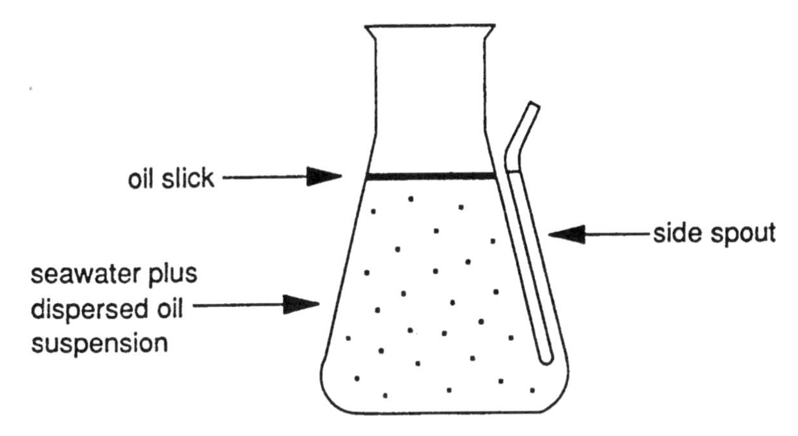

FIGURE 3-7 Schematic representation of the glassware used in the swirling flask test. Water samples containing dispersed oil are collected by pouring through the spout attached at the bottom of the flask.
SOURCE: A. Venosa, Environmental Protection Agency.

of six oil-dispersant combinations. The most common method for quantifying dispersion effectiveness in the swirling flask test is measurement of the absorbance of long-wave ultraviolet light (e.g., averaging the absorbance at 340 nm, 370 nm, and 400 nm) by methylene chloride extracts of the aqueous samples collected after the settling period. Some investigators, however, have concluded that gas chromatographic analysis of these extracts is preferable, because this measurement is less sensitive to interference by dispersants and some oils have very low absorbance at the wavelengths of interest (Fingas et al., 1995c; Blondina et al., 1997). Others contend that the increased time required for analysis by gas chromatography coupled with its much lower precision make this alternative less attractive (Sorial et al., 2004a). Measurement of dispersed oil concentration by absorbance requires the use of appropriate calibration procedures, but with the exception of oils with extremely low absorbance at the target wavelengths, agreement between the two methods is very good (Fingas et al., 1995c; Fingas and Ka'aihue, 2004b). This difference in the method used to quantify the concentration of dispersed oil is only one of several ways in which the swirling flask test used by EPA to evaluate products for inclusion on the NCP Product Schedule (EPA, 2003) differs from the ASTM standard method (ASTM, 2000). Other important differences between these two versions of the swirling flask test include the use of synthetic seawater in the EPA version of the test versus sodium chloride in the ASTM test, use of a higher DOR in the EPA test (1:10 vs. 1:25 for the ASTM standard), and specification of a 0.75-inch orbital diameter for the gyratory shaker used in the EPA test versus a 1-inch orbital diameter in the ASTM standard.

Another modification of the swirling flask test was introduced to overcome perceived limitations due to low energy input. This modification is known as the baffled flask test (Venosa et al., 2002; Sorial et al., 2004a,b). The baffled flask test uses a modified 150-mL trypsinizing flask that contains a stopcock near the bottom of the flask (Figure 3-8). These flasks have four baffles (i.e., indentations in the glass) at the bottom of the flask that increase turbulence during mixing by preventing development of a vortex due to the swirling motion of the gyratory shaker. The baffled flask

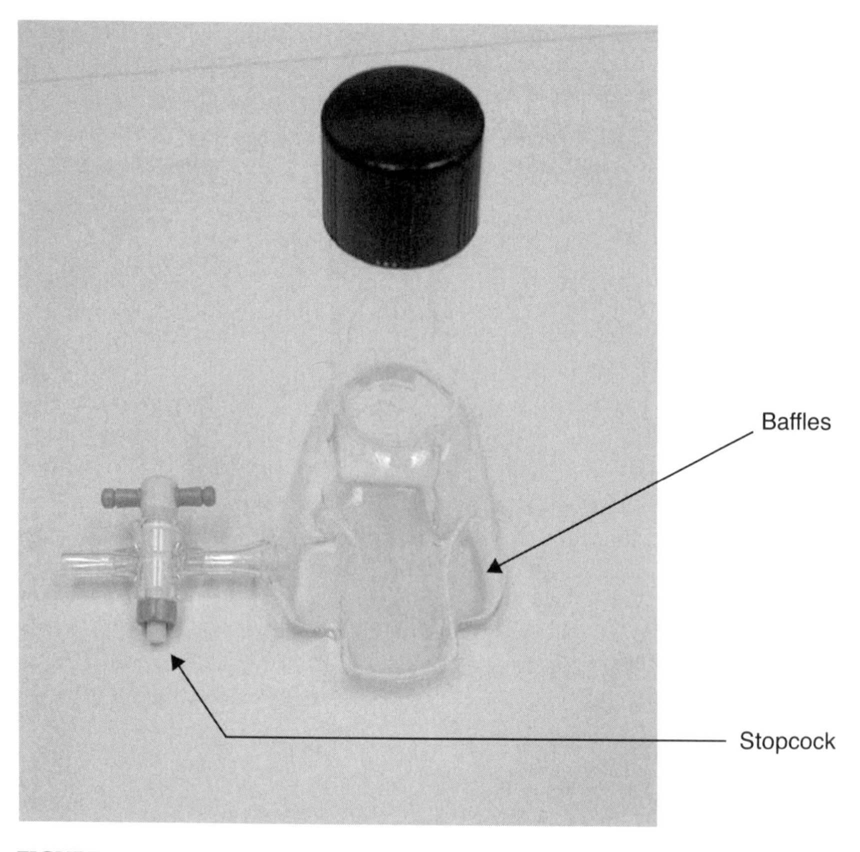

FIGURE 3-8 Photograph of the glassware used in the baffled flask test. This flask has four glass baffles at its base and a stopcock that allows collection of water samples containing dispersed oil by draining rather than pouring.
SOURCE: A. Venosa, Environmental Protection Agency.

test resulted in an average of four to five times more dispersed oil than the swirling flask test when eighteen dispersants were tested with two crude oils (South Louisiana and Prudhoe Bay). More importantly, the baffled flask test was much more precise than the swirling flask test (Sorial et al., 2004a). Dispersion effectiveness can be measured using the same methods as have been used for the swirling flask test, but measurement of the absorbance of methylene chloride extracts between 340 and 400 nm is preferred. Recently, the energy dissipation rates in the swirling flask and baffled flask tests were compared using a Hot Wire Anemometer to characterize the turbulence characteristics (e.g., the velocity gradient, *G*, and the energy dissipation rate per unit mass) of both systems (Kaku et al., 2002). Flask average energy dissipation rates in the swirling flask were about two orders of magnitude smaller than those in the baffled flask, and it was concluded that the turbulence in the baffled flask more closely resembled the turbulence occurring at sea during breaking waves.

Advantages and Disadvantages of Bench-Scale Testing

Bench-scale tests can be very useful for determination of the chemical effectiveness of oil dispersants. Because they are rapid and relatively inexpensive, bench-scale tests can evaluate a wide variety of experimental conditions in a relatively short period of time. As such, they are ideally suited for studies that are fundamentally empirical in nature (e.g., determination of the effectiveness of various dispersant-oil combinations, salinity or temperature effects on chemical effectiveness, relationships between oil composition or weathering and dispersant effectiveness). The relative ease with which treatments can be replicated independently in bench-scale studies is conducive to determining the statistical significance of any observed treatment effects, and statistically significant interactions among treatment factors can be identified using properly designed experiments. In addition, the use of small closed experimental units makes it relatively easy to perform mass balances (although this is often not done) for quality control purposes.

A major disadvantage of bench-scale testing is that it is difficult to scale the results to predict performance in the field because the test conditions do not simulate field conditions, especially energy regimes and dilution due to horizontal and vertical advection and turbulent diffusion. Scaling is difficult because the sensitivity of the response (e.g., dispersion efficiency) to variations in the test conditions is not well understood. In addition, although there has been some recent work in this area (Fingas, 2004a; Kaku et al., 2002), the energy input is rarely measured in common effectiveness testing systems. The generally poor correlation among performance estimates that are provided by different bench-scale systems may be due, at least in part, to the poor characterization of treatment conditions in these systems.

To be most useful, future bench-scale effectiveness testing should incorporate the following modifications. First, energy dissipation rates should be determined for each system over a range of operating conditions. This will be accomplished more easily in some systems than in others, but this parameter has a very large effect on chemical effectiveness and should be characterized for proper interpretation of the results, especially when an objective is comparison with other experimental systems. Second, the chemical effectiveness should be determined over a range of energy dissipation rates. The strong, and possibly nonlinear, dependence of effectiveness on energy dissipation rate implies that measurement at a single condition will be less useful than determining the relationship between these variables. The wide range of energy dissipation rates that can be experienced at sea reinforces the importance of understanding the relationships between energy input and chemical effectiveness. Finally, the definition of chemical effectiveness should include measurement of the dispersed-oil droplet-size distribution in addition to its concentration. Careful determination of the relationships between energy input and droplet-size distributions for a variety of oils that differ in physical and/or chemical characteristics will provide the information that is necessary to determine whether a general predictive model of chemical effectiveness can be developed.

Wave Tanks

Role and Objectives of Wave-Tank Testing

Wave-tank tests are expensive and messy, but when carefully done, they can bring greater realism to the study of dispersants compared to bench-scale tests. As described above, laboratory studies measure only chemical effectiveness; effectiveness tests conducted in wave tanks have the potential to also include some level of operational effectiveness. In particular, dispersant application equipment that produces dispersant droplets with size distributions and impact velocities that are similar to those encountered in spill-response operations can be used in tank tests. The physical characteristics of most wave tanks, however, imply that the encounter probability of the dispersant with the oil slick will be higher than can be achieved during a real spill response. So, wave-tank tests provide upper limits on operational effectiveness. In addition to the added realism provided by the ability to include some aspect of operational effectiveness in the study design, the mechanism by which energy is provided to the dispersant-treated oil slick in wave-tank studies (i.e., waves) is more similar to the mechanism that operates in the sea surface than can be accomplished in any bench-scale effectiveness test. As discussed earlier, there appear to be at least two mechanisms by which dispersed oil droplets can be generated: the dynamic pressure force of turbulent flows dominates at high Reynolds numbers and results in formation of relatively large droplets, whereas viscous shear due to small turbulent eddies dominates at low Reynolds numbers causing small droplets to form (Li and Garrett, 1998). Hence, mechanistic similarity might be important (i.e., energy dissipation rate alone might not be an adequate scaling factor for dispersant effectiveness). Of course, all tanks have walls. Therefore, no tank test will ever be completely realistic, because they cannot adequately incorporate the hydrodynamic effectiveness component. Nevertheless, given the costs and parameter-control difficulties associated with field tests, and the subjective nature of much of the data that can be collected, wave-tank tests are an important tool that can be used to tie the artificialities of laboratory studies to the operational realities of dispersant use in spill response. As such, wave-tank tests should be judged primarily on the basis of the additional realism—over laboratory studies—that is incorporated into their test design while remaining sufficiently controlled to allow replication and collection of quantitative data.

Before dispersants can be accepted by an informed public as a potential primary tool in oil spill response, several important issues need to be thoroughly investigated. In each of these areas, well-designed tank tests should be capable of furthering knowledge considerably.

(1) *Structural Effects.* Weathering is not uniform throughout oil slicks. Photooxidation and evaporation, especially for certain high wax-content oils, can result in formation of a highly viscous "skin" that may provide significant resistance to penetration of chemical dispersants (Berger and MacKay, 1994; Payne and Phillips, 1985a,b). Effectiveness tests that use oil that has been artificially weathered by evaporation of volatile components from a well-mixed bulk phase may overestimate the operational effectiveness (i.e., dispersant penetration of the floating oil) by underestimating the resistance provided by the viscous film that could be encountered by a dispersant droplet that contacts oil that weathered as a floating slick under natural sunlight. Thus, wave-tank tests that can simulate oil weathering as it would occur at sea (i.e., as floating slicks) should be conducted. These studies should also investigate the evolution of the physical-chemical characteristics and the operational dispersibility, as oil weathers in a slick.

In this regard, the formation of water-in-oil emulsions is particularly important in inhibiting dispersant effectiveness. To date, large wave tanks have not been used to examine the performance of dispersants on water-in-oil emulsions generated from weathering of oil on the sea surface. Ideally, these emulsions would be generated in adjacent wave tanks or other systems that can provide continuous mixing of oil and water for hours to days. The effectiveness of dispersants on these blended emulsions could then be tested under more realistic field conditions. The rheological and chemical properties of the test emulsions should be characterized and compared to data from emulsified oil samples collected during actual oil spills. The dispersibility of the artificially generated emulsions should be tested over a range of temperatures, including cold, subarctic conditions. If this approach is successful, it could be expanded to investigate dispersant effectiveness on water-in-oil emulsions in the presence of ice.

(2) *Dispersant Application.* Dispersant application efficiency is affected by dispersant droplet size and velocity. If this aspect of operational effectiveness is to be investigated in wave tanks, the dispersant application system should simulate the droplet-size distributions and impact velocities that are characteristic of specific application methods (e.g., aircraft, helicopter, vessel). These parameters should be measured to verify that the desired characteristics have been achieved. The dispersant distribution over the target area also should be characterized at some point during these tests. In at least one instance, plastic sheet walls surrounding the tank were used to capture drifting spray, and trays were set up within the target area to measure the dosage that was applied to the oil (S.L. Ross, 2002). Although it is unlikely that the characteristics of real dispersant application systems can be accurately reproduced in a wave tank (even a very large one), measurement of effectiveness as a function of dispersant droplet-size distributions and impact velocity may provide information that can be used as input to dispersant effectiveness models.

(3) *Mixing Energy.* As described previously, mixing energy is one of the most important factors determining dispersant efficiency. Many oils will physically disperse even in the absence of chemical dispersants if sufficient mixing energy is provided. As with effectiveness tests at laboratory scales, mixing energy should be measured as a routine part of system characterization, and effectiveness should be measured over a range of mixing energies that span the range that can be realistically expected in the environment of interest. The wave energies used in the experimental system should be scalable to actual sea states.

(4) *Coalescence and Resurfacing of Dispersed Oil Droplets.* In the past decade there have been several studies that looked at the effects of dispersant stripping, droplet coalescence, and resurfacing of dispersed oil (Fingas et al., 2002a; Bonner et al., 2003; Sterling et al., 2004c). The extent

to which this occurs will depend to a large extent on the hydrodynamic effectiveness of dispersion (i.e., the relative rates of coalescence and dilution of dispersed oil droplets by turbulent diffusion) and will exert a strong influence on the ultimate fate of the dispersed oil. The coalescence rate depends on the number concentration of dispersed oil droplets (Sterling et al., 2004c), which will decrease as the dispersed oil plume spreads and mixes with surrounding seawater. As described previously, however, hydrodynamic effectiveness cannot, in general, be investigated in the laboratory or in wave tanks, because these are closed systems with little or no dilution potential, and coalescence will be promoted by providing mixing energy over a prolonged period of time (i.e., by increasing the frequency of droplet-droplet collisions). The relative role of coalescence may be significantly reduced in very large wave tanks where dilution more closely approximates natural conditions. The extent to which coalescence and resurfacing will occur in the field, however, can only be fully investigated in field studies or by incorporating coalescence into a comprehensive dispersed oil fate and transport model. Coalescence kinetic parameters can be estimated in the laboratory (Sterling et al., 2004c). The effects of temperature and ice on dispersed oil droplet size, coalescence, and resurfacing also should be investigated to evaluate the range of conditions under which dispersants are likely to be effective, and these investigations would probably be most realistic in very large wave tanks where dilution more closely approximates natural conditions.

Description of Wave Tanks Available for Mesoscale Dispersant Testing

This section provides brief descriptions of some facilities that are available for testing of dispersant effectiveness in wave tanks. These descriptions focus primarily on the physical facilities and the tools available for measuring experimental conditions and results. Large tanks or facilities created to allow complex inter-comparative studies are discussed first, while smaller and simpler tanks are included for completeness.

Oil and Hazardous Materials Simulated Environmental Test Tank (OHMSETT) The largest test tank available in the world for dispersant testing is the Oil and Hazardous Materials Simulated Environmental Test Tank (OHMSETT), operated by the Minerals Management Service and located on the grounds of the Naval Weapons Station Earle in Leonardo, New Jersey (Figure 3-9). This facility was originally designed for testing mechanical oil recovery equipment, such as booms and skimmers. In recent years, it has been modified to accommodate dispersant testing, including the ability to chill the seawater in the tank to arctic temperatures and modifying the seawater filtration system to improve removal of dis-

FIGURE 3-9 Aerial view of the OHMSETT test tank facility.
SOURCE: J. Mullin, Minerals Management Service, http://www.ohmsett.com/.

persed oil and dissolved dispersant (J. Lane, U.S. Minerals Management Service, Herndon, Virginia, written communication, 2005).

The OHMSETT facility includes an aboveground, concrete tank that is 203 m long, 20 m wide, and 3.4 m deep. Six viewing windows are located at intervals along one side to allow for underwater observations. Brackish water is pumped from a nearby bay, filtered, and the salinity adjusted by the addition of salt with major ion composition similar to sea salt. The tank is usually filled to a depth of 2.4 m, giving a working volume of approximately 9,700 m^3. The tank is spanned by three movable bridges, which can move along the tank at speeds up to 3.3 m/s. In addition to the administrative building alongside the tank, there is a multistory control complex at one end of the tank affording a complete view of the facility. At the south end of the tank is a paddle-type wave generator, capable of producing either smooth or cresting regular waves. At the north end there is an artificial "beach" that can be raised or lowered to either absorb or reflect wave energy, which allows users to produce waves with specific characteristics (e.g., long, even swells or harbor chop). Oil for dispersant tests has typically been added to the water surface in an approximately 10,000 ft^2 area within containment booms. This represents approximately 23 percent of the available surface area of the tank. Note, however, that the entire volume of the tank is potentially available for dilution of the dispersed oil plume.

In dispersant tests that have been conducted at OHMSETT—beginning in March 2002—the test oil has been applied to the water surface through a manifold mounted to the leading edge of the main bridge (Figure 3-10). Oil has been applied while the main bridge advanced at a speed of 0.5 m/s, and the dispersant has been sprayed on the resulting oil slick from a nozzle array hanging below the trailing edge of the same bridge (Figure 3-11). For these conditions, the time interval between application of the oil and the dispersant is approximately 10 seconds. The short time period between application of oil and dispersant is the basis of some criticism of cold-water tests that were conducted at this facility, because the oil was heated to allow it to be pumped through the oil distribution manifold, and some suggest that 10 seconds is not enough time for the floating oil to cool to the temperature of seawater (PWSRCAC, 2004). To address this potential problem, MMS has funded research to investigate the cool-

FIGURE 3-10 OHMSETT oil distribution system.
SOURCE: S.L. Ross, 2002.

FIGURE 3-11 OHMSETT dispersant spray bar in operation.
SOURCE: S.L. Ross, 2002.

ing rate of heated oil in contact with cold seawater and has redesigned the oil-distribution manifold to allow application of cold, highly viscous oil in future cold-water tests (J. Mullin, U.S. Minerals Management Service, Herndon, Virginia, written communication, 2005). Pending peer review of this research, or a repeat of the cold-water tests using an improved oil-distribution system that does not require heating the oil, OHMSETT test results should be used with caution to gauge the effectiveness of chemical dispersants in cold water.

The large size of the OHMSETT tank offers advantages to experimenters wishing to investigate certain aspects of operational effectiveness (e.g., the dispersant application equipment can produce dispersant droplets with realistic size distributions) and hydrodynamic effectiveness (e.g., the facility allows dispersed oil to be transported in a relatively large volume of water). It also permits studies of effectiveness under specialized conditions (e.g., in broken ice). However, the large size of the tank also presents several problems. Primary among these is the high cost of operating a facility of this size (e.g., the cost of chilling 9,700 m^3 of seawater is considerable). This financial constraint often leads to experimental designs that lack sufficient replication to support statistical analysis of the results. Another size-related limitation is the inability to shield the water surface from wind, which can cause the oil slick to drift to one side of the tank over relatively short periods of time and, therefore, requires experimenters to apply the dispersant immediately after application of the oil. As described above, this practice has led to questions regarding the validity of several high viscosity oil tests. In addition, the tank is too large to allow the water to be replaced after each test, and even with improved filtering, some observers are concerned that residual oil and dispersant can affect subsequent tests. The facility operators have determined the maximum dispersant concentration (400 ppm) that can be present in the water without affecting the validity of subsequent effectiveness tests (S.L. Ross, 2000), and to date, this concentration has not been exceeded in sequential tests. The presence of dispersed oil from previous tests, however, affects the water clarity (limiting the visibility of dispersed oil plumes) and precludes determination of the size distribution of dispersed oil droplets in the water column during subsequent tests when the water is not adequately filtered between successive runs. Finally, the size of the OHMSETT tank and its associated equipment is likely to increase the difficultly of closing mass balances through collection of non-dispersed surface oil, measurement of the concentration of dispersed oil droplets in the water column, and quantification of the oil that escaped the boomed test enclosure or adhered to the boom itself. The addition of a secondary containment boom outside the north end of the 10,000 ft^2 experimental area has significantly improved collection of surface oil that splashes out of the test enclosure (it is then included with the other non-dispersed oil collected from the water surface within the test area), but quantifying the oil that adheres to the boom itself remains difficult. To date, dispersion efficiencies have been calculated by comparing the volume of surface oil recovered from dispersant-treated slicks to that recovered from control slicks that are not treated with dispersants, but complete mass balances have not been performed at this facility.

EPA/Department of Fisheries and Oceans Wave Tank at the Bedford Institute of Oceanography A new wave tank for investigation of dispersant effectiveness was recently built at the Bedford Institute of Oceanography (BIO) in Halifax, Nova Scotia, with joint funding from the EPA and the Department of Fisheries and Oceans (DFO) Canada (Figure 3-12). Although this facility was designed specifically for testing dispersant effectiveness and evaluating their effects, experiments involving dispersants and oil have not yet been conducted at the time of this writing.

The BIO/EPA wave tank is 16 m long, 0.6 m wide, and 1.2 m deep (total volume of 8.2 m^3 when filled to the typical level). The volume of

FIGURE 3-12 Photo of the EPA/DFO wave tank at BIO, Halifax, NS, Canada.
SOURCE: K. Lee, Fisheries and Oceans Canada, Centre for Offshore Oil and Gas Environmental Research.

seawater in this tank is small enough that it can be replaced relatively quickly between tests, reducing concerns about the build-up of dispersant or dispersed oil concentrations between runs. A disadvantage of the small tank volume is that it precludes investigation of hydrodynamic effectiveness (e.g., dilution of the dispersed oil plume by turbulent mixing). The facility has a flow-through capability that will enable it to simulate some aspects of the dilution that can occur in open water, but this capability was included primarily to allow chronic toxicity studies to be conducted under more realistic exposure conditions. A weakness of simulating dilution due to advection and turbulent diffusion under a slick at sea by inducing flow of clean seawater through the tank is that, as described in Chapter 4, the at-sea rate is very scale dependent. Loss of oil to the walls of the tank will be minimized by a bubble curtain, which is created by forcing compressed air through holes in a copper tube that is submerged about 5 to 7.5 cm below the water surface adjacent to the walls of the tank. The effectiveness of this approach, especially at higher wave energies, has not yet been determined. In addition, the effect of turbulence created by the bubble curtain on dispersion effectiveness at low wave energies should be carefully evaluated before this system is used extensively. The design of this tank enables it to produce a wide range of waves, including breaking and nonbreaking waves, at energy levels that are typical of sea surface conditions. The wave generator is capable of producing regular waves of varying period and repeatable breaking waves. Another objective of this facility was to allow measurements that will facilitate mass balance calculations; protocols are being developed for this purpose.

Shoreline Environmental Research Facility The Shoreline Environmental Research Facility (SERF), located near Corpus Christi, Texas, contains nine wave tanks, each of which is 33.5 m long by 2.1 m wide by 2.4 m deep (Figure 3-13; Page et al., 2000a). Each tank is equipped with a computer-controlled wave generator that can produce variable wave patterns and seawater inlets and outlets that allow the user to vary the water level in the tank to simulate tides. The SERF wave tanks have the ability to simulate nearshore environments by constructing sand beaches, including a flat back-beach area just above the high-tide line. The tanks can be operated with a high-tide water depth of 2.0 m and a tidal range of about 0.6 m.

FIGURE 3-13 Aerial view of SERF wave tanks.
SOURCE: J. Bonner, Texas A&M University-Corpus Christi.

The SERF is unique among wave-tank facilities in two ways. First, and most obviously, it is specifically designed for simulating nearshore environments, which may contain high concentrations of suspended solids due to resuspension of the shoreline sediment. Therefore, it can be particularly useful for evaluating the ability of dispersants to prevent oil contamination of shorelines. Second, and more importantly, it includes multiple identical wave tanks so independent replication of treatments is much simpler than in facilities that contain a single tank. Although not unique, the capability for continuous flow of clean seawater through the tank allows dilution of the dispersed oil plume to be considered in the experimental design. Finally, the SERF testing protocols have been developed over a period of several years with the objective of closing oil mass balances. To this end, investigators at the SERF measure oil concentrations in several compartments, including the water surface, the water column, the shoreline sediments, and the tank walls (Bonner et al., 2003). Although it may not be possible to account for 100 percent of the added oil, the measurements required to perform mass balances provide a much more detailed picture of the dispersed oil fate than do measurements of only one compartment.

S.L. Ross Wave Tank A small, indoor wave tank is available at the S.L. Ross facility in Ottawa, Ontario, Canada (Figure 3-14). This tank is 10 m long, 1.2 m wide, and 1.2 m deep; it is usually operated filled with 0.85 m of 32 percent salt water (total volume = 10.2 m^3). A wave-generating paddle is located at one end, and a wave-dissipating beach is at the other. A submerged air diffuser creates a bubble curtain that contains oil within a rectangular region in the tank even in the presence of waves. Dispersant is applied through flat-fan nozzles—similar to those used in full-scale, boat-based dispersant application systems—from an overhead spray boom that is mounted above the center of the tank. The amount of dispersant that is applied is measured by collecting the spray in a tray positioned above the water surface at one edge of the oil containment zone (S.L. Ross, 1997).

SINTEF Flume The SINTEF facility in Norway has an elliptical flume that has been used for oil weathering and dispersion studies. The flume has a circumference of 9 m, with a 4-m long major axis (Figure 3-15). The tank is 0.5 m wide and is operated at a water depth of 0.4 m (total volume of 1.75 m^3). The flume is equipped with a wave generator, submerged pumps that circulate the water around the elliptical track, fans that can simulate surface wind, and a UV lamp for photooxidation studies. This facility has been used primarily for oil weathering studies (Daling et al.,

FIGURE 3-14 S.L. Ross wave tank, Ottawa, Ontario, Canada.
SOURCE: S.L. Ross and MAR Incorporated, 2003; courtesy of S.L. Ross and MAR Incorporated.

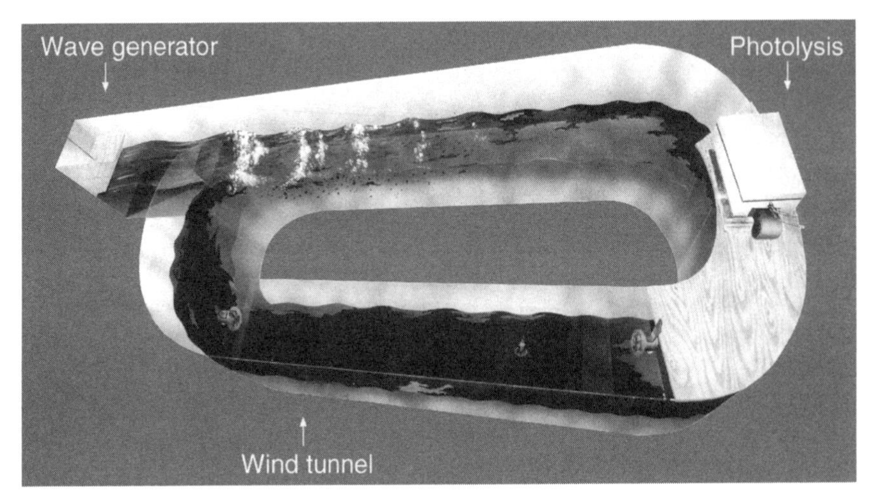

FIGURE 3-15 Schematic diagram of SINTEF hydraulic flume. The flume has a circumference of 9 m, with a 4-meter long major axis.
SOURCE: P. Daling, SINTEF.

1998), but it has been used to study the dispersibility of heavy bunker fuel oil (Fiocco et al., 1999).

The Cedre Polludrome The configuration of the Cedre Polludrome in France is similar to the hydraulic flume at SINTEF, but it is larger (Figure 3-16) (Guyomarch et al., 1999c). The Polludrome flume is 0.6 m wide and is operated with a 1 m water depth (total volume is 10.5 m^3). Like the SINTEF flume, the Polludrome is equipped with an adjustable frequency wave generator, a fan to produce wind across the water surface, pumps for generation of currents, and UV lamps that allow experimenters to simulate photooxidation processes. In addition, the Polludrome is connected to a large storage tank that can be used to pump water into and out of the flume to simulate tides. Finally, the Polludrome has a long straight section that extends beyond the elliptical flume—in line with the wave generator—in which a shoreline can be constructed. The Polludrome has been used for a number of dispersant studies, particularly with higher viscosity oils where multiple dispersant applications can be evaluated (Guyomarch et al., 1999c).

Design of Effectiveness Tests in Wave Tanks

The primary advantage of wave-tank studies over laboratory-scale tests is the ability to investigate some components of operational effectiveness and introduce the energy that drives formation of small oil drop-

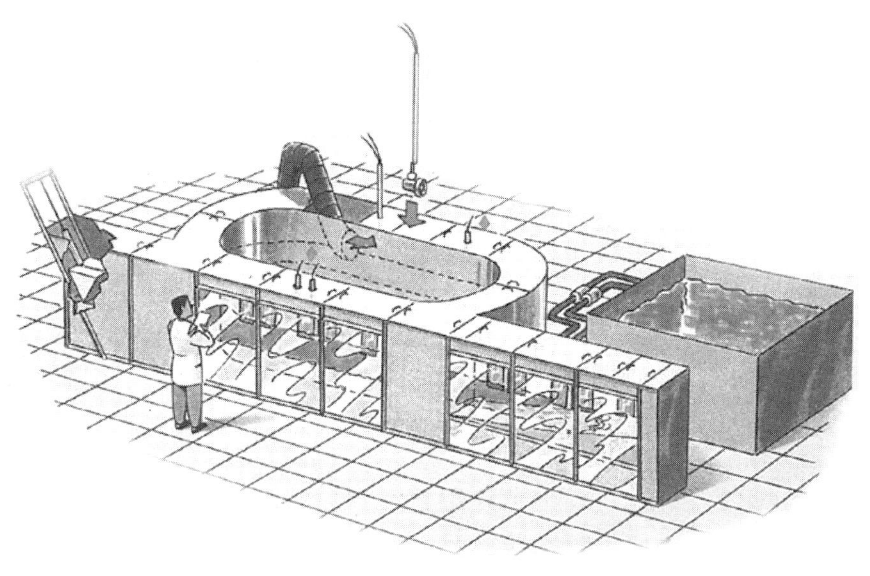

FIGURE 3-16 Schematic diagram of the Cedre Polludrome.
SOURCE: Guyomarch, et al., 1999c; courtesy of J. Guyomarch, Cedre.

lets through a mechanism that is similar to that which occurs at sea (i.e., waves). Whenever possible, the design of mesoscale dispersant effectiveness tests, including hydraulic flumes and wave tanks, should incorporate these factors.

One of the major factors affecting operational effectiveness during spill response operations is the patchy distribution of oil slicks that results from Langmuir circulation and related near-surface transport phenomena, but this is difficult to simulate in wave tanks. A second factor is the interaction of dispersant with floating oil. This requires that the experimental design include dispersant-application equipment that generates realistic droplet-size distributions and impact velocities and that the physical characteristics of the floating oil match those that are expected to exist *in-situ* as closely as possible. Thus, wave-tank tests should use oil that is weathered in a realistic manner, preferably on the water surface in the presence of waves and at a temperature that is representative of the environment of interest. Penetration of the oil by the dispersant may be affected by the viscosity of the oil, especially in the thin film in contact with the oil-air interface, which will depend on the extent to which the oil has evaporated and formed a water-in-oil emulsion. Similarly, the dispersant properties at the time of application to the floating slick should be representative of those that would be expected to prevail during a spill response operation (e.g., if cold-water dispersion is under investigation, the dispersant should be applied at a temperature that is similar to the expected ambient temperature of the application vehicle, because the viscosity of the dispersant will affect the size of the droplets that are formed during spraying; Byford et al., 1983).

Many factors that affect chemical effectiveness have been investigated in laboratory-scale experimental systems, including the effects of water characteristics (e.g., salinity, temperature, and suspended sediment concentrations) and oil composition on dispersion effectiveness. These also can be investigated in wave-tank tests if one has reason to believe that there may be an interaction between these factors and the mechanism through which energy is provided to produce droplets, but at a minimum, they should be controlled or measured. The ability to reproduce the mechanism of droplet formation is one of the main advantages of wave-tank tests over those conducted in the laboratory. Therefore, wave-tank tests should measure and correlate the turbulent energy dissipation rate used to those that occur in the real world. Because waves produced by local wind are expected to be the main source of turbulent energy to disperse the oil in open coastal waters, wave-tank tests should generate waves that are controlled, well characterized, and reproducible (Bonner et al., 2003). Because wave energy in the sea surface varies over a wide range in short time periods (Delvigne and Sweeney, 1988; Agrawal et al., 1992), it is a parameter that should be investigated. In addition, the spatial variation in turbulent shear should be characterized, especially in larger wave tanks, when hydrodynamic effectiveness is under investigation.

If one is interested in investigating oil dispersion in relatively narrow estuaries or rivers, the current and bottom friction will be additional major sources of turbulent energy generation and dissipation. In this case, the wave tank tests should reproduce the expected estuarine and riverine flow fields. The concept of hydraulic radius, instead of water depth, should be used in scaling the flow field of the wave tank to that in estuaries and rivers (Chow, 1988).

In addition to quantifying the energy dissipation rate, the fraction of added oil that becomes entrained in the water column should be measured in wave-tank studies. This is accomplished by either measuring the amount of oil remaining on the water surface after mixing in the presence of dispersant (Brown et al., 1987; Brown and Goodman, 1988; Louchouarn et al., 2000; Belore, 2003; Bonner et al., 2003) and/or by measuring the oil concentration in the water column (Brown et al., 1987; Brown and Goodman, 1988; Bonner et al., 2003). Both of these techniques suffer from limitations. Measurement of surface oil estimates dispersion effectiveness by difference and, therefore, measurement errors that lead to incomplete recovery (including transport to compartments that are not explicitly considered, such as the atmosphere and the walls of the tank) are considered to represent dispersion (Fingas and Ka'aihue, 2004c). These losses are not expected to be the same in dispersed and undispersed oil slicks; so, they cannot be estimated using control treatments. Measurement of oil concentrations in the water column is complicated by the heterogeneous distribution of oil in a chemically dispersed plume, which necessitates collection of a large number of samples with high resolution in space and time. Analysis of oil concentrations with appropriate spatial and temporal resolution requires a method that can provide results in real time, such as *in-situ* fluorometry, but this method should be carefully calibrated and has been criticized as being subject to large systematic errors (Lambert et al., 2001a). Some attempts have been made to close mass balances during dispersant effectiveness tests in wave tanks (Brown et al., 1987; Brown and Goodman, 1988; Bonner et al., 2003), and although none have been completely successful (Fingas and Ka'aihue [2004c] report that oil recovered after wave-tank studies has ranged from about 10 to 100 percent of that added, with recent studies being in the range of 50 to 75 percent), this exercise provides useful information regarding the fate of the oil and the uncertainty in the estimates of dispersion effectiveness. Therefore, mass balances should be attempted in all wave-tank studies of dispersant effectiveness.

In addition to measuring the concentration of dispersed oil, the droplet-size distribution should also be measured. The size and density of the dispersed oil droplets will determine their rise velocity and, therefore, whether they will be stably entrained in the water column under ambient mixing conditions or will eventually float to the surface and reform a floating slick. Efficient chemical dispersion of oil should result in a high concentration of oil droplets with a volume median diameter less than about 50 μm (Byford et al., 1984; Daling et al., 1990a; Lunel, 1995b). Droplets that are larger than this are likely to resurface if the mixing energy is removed or significantly reduced.

Well-designed experiments using wave tanks have an important role in the study and quantification of factors controlling dispersant effectiveness. With more realistic mechanisms for energy input and rigorous measurements, it is hoped that such tests can be used to develop better predictive models of dispersant effectiveness.

Field Studies

Objectives of Field Studies

Historically, one of the major motivations for conducting full-scale sea trials was skepticism of the validity of laboratory and mesoscale tank tests. These smaller-scale tests are frequently criticized for inaccurately simulating the processes that contribute to dispersion of oil slicks at sea. In particular, uncertain or improper scaling of the laboratory systems (e.g., oil-water ratios, mixing energy) relative to conditions at sea and the effects of system boundaries (i.e., wall effects) on the observed effectiveness are commonly identified as detracting from the realism of laboratory systems and wave tanks. A perceived advantage of full-scale field trials is that they are the best representation of reality that can be achieved while maintaining some degree of control over the design of the experiment. Although this control is desirable because it allows the experimenters to limit or, at least, to identify and measure the uncontrolled variables, it also introduces artificiality into the test.

From a more fundamental perspective, the motivation for studying dispersant effectiveness in field studies derives from the opportunity to study phenomena that cannot be addressed at the smaller scale of laboratory and wave-tank systems. These include, for example, greater opportunity to investigate operational effectiveness issues such as the use of real application equipment (e.g., aircraft) to apply dispersants to oil slicks under real conditions (e.g., patchy oil distribution caused by Langmuir circulation and eddies of various sizes) resulting in realistic encounter rates. Similarly, field studies may present the only opportunity to investigate the hydrodynamic effectiveness of chemical dispersion (i.e., dilution of the dispersed oil plume due to horizontal and vertical diffusion resulting from realistic currents and eddies). Both of these processes, however, require relatively large experimental oil spills to be sufficiently realistic. For a variety of practical reasons, most planned field studies involve small quantities of oil (e.g., 20 to 50 tonnes [roughly 5,000 to 13,000 gallons]) relative to what is released during real oil spills (e.g., see the case studies presented in text boxes and the subsection on the Gulf of Mexico dispersant applications in this chapter; only the M/V *Blue Master* spill [16 tonnes] was comparable in size to most field studies; other spills ranged in size from about 320 tonnes [the Poseidon pipeline spill] to 87,000 tonnes [the *Sea Empress*]). Small spills, even when studied under field conditions, will result in operational and hydrodynamic effectiveness that is better than could be achieved under more complex response conditions. That is, the dispersant encounter rates would be too high and dilution of the dispersed plume would be too fast to extrapolate directly for prediction of performance or effects of real oil spills (see discussion of surface transport in Chapter 4 for more details). In addition, although field studies allow dispersant effectiveness to be investigated under realistic conditions, only a few realizations of all possible conditions can be specifically tested due to financial and logistical constraints, and these may not be the conditions of most interest. Instead, the conditions that can be investigated are those that prevail at the time the study is conducted, and investigators have only limited control over what those conditions will be.

Additional justifications for conducting at-sea trials of dispersant effectiveness include that they can provide opportunities to develop and test instrumentation for monitoring dispersion effectiveness (e.g., surface oil thickness and aerial extent, water column concentrations of dispersed oil), they can be used to train spill response personnel, and they can be used to verify dispersed oil fate and transport models. Although all computer models are simplifications of the real world and, therefore, should not be expected to exactly simulate the complex behavior of oil in the environment, the underlying conceptual models should incorporate the major oil transport processes and the mechanisms that govern its fate. Carefully executed field studies can inform these conceptual models by testing the suspected cause-and-effect relationships that control dispersant effectiveness. In addition, field studies can be used to calibrate model parameters by providing measured dispersed-oil concentration distributions for specific well-characterized initial and boundary conditions that can be compared to model output. Furthermore, field studies can be used to validate model output to evaluate the reliability of the model predictions.

Conversely, design of chemical dispersion field studies should be guided by modeling, especially the expected transport of the surface and dispersed oil plumes. For example, models can be used to identify sampling locations and determine the required sampling frequencies. Furthermore, oil concentration predictions will assist in specifying sampling and analytical methods that will be used (e.g., sample size affecting detection limits and the dynamic range of expected concentrations and phase—dissolved vs. oil droplet—of the oil in the collected samples).

Design of Field Studies

A full-scale field trial can be very costly (e.g., potentially in excess of U.S. $500,000). The major costs include permitting during the planning phase (Payne and Allen, in press), mobilizing the vessels, aircraft, and personnel that are required to carry out the study during the field exercise itself, and analytical chemistry costs associated with measuring oil concentrations and fate (dissolved vs. particulate) in field-collected samples. The experiments are usually carried out far from populated areas, which increases the travel time to the site, further increasing time and costs. Once committed, the experiment is at the mercy of the prevailing weather, which will dictate whether any work can take place at all. For example, in a recent field trial conducted in the United Kingdom, experiments could not be performed on the first scheduled day due to excessively high wind speeds, which made the small boat operations that were required for monitoring unsafe, whereas on the second day, the experiment was delayed due to insufficient wind speed, such that poor dispersion of the target heavy fuel oils (IFO 180 and IFO 380) was expected (Lewis, 2004). Weather contingencies of this sort can cause huge costs overruns.

In addition to the high costs of field studies, permitting and legal issues can be major impediments to conducting field studies, and they are a major reason that no field studies have been conducted in U.S. waters since 1979. Although there are published guidelines for obtaining an EPA permit for planned spill experiments (EPA, 2001), those requirements can be quite onerous and include a requirement that all parties assume financial and legal responsibility for any unintended consequences of the study. In addition to the EPA, approval may be required from several other federal, state, and county agencies (Payne and Allen, 2004; in press). In some cases, it can be difficult to identify relevant permitting requirements and obtain approval from these agencies.

Although they are often considered to be the best representation of "reality," field trials are also subject to limitations. A major limitation is that a very limited data set can be obtained from any one trial. As such, the objectives should be clearly defined and reliable procedures should be established to ensure that the required results are achieved. Poor experimental design will produce results that are difficult to interpret unambiguously.

Design of field studies should involve principles similar to those used in the design of any experiment. In particular, a primary objective should be to obtain an unbiased estimate of the variation that exists between two experimental units (i.e., oil slicks) that are treated identically to allow evaluation of the statistical significance of any differences that are observed between experimental units that are subjected to different treatments (e.g., dispersant-treated slicks versus untreated control slicks). This objective requires treatments to be independently replicated and randomly distributed or interspersed throughout the experimental domain (Box et al., 1978; Hurlbert, 1984; Montgomery, 1997; Ruxton and Colegrave, 2003). Independent replication of treatments requires that they be conducted in independent experimental units. It is not sufficient to collect multiple samples from the same experimental unit; this "pseudoreplication" only serves to characterize the spatial and/or temporal heterogeneity of the experimental unit, it cannot characterize the degree of variation that exists between experimental units independently of treatment (Hurlbert, 1984; Ruxton and Colegrave, 2003). Although alternative designs (e.g., BACI—before-after-control-impact designs) may be appropriate when replication is not possible, such as studies involving spills of opportunity, these are not appropriate for planned field studies. Even when BACI-type experimental designs are required, modifications that include independent replication of controls are recommended (Underwood, 1994). When field studies are conducted over several days, the experimental domain should be interpreted to include both spatial and temporal dimensions. Therefore, independent replicates for specific treatments should be performed on different days, which will presumably sample a range of weather conditions.

In reality, financial and technical constraints limit the degree to which the design of field studies can comply with these principles of sound experimental design. Financial constraints limit the scope and duration of field studies and, consequently, the ability to independently replicate treatments and intersperse the replicates over space and time. This is further complicated by the vagaries of weather, which may increase the within-treatment variance of properly interspersed replicates and obscure the ability to detect statistically significant between-treatment differences. Paired experimental designs may be useful in reducing the effects of weather, but the apparently nonlinear response of dispersion effectiveness to mixing energy (Fingas et al., 1994; Fingas et al., 1996a) may reduce the benefit of this approach, because the difference between the dispersant treatment and the control might vary with energy level.

In addition to the inability to control weather and, therefore, to set mixing energy as an independent variable, field studies are subject to an additional important technical limitation: the inability to quantitatively measure effectiveness for use as an endpoint in statistical comparisons of treatments. Dispersant effectiveness in sea trials has been monitored by measuring surface oil and dispersed-oil concentrations in the water column, but neither method produces satisfactory results. Surface oil is commonly monitored remotely using aircraft-mounted side-looking airborne radar (SLAR) and ultraviolet (UV) and infrared (IR) line scanners. SLAR and UV scanners are not sensitive to oil thickness, and although IR scanners can distinguish between thick and thin oil slicks, they cannot measure the oil thickness (Goodman and Fingas, 1988; Lewis et al., 1995a,b; Lewis and Aurand, 1997). As a result of these limitations, the volume of surface oil cannot be measured, and therefore, the effectiveness of dispersion cannot be quantified using these methods. In fact, treatment of floating oil with a chemical dispersant may cause the thick part of slick, which can be detected by IR scanning, to increase in area due to decreased oil-water interfacial tension (Goodman and MacNeill, 1984; Goodman and Fingas, 1988). More advanced remote-sensing technologies, such as microwave radiometry (Schroh, 1995) and laser-ultrasonic detection (Choquet et al., 1993; Brown et al., 2000), can be used to estimate the volume of floating oil, but interpretation of these data is sometimes difficult (Lewis and Aurand, 1997). For example, successful use of the laser-ultrasonic technique, known as Laser Ultrasonic Remote Sensing of Oil Thickness (LURSOT), from an aircraft has not yet been reported (Brown et al., 2000).

Measurement of dispersed oil in the water column can be more difficult. Early studies attempted to measure dispersed oil concentrations by collecting grab samples at various locations and times following application of a chemical dispersant, but the heterogeneity of dispersed oil distribution is too great to obtain meaningful results with the limited number of grab samples that can be collected and analyzed (Brown et al., 1987). Real-time measurement of dispersed oil concentrations using continuous fluorometry allows collection of a more dense data set that improves the ability to characterize the concentration distribution of dispersed oil along a transect through the slick, but these measurements are still limited by the inability to collect samples simultaneously at multiple positions within the dispersed oil plume (Lunel, 1995a; Lewis, 2004). Furthermore, careful calibration of fluorometers is necessary to obtain quantitatively useful results (Lambert et al., 2001a). Although calibration methods vary greatly among investigators, the most reliable method appears to be collection of water samples directly from the fluorometer effluent. Even when calibrated appropriately, however, *in-situ* fluorometry is subject to interferences that can affect its quantitative reliability. Because the nature of the dispersion and dilution processes results in a dispersed oil plume that is heterogeneous in space and time, unambiguous quantitative interpretation of dispersed oil concentrations for the purpose of estimating mass balances is difficult. In this respect, continuous release of oil from a fixed point into a current may be a more effective experimental design (Lunel, 1994b, 1995a), but this design lacks many of the elements of realism that are sought by field studies. For example, operational effectiveness will be unrealistically high due to application of dispersant from a boom mounted close to the oil discharge position, and the short time period between oil discharge and dispersant application allows for no weathering and limited spreading of the slick. Also, the hydrodynamic effectiveness will be artificially high, because this experimental design lays down a very narrow (initially 1-m wide) oil slick. Discharge of the oil into a current is also likely to increase the hydrodynamic effectiveness by increasing the horizontal turbulent diffusion coefficients. So, although this experimental design is useful for many purposes, it will almost certainly overestimate the effectiveness of dispersion relative to a real spill response operation.

The inability to make measurements that are adequately quantitative in sea trials has led some investigators to rely solely on visual observation (Lewis, 2004), which is purely qualitative and very sensitive to viewing conditions (e.g., position of sun relative to viewer, cloud cover, viewing angle). The reliability of data collected by visual observation would be improved by using "blind" observation techniques, in which the observers are not informed of the treatment that is applied to experimental slicks. This would require treatment of control slicks with formulations that contain the dispersant solvents but lack the surfactants. In addition, the observers must be extremely careful to avoid interacting with other observers when making observations to reduce the potential for nonindependence of the observations. Visual observation is also a central component of the Specialized Monitoring of Advanced Response Technologies (SMART) protocols for monitoring the performance of dispersants in oil-spill response operations. Although qualitative data may provide anecdotal evidence for dispersant effectiveness, which may be suitable for some purposes, it cannot be used to validate fate and transport models.

Review of Past Field Studies

A number of controlled field trials of dispersant effectiveness have been conducted in Canada and Europe since the 1989 NRC review (McDonagh and Colcomb-Heiliger, 1992; Lunel and Lewis 1993a,b; Brandvik et al., 1995, 1996; Walker and Lunel 1995; Lunel 1993, 1994a,b, 1995a,b; Lewis et al., 1995a,b, 1998a,b; Lunel et al., 1995a,b,c; Strom-Kristiansen et al., 1995; Walker and Lunel, 1995; Lunel and Davies, 1996; Fiocco et al., 1999). Many of these studies have been reviewed and summarized (S.L. Ross, 1997; Fingas and Ka'aihue, 2004b), and the proceedings of a two-day symposium on oil-spill dispersant applications in Alaska are also available (Trudel, 1998).

No attempt will be made to duplicate or even briefly cover the findings presented in these documents. Instead, several of the most significant lessons learned—specifically with regard to applications and dispersant-treated oil behavior—will be briefly highlighted in the following paragraphs.

It is now known that oil spills are composed of thick slicks (usually thicker than 1 mm) that contain most of the oil volume (the rule-of-thumb is that 90 percent of the oil volume is contained in 10 percent of the area), and that these patches are surrounded by thinner sheens (about 1 to 10 μm or 0.001 to 0.01 mm) (S.L. Ross, 1997). This combined thick and thin slick spreading is of great importance with regard to dispersant effectiveness. From field trials and actual dispersant treatment of accidental oil spills, it is now generally accepted that the one pass concept for dispersant application is not appropriate for dealing with the thicker part of spills, and that the multi-pass approach (as has always been used in United Kingdom) is the only way to completely dose the thicker portions of marine spills (Lunel et al., 1997b).

Daling and Lichtenthaler (1987) compared the results of laboratory effectiveness tests with the results from several small field trials. They showed that the correlation between effectiveness measured using the three different laboratory test systems and between field and laboratory tests was poor. There was, however, fairly good correlation between the mean results for the different dispersants from the three laboratory tests and field tests. That is, dispersants that performed poorly in the laboratory also performed poorly in the field, but the lab tests were not able to predict the dispersibility of a specific oil by a specific dispersant under defined conditions at sea with any satisfactory level of accuracy. The results of more recent comparisons of laboratory effectiveness data and field trials are shown in Table 3-3, which demonstrate that the field effectiveness was generally lower than values obtained in the laboratory (Fingas and Ka'aihue, 2004b). The higher effectiveness in laboratory studies may indicate that the energy levels were higher in the laboratory tests than in the field studies, which is contrary to what was thought in previous years (Lunel, 1994a).

Based on the monitoring results from field studies and actual spills, it can be concluded that it is difficult to estimate average concentrations under treated slicks because of the significant heterogeneity both horizontally and with depth into the water column (Brandvik et al., 1995; Lewis et al., 1998b). Figure 3-17 shows the horizontal and vertical distribution of total petroleum hydrocarbons from small test spills as determined by UV/fluorescence before and after dispersant treatment. Before treatment, the maximum concentration in the surface waters (<0.5 m) was less than 1 ppm, but during treatment, this increased to nearly 6 ppm

TABLE 3-3 Comparison of Laboratory and Field Dispersant Effectiveness Results (from Fingas and Ka'aihue, 2004b)

			Effectiveness Results in Percent					
Oil type	Dispersant	Field Test	SF GC	SF CA	IFP	WSL Lab 1	WSL Lab 2	Exdet
Medium fuel oil	Corexit 9527	26	54	50	91	42	42	67
Medium fuel oil	Slikgone NS	17	49	46	94	29	23	50
Medium fuel oil	LA 1834/Sur	4	2	2	50	16	11	38
Forties crude	Slickgone NS	16	47	65	95	28	25	60
Forties crude	LA 1834/Sur	5	2	61	61	15	12	53
Correlation with field test (R^2)			0.89	0.7	0.54	0.87	0.94	0.41
Ratio lab test/field test			0.4	0.35	0.19	0.56	0.62	0.27

NOTE: SF = Swirling Flask, GC = Analysis by Gas Chromatography, CA = Colorimetric Analysis, IFP = French Institute for Petroleum Test, WSL = Warren Springs Laboratory Test.

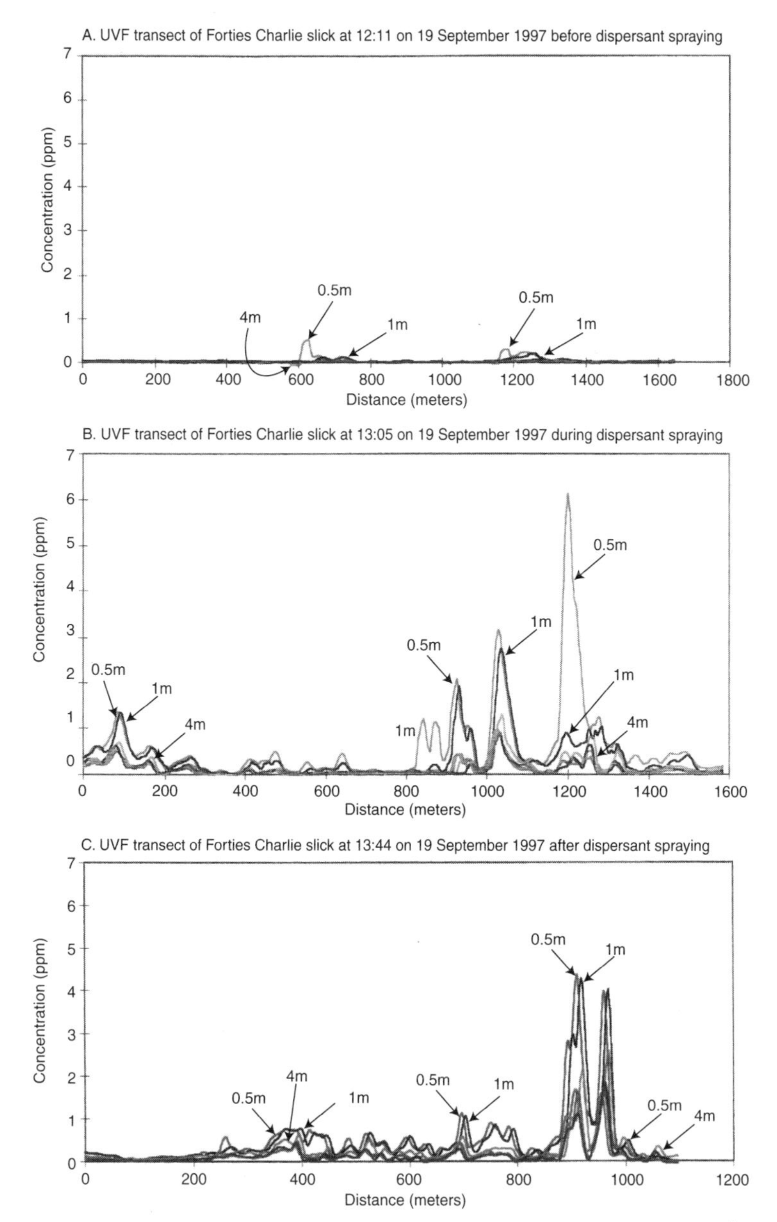

FIGURE 3-17 Dispersed oil concentrations under an approximately 27 m^3 surface slick of Forties crude oil (a) before, (b) during, and (c) after spraying with 2,250 liters of Corexit 9500 during the 1997 North Sea field trials. Sampling depths for the major peaks in dispersed oil concentrations are labeled, and in many, but not all cases, the 0.5 and 1.0 m depths were very similar.
SOURCE: Modified from A. Lewis, et al., 1998b; courtesy of AEA Technology.

with lesser concentrations at depth. After approximately 45 minutes, concentrations at depth also increased, but generally to only 1–2 ppm.

The heterogeneity of the dispersed oil plume makes it difficult to obtain reliable estimates of the mass of dispersed oil beneath a treated slick in field studies, which is essential to achieving closure of a mass balance. Fingas and Ka'aihue (2004b) consider estimation of a mass balance to be among the most important factors in obtaining reliable effectiveness estimates from field studies, but they conclude that no field trials have achieved mass balance closure, which is difficult even in more controlled tank tests where up to 70 percent of the oil may be missing from the final mass balance. Other factors considered by these authors to be critical for obtaining reliable results from field studies include the use of proper con-

BOX 3-3
Case Study: North Sea Trials

Spilled Oil Type/Volume/Conditions: A series of experimental spills in the North Sea were conducted in 1990, 1993, 1994, 1995, and 1997 for various dispersant applications. Test oils included Forties Blend crude oil, Troll crude oil, Alaska North Slope crude oil, IFO-180, and a 50:50 blend of medium fuel oil and gas oil (MFO+GO) with an API gravity of 22. Spill volume was typically 15–50 m^3.

Physical and Biological Setting: Open-water setting in water depths greater than 90 m. Tests usually conducted in the summer, with water temperatures of 15° C (roughly 59° F), winds 5–10 m/s.

Dispersant Application:

1990 Test—Objective was to have a steady-state oil discharge so that replicate measurements could be made of the dispersed oil concentrations under the treated slicks to better quantify dispersant effectiveness. Four continuous releases of 50 liters per minute of MFO+GO, with dispersant application 12–15 minutes later at a dispersant:oil ratio of 1:20, using OSR-5, Slickgone NS, and 1100X, and no dispersant as a control. Winds were up to 14 m/s.

1993 (May) Test—Objective was to determine dispersant effectiveness on a medium fuel oil with low concentrations of light ends, so evaporation would not be significant. Winds were 11–22 miles per hour (roughly 17–35 kilometers per hour). Single releases of 20 m^3 of MFO+GO, with application of 2 tonnes (roughly 588 gallons) of Dasic Slickgone NS 1.5 hours after release at a dispersant:oil ratio of 1:10 during 10 spray runs. A second application of 2 tonnes was conducted in the afternoon on the remaining oil. There was a similar untreated oil release as a control.

trol slicks (i.e., not treated with dispersant), the need for remote sensing, and the use of proper analytical procedures in the field (e.g., fluorometer calibration) and during laboratory analysis of samples collected in the field. In their survey of field trials completed before 1990, the effectiveness estimates averaged around 33 percent, but inclusion of more recent trials decreases the average effectiveness to only 16 percent (Fingas and Ka'aihue, 2004b). This is somewhat surprising, because better experimental designs and methods were used in the more recent tests, but the effectiveness of recent tests may have been reduced by the use of heavier oils.

A large number of at-sea trials were conducted between 1993 and 1994 (see Box 3-3 on the North Sea trials), and Table 3-4 presents a summary of the dispersant efficiency data for different oils tested and the different

1995 (August) Test—Objective was to test effectiveness of two application methods on an crude oil: helicopter bucket versus boat-mounted spray arm; and to calibrate aerial remote sensing sensors with ground truth data on the surface oil slicks. Single releases of 15 m^3 of topped Troll crude oil with two applications of Corexit 9500 at a final dispersant:oil ratio of 1:20 on the thick oil slicks 2 hours post release. Winds were roughly 17–25 kilometers per hour.

1997 (September) Test—Objectives were test the degree to which multiple applications of dispersants (Corexit 9500 and Dasic Slickgone NS) can disperse high viscosity emulsions formed after 1–2 days at sea and the dispersibility of heavy fuel oils.

Monitoring Results:

Effectiveness: In all tests, remote sensing aircraft equipped with SLAR, video, ultraviolet, and infrared cameras were used to track the behavior of surface slicks. Oil concentrations in the water column were monitored using field fluorometers towed through the slicks at multiple transects at different depths and distances downcurrent. The field fluorometers were calibrated using discrete samples.

1990 Results—For slicks of a medium fuel oil treated with OSR-5 and Slickgone NS (Type III dispersants), the slick length was reduced compared to the control slick. Using the multiple transect data, it was estimated that, 12–15 minutes after dispersant application, OSR-5 dispersed 21–42 percent of the oil, Slickgone NS dispersed 11–27 percent, 1100x dispersed 6–17 percent, and there was 0–3 percent natural dispersion.

1993 Results—For slicks of a medium fuel oil treated with Dasic Slickgone NS, the surface area of the slick initially increased in area due to spreading by 5 hours after treatment, then by 9 hours after treatment reduced in surface area, compared to the control slick. The first dispersant application had a temporary effect of lowering the water content and viscosity of the surface oil, however, within 4 hours, the water content and viscosity of the treated oil was the same as the control slick. The oil concentrations under the slick prior to dispersant application were about 1 ppm and only extended down to 1 m. After dispersant application, oil concentrations were typically 1–10 ppm down to 5 m, with a maximum concentration of 25 ppm. There was a 16-fold increase in the volume of dispersed oil under the treated slick compared to the control.

1995 Results—For a slick of stabilized Troll crude oil treated with Corexit 9500 2 hours after release, no thick oil patches were observed 15 minutes (for helicopter application) and 30 minutes (for boat application) after dispersant application. The oil increased from 10 percent water content immediately after release to 20–70 percent water content at the time of dispersant application two hours later. The oil viscosity increased from 20 cSt to 100–1,000 cSt after 2 hours, and 3,000 cSt (control slick) at 6 hours after release. After about 1.5 days, the control slick reached 7,000 cSt, the vis-

TABLE 3-4 Summary of Dispersant Efficacy Data from 1993–1994 Sea Trials

Energy Regime	Wind Speed (m/s)	Date	Oil Dispersant	Percent Dispersed (mean)	Standard Deviation
Low	3	7/9/93	MFO	0.8	0.7
Low	5	8/19/94	MFO-Slickgone NS	8	4
High	10	7/9/93	MFO	2	0.7
High	7	8/22/94	MFO	4	2
High	7	8/25/94	Forties	5	3
High	10	7/9/93	MFO-1100X	10	4
High	10	7/9/93	MFO-Slickgone NS	17	6
High	6	8/23/94	MFO-Slickgone NS	16	7
High	6	8/25/94	Forties-Slickgone NS	16	6
High	7	8/22/94	MFO-Corexit 9527	26	10
High	10	7/9/93	MFO-OSR5	30	7

	Percent Dispersed		
Energy Regime	MFO	MFO-Slickgone NS	Ratio of Chemical Dispersion to Natural Dispersion
Low	0.8	8	10
High	3	17	6

SOURCE: Data from Lunel et al., 1995b and S.L. Ross, 1997.

cosity limit for dispersibility of Troll crude oil. Oil concentrations in the water under the control slick were 0.05–0.3 ppm, versus those under the helicopter-treated slick which were 10–20 ppm after 0.5 hours, and 1–3 ppm after 1.5 hours. Under the treated slicks, the oil concentrations were uniform to 8 m depth.

1997 Results—Emulsified crude was rapidly dispersed after two days at sea, even though the Forties crude had a viscosity of 4,000–4,500 cP and the Alaska North Slope crude had a viscosity of 15,000–20,000 cP. Corexit 9500 partially dispersed the IFO-180 after 4 hours at sea, though there was little effect on the heavy fuel oil when the viscosity exceeded 20,000 cP. These field experiments demonstrated that emulsified oils with viscosities up to 20,000 cP can be effectively dispersed, extending the window of opportunity for dispersant use.

SOURCE: Summarized from Brandvik et al. (1996), Lunel (1995a), Lunel et al. (1997b), and Lewis et al. (1998a). *Effects:* Not assessed.

energy regimes encountered. There is a clear ranking in percentage of oil that the different formulations successfully dispersed into the water column in the field as the encountered energy regime increased; however, it should be noted that the overall percent dispersed values were relatively low. Although this ranking had been well documented for laboratory tests, these data were the first set from field trials where the ranking could be quantified. The tested dispersants increased the rate of dispersion by 6 to 10-fold compared with natural dispersion in the case of a medium fuel oil and 3-fold in the case of Forties crude oil. Comparison of the dispersion data for the low-energy regimes (0–5 m/s wind speed [roughly 0–10 knots]) with the higher-energy regimes (6–10 m/s wind speed [roughly 12–20 knots; 22–37 kilometers per hour; 13–22 miles per hour]) shows that natural entrainment is enhanced through the use of dispersants by about the same factor in low-energy regimes (10-fold) and in higher-energy regimes (6 to 10-fold).

Dispersion effectiveness measured in the 1997 AEA North Sea field trials (Box 3-3), which used Forties blend and Alaska North Slope crude oil weathered on the water surface for 45 and 55 hours, respectively, was much greater than the effectiveness that was observed in the 1993–1994 studies that were described above. In these studies, the naturally formed water-in-oil emulsions were completely dispersed (Lewis et al., 1998a,b; Fiocco et al. 1999). The naturally emulsified Forties oil had a viscosity of 4,000–4,500 cP with a water content of 50 percent by volume, whereas the emulsified ANS oil had a viscosity in the range of 15,000–20,000 cP and a water content of 20–30 percent by volume. The emulsified Forties oil was rapidly and totally dispersed after multiple aerial applications of either Corexit 9500 or Dasic Slickgone NS. The rate of dispersion of the emulsified ANS resulting from four aerial applications of Corexit 9500 to the thicker oil patches appeared to be slightly slower than that observed for the weathered Forties oil, but it was also totally dispersed. More variable and less effective dispersion was observed after aerial application of Corexit 9500 to 4- to 23-hour weathered and partially emulsified IFO-180 fuel oil with viscosities ranging from 5,000 to 12,000 cP and water content ranging from 20 to 30 percent. Emulsified IFO-180 fuel oil with a viscosity above 20,000–30,000 cP was not dispersible to any degree.

One of the direct charges to the committee was to address "how laboratory and mesoscale experiments could inform potential controlled field trials and what experimental methods are most appropriate for such tests." The body of work completed to date has provided important, but still limited understanding of many aspects of the efficacy of dispersants in the field and behavior and toxicity of dispersed oil. Developing a robust understanding of these key processes and mechanisms to support decisionmaking in nearshore environments will require taking dispersant research to the next level. Many factors will need to be systematically varied in settings where accurate measurements can be taken. It is difficult to envision the proper role of field testing in a research area that has yet to reach consensus on standard protocols for mesocosm testing. The greater complexities (and costs) of carrying out meaningful field experiments suggest that more effort be placed, at least initially, on designing and implementing a thorough and well-coordinated bench-scale and mesocosm research program. Such work should lead to more robust information about many aspects of dispersed oil behavior and effects. When coupled with information gleaned through more vigorous monitoring of actual spills (regardless of whether dispersants are used effectively in response), this experimental work should provide far greater understanding than is currently available. Upon completion of the work discussed below, the value of further field-scale experiments may become obvious. In any case, such field-scale work would certainly be better and more effectively designed and executed than is currently possible. **Future field-scale work, if deemed necessary, should be based on the systematic and coordinated bench-scale and wave-tank testing recommended in this report.**

Effectiveness Testing Using Spills of Opportunity

In the arena of public opinion, no test can hope to have the positive impact of an actual success in using dispersants during a real spill. There are several areas around the country where the volume of crude oil traffic is so large that small spills are somewhat common. In these areas, it might be possible to develop a plan for using dispersants as a first-strike tool to respond to a small spill that would ordinarily be cleaned up mechanically. Resistance from the responsible party due to concerns about additional liability that could result from curtailment of mechanical response activities in the area of the dispersant field study may make it difficult to obtain authorization for this type of research.

The principles of field study design as discussed above should be considered with regard to the additional limitations imposed by spills of opportunity, and many of these have been incorporated into draft Spill of Opportunity Contingency Study Plans that have been or are being prepared for several RRTs in different parts of the country. The most formalized of these documents is the *Texas General Land Office "Spill of Opportunity" Dispersant Demonstration Project Description* (Aurand et al., 2004). The primary objectives are to evaluate the operational efficiency of dispersant application and monitoring under realistic spill-response conditions, assess the fate of the dispersed oil plume, and evaluate the interaction of the dispersed oil plume with sediments in shallow estuarine waters. Should it be approved and incorporated into a spill response plan (including identification and pre-placement of sampling equipment, and stand-by contracts for personnel), it will provide considerable advantage in marshaling all the components necessary to adequately sample and monitor the results from dispersant applications when and if they occur in the designated areas. Following the lead in Texas, industry and federal agencies should develop and implement detailed plans (including preposition of sufficient equipment and human resources) for rapid deployment of a well-designed monitoring plan for actual dispersant applications throughout the United States.

There were two instances in U.S. waters over the last 17 years where ad hoc spill-of-opportunity studies were conducted during real spill events. During the September 1987 *Pac Baroness* oil spill off Point Conception, California, the effectiveness of treatment of a 100 meter by 700 meter portion of the slick with 41 gallons of Corexit 9527 by helicopter was documented (Payne et al., 1991c). Continuous subsurface UV fluorescence measurements and grab samples of water from beneath the slick were also obtained from a support vessel before and after dispersant application. Unfortunately, the results of the tests were equivocal because the slick was very thin in the treated area and only a small portion of the slick was treated. In addition, 15- to 20-knot (roughly 27 to 37 kilometers per hour) crosswinds caused significant breakup and dispersion of the surface slick in both treated and untreated control areas. SLAR data did not show definitively that any changes occurred because the resolution of the technique from the observation altitude 5,000 ft (roughly 1,500 m) was not sufficient to observe changes in the small treated area. The aerial UV scans suggested that changes occurred in the treated slick, but the *in-situ* UV fluorescence measurements and subsequent chemical analyses did not indicate that significant dispersant-enhanced entrainment occurred.

The lessons learned from this study led to development of detailed plans for investigating dispersant effectiveness at future spills of opportunity. The recommendations included:

- Detailed plans for different coastal regions should be prepared in advance.
- During execution of the plan, target areas for dispersant application should be identified by smoke bombs and surface buoys or drogues.
- The dispersant should be applied into the wind to minimize drift away from the target area, and two surface vessels should be used in addition to a helicopter observation platform for documentation of dispersant application effectiveness. If possible, the vessels should be perpendicular to and along the dispersant flight line to document dispersant drift away from the target area.
- Both videotape and 35 millimeter (or digital) photography should be used to document the experiment.
- Water-column oil concentrations should be measured using *in-situ* UV fluorescence and chemical analyses should be completed on grab samples collected from the output of the fluorometer as well as more traditional water sampling equipment at different depths.
- For large areas, remotely monitor the slick using SLAR at 5,000 to 7,000 ft (roughly 1,500 to 2,100 m) (which is useful under all weather conditions) and IR/UV at 400 ft (roughly 120 m) (effective only in clear weather). Other remote sensing techniques may also be more appropriate.

Utilizing the lessons learned from the *Pac Baroness* study, additional spill-of-opportunity dispersant trials were undertaken at the *Mega Borg* fire and oil spill off Galveston, Texas, in 1990 (Kennicutt et al., 1991; Payne et al., 1993). The ship's cargo was a light Angola Planca crude oil (API gravity = 38; viscosity = 4.58 cSt at 30° C [roughly 86° F]). Dispersant effectiveness was monitored by concurrent observations from the command control aircraft and dispersed oil concentrations were monitored using UV fluorescence continuously at a depth of 4 m along transects through the slick and a discrete water sampling program.

The distribution of dispersed oil droplets was very heterogeneous and reflected the patchy distribution of oil on the water surface before dispersant application. Maximum concentrations of dispersed hydrocarbons in the center of the treated zone were 22 mg/L for total aliphatics (primarily dispersed droplets) and 5.4 μg/L for total aromatics 60 to 90 minutes after dispersant application. Elevated levels were generally limited to the upper 1–3 m of the water column. Concentrations in the upper 1–3 m of the untreated control zones were significantly lower (1.2–3.9 mg/L and 0.8–1.7 μg/L for total aliphatic and aromatic hydrocarbons, respectively). The dispersed aliphatic hydrocarbon concentrations at a depth of 9 m in the treated and control areas were similar (2.5–2.7 mg/L), suggesting that they represented a background, steady-state concentration of very fine, physically dispersed oil droplets that were formed by natural dispersion of the slick during the six days before the dispersant tests began. The ratio of the concentrations of aliphatic to aromatic hydrocarbons showed no evidence of significantly enhanced dissolution of lower- and intermediate-molecular-weight aromatics as a result of chemical dispersion. If such dissolution had occurred, however, it is possible that the dissolved-phase polynuclear aromatic hydrocarbons (PAH) were lost to evaporation directly from the 29° C (roughly 102° F) seawater in the upper mixed layer before the water samples were collected.

One of the major disadvantages identified in both of these ad hoc spill-of-opportunity studies was that many of the resources (boats, aircraft, response personnel, etc.) necessary to assist with the execution of the program were tied up with response activities. Also, radio communications among all the operating platforms (observation aircraft, directional aircraft, dispersant application aircraft, sampling and observation boats, and Unified Command personnel) were difficult at best, and often nonexistent during the field operations. Finally, both spill-of-opportunity studies were relatively far from land (16–25 miles [roughly 25–40 kilometers]) and refueling of the observation/command control aircraft coordinating the dispersant trials was problematic.

As with the planned trials discussed earlier, the measured subsurface oil concentrations were extremely patchy, and there was no way to integrate or average the concentrations over time and space to even begin to approach a percent dispersed oil calculation. Finally, during spill-of-opportunity studies, the oils may not be amenable to chemical dispersion, or in the case of the *Mega Borg*, the oil may be so light, that it disperses naturally, making comparisons of treated vs. untreated areas tenuous at best.

From the early API tests in 1975 and 1979 to the most recent field trials and measurements completed in 1997, only one well-documented spill in which modern dispersants have been used has been studied in an efficient and controlled manner (Lunel et al., 1997a; Lunel, 1998). That was the *Sea Empress* oil spill in Milford Haven, UK, in 1996 (see case study in Box 3-4).

In future spill-of-opportunity tests, it is recommended that both dissolved-phase and particulate oil droplets be sampled (Payne et al., 1999; Payne and Driskell, 2003) so that measured concentration data can be used to validate computer-model predictions of these phases and so that the

BOX 3-4
Case Study: *Sea Empress,* Milford Haven, Wales, United Kingdom

Spilled Oil Type/Volume/Conditions: T/V *Sea Empress* grounded outside of Milford Haven on 15 February 1996, releasing a total of 72,000 tons (roughly 23 million gallons) of Forties Blend crude oil over a period of six days. Forties Blend is a relatively light crude oil, with an API gravity of 40°, viscosity of 3.88 cSt at 20°C, and pour point of –3° C. It readily forms emulsions of up to 60–70 percent water.

Physical and Biological Setting: The outer coast consists of exposed steep cliffs with pocket beaches of sand and gravel. Water depths are greater than 20 m at 1 km offshore. Milford Haven is a sheltered bay with extensive tidal flats and marshes as well as beaches and rocky shores. There were sustained winds of 15–40 knots (roughly 27–74 kilometers per hour) throughout the spill. Offshore islands are important seabird sanctuaries with internationally important populations of puffins, guillemots, gannets, and Manx shearwaters. The area includes one of three marine nature reserves in the UK, two nature preserves, and 35 sites of Special Scientific Interest. It has popular tourist beaches and a local fishing industry.

Dispersant Application: A total of 445 tonnes of seven different dispersant products were applied over seven days, mostly by DC3 spray aircraft. Spotter planes were used to direct the spraying to the thickest parts of the fresh oil releases each day. The first test application conducted 18 hours after the initial release of 2,000 tonnes was not effective, based on visual observations. The reason for the delay in application was that the initial oil release was carried by the ingoing tide into the Milford Haven which was an area where dispersant use was not approved. Oil moving back out of Milford Haven on the subsequent outgoing tide had weathered and was not amenable to dispersant by the time it was targeted. An oil-weathering model was used to predict that, after 12–18 hours, 40 percent of the oil would have evaporated and a 70 percent water-in-oil emulsion would have formed. A second test using both dispersant and a demulsifier 36 hours after the initial release was partially effective. On day 4, there was another release of 2,000 tonnes, where dispersant application on the fresh oil was determined to be highly effective, thus full-scale dispersant application was approved. There were continued daily releases of 5,000–20,000 tonnes of oil during a three-hour period of the falling tide. These slicks formed very discrete targets with limited spread of the oil. Each day, the thickest parts of the fresh oil slicks were repeatedly sprayed until they had been dispersed, then larger patches of more weathered oil offshore were sprayed. The last oil release occurred on day 7, and dispersant applications were terminated on day 8 when it was determined that they were no longer effective on the emulsified oil. Dispersant applications were generally restricted to beyond

1 km of the shoreline, to meet the requirement of a minimum of 20 m water depth, though there were exceptions, particularly on one occasion when a patch of oil that had migrated north of the entrance to Milford Haven returned and there was a threat of it being carried in by the tidal movement, The decision to spray the target was made as the consequence of not doing so was not an acceptable trade-off.

Monitoring Results:

Effectiveness: There was an extensive monitoring program to document the effectiveness of each dispersant application consisting of visual observations from spotter aircraft, SLAR imagery, visual observations from boats, and measurements of oil concentrations in the water column using field fluorometry. The dispersants were most effective on the fresh oil releases, as indicated by plumes of dispersed oil in the water and large reductions in surface slicks. It was not possible to determine the relative effectiveness of the different dispersant products.

Fluorometry measurements through the water column showed some natural dispersion, with oil concentrations of 3 ppm at 1 m, but less than 0.5 ppm at 4–5 m, indicating the formation of relatively large droplets during natural dispersion that remained in the upper water column (Lunel et al., 1997a). Oil concentrations under treated slicks were typically also 3 ppm but uniformly mixed down to 5 m, indicating the formation of smaller dispersed droplets that were vertically mixed under the strong wind conditions. Fluorometry and visual observations from boats were used to document that dispersant application on emulsified oil did increase the oil concentrations and depth of oil mixing into the water column. The first dispersant application appeared to break the emulsion, whereas subsequent applications increased the concentrations of dispersed oil into the water (Lunel et al., 1997a).

Even with extensive monitoring, it was difficult to determine dispersant effectiveness. A mass-balance approach was used, as follows (Lunel et al., 1997a): (1) 40 percent was estimated to be lost by evaporation, based on a calibrated oil weathering model; (2) 3 percent was recovered at sea; (3) 7 percent stranded on the shoreline; and (4) the remaining 50 percent was assumed to have been dispersed. Experience and modeling was used to estimate that 10 percent of the oil would have naturally dispersed under the spill conditions (Lunel et al., 1997a). Thus, it was estimated that 40 percent of the oil (about 29,000 tonnes) was chemically dispersed. The dispersant-to-oil ratio was calculated to be 1:65, based on use of 445 tonnes of dispersants and the chemical entrainment of 29,000 tonnes of oil.

Effects: The spill impacted 6,900 birds (mostly migrating scoters), and an estimated 5,000 tonnes of oil stranded onshore, resulting in shoreline oiling of 98 km as heavy, 34 km as moderate, and 66 km as light (Harris, 1997; Law et al., 1997). Oil concentrations in the water column below treated slicks were generally 1–10 ppm and uniformly mixed down to 5 m (the maximum depth of measurement) (Lunel et al., 1997a). Within 5 km downcurrent of the grounding site, oil concentrations in the water column were 0.5–0.6 ppm throughout 4 days after termination of dispersant applications; by 12 days after termination, oil concentration were 0.2 ppm or lower. At distances of >10 km downcurrent, oil concentrations were 0.2–1.0 ppm throughout 6 days after termination of dispersant application; by 12 days after termination, they were 0.2 ppm or lower.

There were reported mortalities of shallow sub-tidal and intertidal organisms, with bivalves and urchins washing up by the hundreds in some areas. Wild salmon, other finfish, crab, lobster, and whelk were found to have low levels of PAH but no taint. Intertidal mussels remained contaminated in one bay with heavy shoreline oiling for 19 months after the spill.

SOURCE: Summarized from Harris (1997), Law et al. (1997), and Lunel et al. (1997a).

data can be compared to values typically used in water accommodated fractions (WAF) generated for dispersed oil toxicity evaluations (see Chapter 5).

Monitoring Dispersant Use During Actual Spills

Monitoring of dispersant use means different things to different people. The mental model one has of concepts or definitions is generally associated with their background and stakeholder role. Dispersant-use monitoring can be separated into two basic categories: (1) information collected to help make timely operational decisions; and (2) data gathered for future analyses of fate and effect (Pond et al., 1997). Operational monitoring should provide information on the application platform's spraying parameters and on whether or not oil is being entrained into the water column. This information should be conveyed immediately to those making the decision on whether or not to continue the operation. The second type of monitoring involves collecting data that can be later used to address the fate and effects of the dispersed oil and may also be used to ground truth some of the operational monitoring information (Hillman et al., 1997). In every dispersant application, operational monitoring is done to some degree. Depending on the circumstances, ground-truth information on fate and effect may or may not be required.

Dispersant effectiveness is a phrase that has been interchangeably used to describe how well the product performs both in the laboratory and in field applications. As discussed previously, there are three components that will determine dispersant effectiveness during spill response: operational effectiveness, chemical effectiveness, and hydrodynamic effectiveness. The common usage of "dispersant effectiveness" to describe performance in the laboratory and the field is unfortunate because laboratory-derived effectiveness usually does not equate to effectiveness in field applications (e.g., see Table 3-3). This dual usage has fostered misconceptions and misunderstanding throughout the response community and the public. As described previously in this chapter, laboratory tests generally measure chemical effectiveness, whereas effectiveness in the field is also dependent on operational and hydrodynamic factors. Therefore, a laboratory effectiveness of 60 percent does not mean that 60 percent effectiveness will be obtained in field applications. Depending on many factors, the field effectiveness for a product may range from 0 percent to 100 percent.

Effectiveness of a dispersant application in the field has been defined as "the amount of the oil that the dispersant puts into the water column compared to the amount of oil that remains on the surface" considering the total amount of the oil that was treated (Fingas, 2002a,b; 2003; Lewis, 2004). U.S. Coast Guard, et al. (2001) define effectiveness based upon the amount of oil that the dispersant puts in the water compared to the amount of oil that was in the area treated. NRC (1989) concluded that a mass balance approach has given good effectiveness estimates in a few elaborate field tests, but "it is complicated, requires set-up time, and is not practical in real spills." In field experiments, the release volume is known, the area of the slick can be measured, and the average thickness for this finite area can be calculated. In addition, dispersants are generally applied to the entire test slick; thus mass balance effectiveness estimates may be applicable. In accidental spills, however, only a portion of the total amount of spilled oil is normally treated, the oil thickness of the treated area is unknown and highly spatially variable, and thus the volume of oil in the treatment area is seldom known to any great accuracy in a timely manner. Presently, there is no valid and reliable method of determining slick thickness in the field, and any estimated value may easily be in error by an order of magnitude (Fingas, 2002a,b).

Prior to development and implementation of a monitoring plan, it is imperative that the stakeholders agree on attainable goals and objectives for the monitoring (U.S. Coast Guard et al., 2001). Among these goals and objectives should be a working definition of field dispersant effectiveness and a set of Standard Operating Procedures (SOPs) with data quality objectives. The definition of field effectiveness could parallel the definitions of mechanical recovery (i.e., percent recovery of the entire spill) or *in-situ* burning (i.e., percent of oil burned from a contained area).

The degree and extent of monitoring should be in proportion to the sensitivity of the environment. In general, the more sensitive the environment, the more emphasis should be placed on monitoring. Sensitivity can be assigned based upon environmental and political parameters. Basically the resource trustees and stakeholders want to know how well the response works and the extent of the effects. There is a heightened concern as the sensitivity increases. Factors that have a direct relationship to sensitivity include, among others, nearness to shore, special habitats such as marine sanctuaries and parks, biological and migratory seasonality, size of incident, and nature of the spilled product. Nearness to shore generally involves environments with shallower water, lower dilution rates, higher productivity, fish and shellfish nursery grounds, higher concentrations of wildlife, greater commercial and recreational use, and shorter response times.

As discussed in Chapter 2, it is very advantageous for the resource trustees and stakeholders to pre-identify sensitive areas, determine where and when dispersant use should be discussed, and outline monitoring objectives. Unless otherwise stated, pre-approval agreements generally are based on the assumption that use of dispersants, under specified conditions, will protect sensitive shoreline and water-surface resources without causing significant impacts to water-column and benthic resources, even assuming 100 percent dispersion of the slick.

Operational Monitoring

The primary reason to monitor operational aspects of dispersant use is to determine if the dispersant application is operationally effective (e.g., that the dispersant is being applied to the surface oil targets). The secondary purpose is to estimate the relative effectiveness of the operation (Fingas, 2003). Additional data also are needed to provide documentation on what dispersant was used, how much was used, when and where it was used, and the environmental conditions at the dispersal sites. Because there is no truly quantitative method to determine dispersant effectiveness in the field, the best that can be done is to qualitatively estimate if the dispersants are working (Henry, 2004).

Effective/Ineffective Dispersant Applications

It is assumed that some portion of the dispersant spray will miss the target due to wind drift of the spray or turning pumps on too soon or off too late (see earlier discussion of dispersant use in response to the T/V *Exxon Valdez* spill). Missing the target excessively should be documented in the monitoring report, and controls should be enacted to minimize this to within acceptable limits. An experienced trained observer is the best

TABLE 3-5 Guidelines to Assist in Determination of Effective/Ineffective Application

Possible Dispersant Action	Possible False Positives	Possible False Negatives
Difference in appearance between treated and untreated slick.[a,b]	Suspended solids or algal blooms may resemble dispersed oil.[a,b]	Dispersion may not be instantaneous, may take several minutes to a few hours to show dispersed plume.[a,b,c,d]
Appearance of plume can range from brown to pale yellow.[a,b,c,d]	Boat wakes through oil may appear as dispersed paths.[a]	Visible cloud or plume not observed, water may be naturally murky.[a,b,c]
Changes in area and thickness of the oil.[c]	Dispersants may have a herding effect on thin oil. May also be seen as lacing.[b,d]	Oil may be dispersing under the slick and not seen.[b,d]
Higher fluorometer readings of dispersed oil in application area vs. background or non-treated slick area.[a,b,c,d]	Rapidly dissipating whitish plume may be caused by dispersant alone (missed target).[d]	
	After initial visual assessment, some dispersed oil may resurface.[d]	

[a]USCG et al., 2001.
[b]NOAA, 1999.
[c]ExxonMobil, 2000.
[d]Fingas, 2003.

way to assess if the dispersant operation is effective or not (Lewis and Aurand, 1997; U.S. Coast Guard et al., 2001; Fingas, 2003; Goodman, 2003; Henry, 2004). Even though there are difficulties with the interpretation of fluorometer data (Fingas, 2003; Goodman, 2003), the addition of confirmation fluorometer readings will help substantiate visual observations that there has been an increase in the amount of oil entrained into the water column under treated slicks. Table 3-5 contains guidelines to assist in determination of effective/ineffective application.

Special Monitoring of Applied Response Technologies

The protocol used by most if not all U.S. regions for obtaining operational monitoring information for dispersant use and *in-situ* burning is Special Monitoring of Applied Response Technologies (SMART) (U.S.

Coast Guard et al., 2001). The purpose of the dispersant section of SMART is to outline a protocol that rapidly can collect information to assist in real-time decisionmaking during dispersant applications (Barnea and Laferriere, 1999). SMART only outlines how to determine if the dispersant application is working, but provides no guidance on how to determine a percent dispersant effectiveness. It relies heavily on personnel being trained using job-aids developed to support SMART (Levine, 1999). For much of coastal and offshore waters of the United States, the resource trustees and stakeholders have designated selected areas as pre-approved for dispersant use. All pre-approved areas have a stipulation that requires use of the SMART protocols for operational monitoring, if operationally feasible. In an effort to better document effectiveness, field portable equipment has now been prepared and staged within various RRTs for immediate deployment in the event of a spill (Gugg et al., 1999; Barnea and Laferriere, 1999; Henry et al., 1999; Henry and Roberts, 2001). Some pre-approvals indicate that SMART will be used for fate and effects monitoring; however, SMART specifically states it "does not monitor the fate, effects, or impacts of dispersed oil" (U.S. Coast Guard et al., 2001). The SMART protocol contains three tiers of monitoring:

Tier I is visual monitoring by a trained observer, preferably using an aircraft separate from the "spotter" aircraft directing the dispersant application (U.S. Coast Guard et al., 2001). The protocol recommends documentation via forms, photography, and videotape. Tier I monitoring may be enhanced through the use of remote sensing instruments, such as infrared thermal imaging, if data are available in real-time. The purpose of Tier I is to visually assess if the operation is working and rapidly report the findings to the decisionmakers. Typical observations include: (1) that the dispersant spray hit the slick; (2) a reduction in the amount of oil on the water surface after dispersant treatment; (3) a change in the appearance of the treated slick; and (4) the presence of a milky or cloudy plume in the water column.

Tier II includes Tier I monitoring and adds an on-water component. From a vessel, water samples are analyzed via continuous flow fluorometer collecting water at a 1 m sampling depth. The protocol recommends comparing fluorometer measurements from three general areas: (1) background water outside the spill area, (2) below the surface oil slick before dispersant application, and (3) an area where the oil slick has been treated with dispersants. The purpose of Tier II is to confirm whether or not oil is being entrained into the water column (Barnea and Laferriere, 1999). A few water samples are collected for later laboratory analysis to validate and possibly quantify the fluorometer measurements.

Tier III is presented as "Additional Monitoring" to collect information on transport and dispersion of the oil into the water column. It follows Tier II procedures but adds multiple depth fluorometer sampling of selected transects and provides for collection of additional environmental parameters, such as water temperature, conductivity, dissolved oxygen, pH, and turbidity.

The SMART protocol includes collection of water samples to validate and quantify the fluorometer readings. Calibration methods and techniques are discussed in Lambert et al. (2001a,b) and Fingas (2002a,b). The validation method can estimate the quantity of "oil" in the water column, but the data cannot be used to differentiate between that part that is dissolved and that part that is in droplets. Fingas (2003, 2004a) discussed the precautions and proper use of fluorometry in the field. His comments on field techniques include awareness of possible contamination using Tygon tubing and maintaining the sampling probe in waters undisturbed by the vessel (in front of or outside the bow wave).

The Alyeska Ship Escort Response Vessel System (SERVS) has developed dispersion monitoring guidelines that are similar to the SMART protocol, but the primary goal of the Alyeska/SERVS protocol is to provide real-time assessment of the environmental effects of dispersion (Hillman et al., 1997). The Alyseka/SERVS protocol relies on aerial monitoring as the primary tool for monitoring dispersant effectiveness and effects with additional support provided by collection of water samples and *in-situ* fluorometry. This protocol is not intended to provide quantitative estimates of dispersant effectiveness, real-time estimates of water-column dispersed oil concentrations, or estimates of oil mass balance. This protocol attempts to monitor the dispersed oil plume by locating the water-column sampling stations and the *in-situ* fluorometry transect relative to drogues that drift with subsurface currents (usually at 2-m depth). Whereas SMART and the Alyeska/SERVS protocols rely on conventional filter fluorometers with a single filter for excitation and another for emission for *in-situ* measurement of dispersed oil concentrations, multiple-wavelength fluorometers and *in-situ* instruments capable of measuring particle-size distributions have been investigated for research use (Fuller et al., 2003). Unfortunately, the performance of these instruments for monitoring oil dispersion at sea has not yet been evaluated (Ojo et al., 2003).

Additional Operational Monitoring

To better document the operation and to possibly provide clues to future questions, several delivery platform and environmental parameters should be recorded. Pre-application documentation should include the name, lot number, and quantity of dispersant loaded on the aircraft or vessel. A sample should be taken, with proper chain-of-custody, from each dispersant lot (to allow for later analysis if verification of product effectiveness is needed). After each sortie, the amount of dispersant remaining onboard should be documented. Also, other data are needed on the platform performance during the application and on the environmental conditions in the application area. Table 3-6 provides guidance on the additional monitoring data or samples to be obtained. Most of the performance data should be automatically recorded on the platform.

Environmental Monitoring

As discussed in Chapter 2, there is a reasonable degree of confidence in the current ability to assess trade-offs, relative to use of dispersants, in offshore waters. In general, offshore waters are considered to be less sensitive to dispersed oil impacts, because of rapid dilution of dispersed oil, than shallower or nearshore environments. But as shallower or nearer to shore waters are evaluated for dispersant use, the sensitivity of the environment and the degree of uncertainty make the assessment more difficult. The database of oil component acute toxicity is much better than the knowledge of the bioavailability of dispersed oil components in the water column. Unfortunately, most of the measurements on concentrations of

TABLE 3-6 Guidance on the Additional Monitoring Data or Samples

Pre-application	Application	Post-application
Name of dispersant	Spray time	Volume of dispersant remaining onboard
Dispersant lot number	Pump rate	
Sample of each dispersant lot number	Speed during application	
Volume of dispersant onboard	Spray height during application	
Platform (aircraft or vessel)	Sample of weathered and neat oil	
Description of spray system	Dispersant applied neat or diluted	
Dispersant pump calibration documentation	Wind speed and direction	
Spray nozzle test documentation	Current speed and direction	
	Air and surface water temperature	
	Cloud cover	
	Surface salinity	
	Wildlife in area (birds, mammals, turtles)	
	Approximate spray width	
	Approximate spray path length	
	Number of passes over same area to achieve adequate dispersion.	
	Sea state	

dispersed oil in open water are from fluorometer readings or from the total extraction of unfiltered water samples. Thus at best, the total concentration of oil components in the water column is known, but not whether the component concentrations reside in the water or in oil droplets is not known. Questions the risk assessors need answers to concerning the dispersed oil include: (1) How are the components of dispersed oil distributed in the water column? and (2) What fractions are in the dissolved phase and what fractions are in droplets or adhered to particulates in the water column? These data cannot be obtained through fluorometry, and Page et al. (2000b) have shown that estimations of oil-component partitioning based upon solubility coefficients alone are not reliable for oil-in-water mixtures. The data can be obtained via discrete large-volume water samples that are collected and filtered immediately to differentiate between components that are truly dissolved and those that are present as dispersed oil droplets (Payne et al. 1999; Payne and Driskell, 2001, 2003). These samples, at a minimum, should be analyzed for dispersed oil droplet and dissolved-phase PAH and total petroleum hydrocarbon (TPH) concentrations in the filtered and unfiltered water. Sample collection should be from several depths and repeated over time. Real time *in-situ* fluorometry data should be used to locate where to take samples and to verify that the discrete samples were taken in the dispersed oil plume. In additional to finite grab examples collected with traditional water-sampling equipment, aliquots of effluent from the fluorometers should also be collected for chemical analysis. Whenever possible, separate fractions for dissolved and particulate/oil-phase components should also be analyzed (Payne et al., 1999; Payne and Driskell, 2001, 2003). Monitoring data, coupled with local transport mechanisms, can be used to validate computer-model predictions, and thus reduce the uncertainty of the fate of dispersed oil components. Ultimately, this will provide decisionmakers with a better tool to assess use of dispersants in sensitive environments.

The trustees of the local resources at risk will determine if other types of monitoring are needed to assess the effects. The extent of monitoring should be based on the sensitivity of the environment and the predicted amount of dispersed oil reaching the resources of concern. The collection and analysis of samples, whether they are sediment, nekton, or benthos, should be conducted so there can be a direct comparison with water-column analytes.

DEVELOPING ADEQUATE UNDERSTANDING OF DISPERSANT EFFECTIVENESS TO SUPPORT DECISIONMAKING

As discussed in Chapter 2 and shown in Figure 2-4, the potential effectiveness of dispersants is a key consideration at several steps in the decision-making process. Significant work has been done to test dispersant products on a range of oil types or refined products under different test conditions (temperature, salinity, etc.). The test protocols were designed to establish a high degree of reproducibility, but were never intended to replicate actual environmental conditions that may be encountered during a spill. However, these kinds of tests are useful to provide guidance on whether or not a test oil is likely to be dispersible under ideal conditions.

The fourth question in Figure 2-4—"Are conditions conducive?"—addresses the range of factors that affect the overall field effectiveness of dispersant application once the oil starts to spread and weather. Currently, predicted dispersant effectiveness for a specific spill event is based on simple models and past experience. In current fate and transport models, dispersant effectiveness is an input value. In the future, it would be desirable to possess the ability to predict dispersant effectiveness over time through the use of a physical-chemical efficiency model. However, additional research is needed to develop the model. **Relevant state and federal agencies and industry should develop and implement a focused series of studies that will enable the technical support staff advising decisionmakers to better predict the effectiveness of dispersant application for different oil types and environmental conditions over time.**

Bench-scale effectiveness tests can provide a valuable tool for investigating the factors and interactions that affect the chemical effectiveness of oil dispersion. A particular strength is their ability to inexpensively and quickly test a large number of conditions. Currently, most bench-scale effectiveness tests incompletely characterize the test conditions and do not systematically vary factors, such as mixing energy, that are known to have a strong influence on the process of oil dispersion. In addition, important response variables, such as oil droplet-size distributions, are not routinely measured. As a result, bench-scale effectiveness tests cannot, in general, provide the type of input that is needed for fate and transport models. **Experimental systems used for bench-scale effectiveness tests should be characterized to determine the energy dissipation rates that prevail over a wide range of operating conditions. Future effectiveness tests should measure chemical effectiveness over a range of energy dissipation rates to characterize the functional relationship between these variables. Finally, evaluation of chemical effectiveness should always include measurement of the droplet-size distribution of the dispersed oil.**

Wave-tank-scale effectiveness tests are particularly useful for investigating factors that cannot be studied in laboratory-scale tests. In addition, the more realistic mechanism of energy input in experiments conducted in wave tanks reduces the sensitivity of results to uncertainties regarding the mechanism of oil-droplet formation and, therefore, scaling of laboratory- or wave-tank-derived effectiveness estimates to sea-surface conditions.

The design of wave-tank dispersant-effectiveness studies should specifically test hypotheses regarding factors that can affect operational effectiveness. These factors include oil properties that are representative of those expected to prevail under spill-response conditions, such as water-in-oil emulsification and the potential for heterogeneity in the rheological properties of the floating oil (e.g., formation of a "skin" that resists dispersant penetration). Dispersant droplet-size distributions and impact velocities should be similar to those that would be expected to be generated by dispersant application methods commonly used in oil-spill response.

Tank tests that determine the ability of mechanical recovery methods to recover oil that has been treated with dispersant but not effectively dispersed, or re-floated oil, should be carried out. A more complete understanding of what limitations the unsuccessful use of dispersants may have on subsequent mechanical recovery methods could greatly reduce concern over relying on operational testing of the dispersant effectiveness in the early phases of spill response.

Energy-dissipation rates should be determined for wave tanks over the range of operating conditions that will be used in dispersant effectiveness tests. The wave conditions used in dispersant effectiveness tests should represent a specific environment of interest. It may be necessary to conduct experiments over a range of energy dissipation rates to adequately represent the environment of interest.

More robust understanding of dispersant effectiveness can be derived from test tanks, if more rigorous protocols are implemented that better quantify the eventual fate of the test oil. **The concentration of oil should be measured in all identifiable compartments to which it could be transferred when dispersant effectiveness is investigated in wave tanks.** This includes, but may not be limited to, the water surface, the water column, the atmosphere, and wave-tank surfaces. Oil mass balances should be reported in an effort to better understand the accuracy of effectiveness quantification. In addition, the droplet-size distribution of the dispersed oil should be measured and reported.

Little is known of the potential leaching of surfactant from floating oil and dispersed oil droplets at realistic oil-to-water ratios and under turbulence conditions that might be encountered in the field. In particular, the effects of surfactant leaching on the effectiveness of oil dispersion and the potential for droplet coalescence should be understood better. **Coalescence and resurfacing of dispersed oil droplets as a function of mixing time should be studied in flumes or wave tanks with high water-to-oil**

ratios (to promote leaching of surfactant into the water column). Periods of wave-induced turbulence should be followed by periods of relative calm to allow droplets to resurface. The surfactant concentration remaining in the resurfaced oil should be measured, and its dispersibility should be measured (by introducing more wave turbulence) to evaluate the ultimate fate of resurfaced oil. Alternatively, oil dispersion should be measured after dispersant is applied and incubated with floating oil under calm conditions to determine the effect of surfactant leaching from a surface oil film on dispersant effectiveness.

Although careful and controlled research in the laboratory or test tank will be important to developing tools to support decisionmaking, the results of dispersant application during real spills will be the most important indicator of whether or not the dispersant application was effective. Field data are essential to a better understanding of the spill-specific conditions that affected the dispersant operation, and they should be used to validate model predictions. **To improve the quality of field data collected during dispersant applications, more robust monitoring capabilities should be implemented. Specific attention should be given to:**

- Developing an environmental monitoring guidance manual for dispersant application monitoring with suggested sampling and analytical techniques, sampling methods, and QA/QC to ensure cost effectiveness and maximum utilization of the data
- Developing a detailed standard operating procedure (including instrument calibrations and data quality objectives) for each sampling and analytical module (SMART is guidance only)
- Developing a definition of field effectiveness
- Measuring dispersed oil droplet and dissolved-phase TPH and PAH concentrations with grab samples of filtered and unfiltered water (these data can then be compared to model predictions and toxicity data for both dissolved and particulate/oil-phase components) as a function of location and time.

4

Transport and Fate

Spilled oil is transported, and its composition and character altered, by a variety of physical, chemical, and biological processes (Figure 4-1). Use of chemical dispersants changes the relative importance of these processes, affecting the fate of the oil, and altering subsequent ecological effects. Thus, it is important to understand the transport and fate of oil with and without dispersant use. A number of comprehensive studies have reviewed these mechanisms including Stolzenbach et al. (1977), Kerr and Barrientos (1979), Huang and Monastero (1982), Payne and McNabb (1984), Payne et al. (1984), Delvigne et al. (1986), Spaulding (1988), Lee et al. (1990), Payne et al. (1991a,b,c,d), Yapa and Shen (1994), ASCE (1996), Reed et al. (1999), Payne and French-McCay (2001), Payne and Driskell (2003), and NRC (1985, 1989, 2003). These mechanisms are reviewed briefly in the first two sections of this chapter, with a focus on how transport and fate influence the subsurface concentration of oil, and how the composition and concentration of surface and entrained oil droplets can be expected to vary with and without application of chemical dispersants. The latter portion of the chapter and Appendix E discuss how the mechanisms are integrated into computer oil spill models that simulate the fate of spilled oil, and how such models are used (or might be used) for purposes of pre-planning, emergency response, and natural resource damage assessment.

TRANSPORT PROCESSES

There are three major modes of transport for spilled oil or petroleum products discussed in the following subsections. The first deals with the

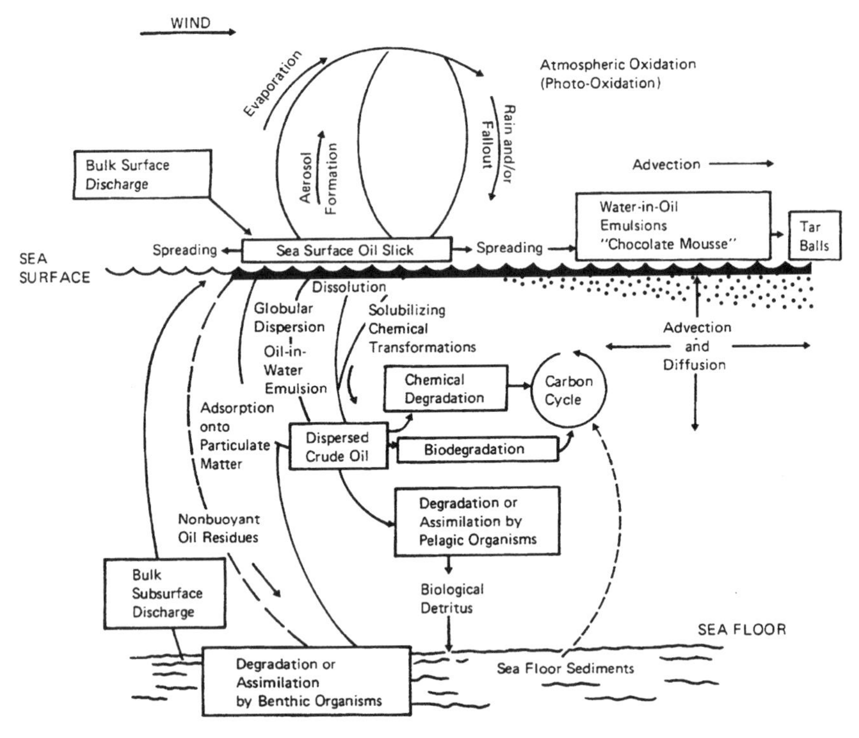

FIGURE 4-1 Major open-ocean oil fate and transport processes.
SOURCE: NRC, 1985.

surface transport of slicks, which is important because the shape, thickness, and location of a slick affect the ability to effectively apply dispersants. The second subsection deals with vertical transport, which is responsible for the initial dilution of dispersed oil. Finally, the last subsection deals with horizontal subsurface transport, which is responsible for the ultimate dilution of dispersed oil.

Surface Transport

Oil spilled directly on a calm water surface spreads radially by gravity and is resisted by inertia, viscosity, and surface tension until the slick reaches a thickness of ~0.1 mm. Fay (1969), Hoult (1972), and others have modeled this spreading under idealized conditions (e.g., instantaneous spill, no wind, no waves). Application of chemical dispersants can temporarily affect this spreading through the phenomenon of herding.

Additional spreading takes place because (1) oil is usually spilled over a period of time and into a moving current, (2) wind, waves, and non-uniform currents diffuse and break up the slick, and (3) droplets periodically disperse and resurface, tending to stretch the plume. This last mechanism has been described by Johansen (1984) and Elliott et al. (1986) and may increase in significance when considering the fate of chemically dispersed oil.

Slick thicknesses were estimated during several well-documented oil spills, usually indirectly by dividing volume/area (Mackay and Chau, 1986; Lunel and Lewis, 1993a,b; Lewis et al., 1995a,b; Walker et al., 1995; Brandvik et al., 1996; Brown et al., 2000). These studies indicate that oil does not spread uniformly, but is irregular in shape and thickness—generally elongated in the direction of the wind and often composed of thick patches (>1 mm) and thinner sheens (<0.01 mm). S.L. Ross (1997) gives a general rule of thumb that 90 percent of an oil spill's volume is contained in 10 percent of its area. Figure 2-5 (in Chapter 2) presents representative descriptions of the wide range of slick thicknesses typical of an oil slick along with an approximation of the estimated volume/unit area for the different thicknesses. The non-uniform characteristics of a slick can be included in models (e.g., Mackay et al., 1980a,b; Lehr et al., 1984), but such models are basically empirical.

Surface spreading has important implications for the operational effectiveness of dispersant application because dispersant delivery systems have finite encounter rates (area coated per unit time) and capacities (total volume of dispersant used; Gregory et al., 1999). As such, dispersants are most effective when they are applied as soon as possible (before the slick has had time to spread and break up), and with the benefit of airborne sensing to identify locations of maximum slick thickness. In particularly thick regions, it is not practical to treat the slicks with a single pass and lacking visual confirmation of dispersion, a multi-pass approach is often used (S.L. Ross, 1997; Lunel et al., 1997b).

Of additional concern is oil that is accidentally released from subsurface blowouts during offshore exploration or production. Here the oil will likely be mixed with substantial quantities of natural gas, which provides the major source of buoyancy. Masutani and Adams (2004) and Tang (2004) describe the spectrum of oil droplet sizes that can be expected as a function of dimensionless exit conditions. The combination of gas and oil forms a buoyant droplet/bubble plume that entrains seawater as it ascends toward the surface. A similar situation, but without the gas, would occur with the rupture of an underwater oil pipeline. Models to describe such plumes have been developed by Yapa and Zheng (1997; 1999) and Johansen (2000) among others. If the oil is released in shallow water (less than roughly 100 m), it will rise as a coherent plume, containing a mixture of gas, oil, and water. Once the plume surfaces, the oil and water will spread radially in a surface layer (Fannelop and Sjoen, 1980). Because of the presence of water, the resulting oil slick will be significantly thinner than those produced by oil spilled directly on the surface. In deeper waters, ambient currents, and potentially density stratification, will cause the gas bubbles and larger oil droplets to separate from the remainder of the plume and ascend as individual (or small groups of) droplets and bubbles (Socolofsky and Adams, 2002). Because droplet rise velocity depends on diameter, the larger oil droplets will reach the surface sooner and closer to their source than the smaller droplets. This fractionation leads to a *substantially* longer (and thinner) plume than would be produced by a surface spill.

Work is being conducted both in the United States and abroad, to assess if and how to chemically disperse oil from a subsurface blowout. In many cases, it is impractical to apply dispersants at the surface because the slick is too thin. However, if the surface slick is subsequently concentrated by Langmuir circulation cells or other convergence mechanisms, dispersants can be applied to the thicker portions. In the absence of such surface convergence, the most effective method would be to apply dispersant within the well (down hole) before the oil can mix with seawater, but this may be difficult, so attention is being paid to schemes that dispense the dispersants directly into the plume. This should be done as close to the seafloor as possible to minimize dilution, and hence achieve the desired dispersant-to-oil ratio (DOR) without bearing the cost and potential environmental consequences of using excessive quantities of dispersant. Some initial concepts for dispersant application to blowouts can be found in Johansen and Carlsen (2002).

Slicks are advected downwind by a combination of wind and waves. Pure advection (without spreading) does not affect the concentration of oil or the effectiveness of dispersants, but it is important for understanding where an oil slick will end up. Many researchers have studied these processes from theoretical and empirical perspectives, and a rule of thumb is that slicks move at approximately 3 percent of the wind speed measured 10 m above the water surface (i.e., the "wind factor" is about 3 percent). For moderate to high sea states, approximately two-thirds of this transport can be attributed to Stokes drift (the fact that near-surface wave orbits in deep water waves do not follow exact circles, as linear theory would suggest, but exhibit a net transport in the direction of wave propagation). The remaining one-third represents the slick moving relative to the water directly underneath it (Lehr et al., 2002). Coriolis acceleration causes the slick to drift ~10–20 percent to the right of the wind in the northern hemisphere, but this effect is often omitted in transport models. Experimental observations support these conclusions, with some suggestion that the wind factor and deflection angle decline with wind speed (Youssef and Spaulding, 1993).

Vertical Transport

Dispersion of a surface slick, whether caused naturally or through application of chemical dispersants, results in the formation of droplets that are entrained into the water column and transported with the subsurface currents. The importance of vertical transport is clearly seen by a simple calculation for illustrative purposes: a surface slick that is 0.1 mm thick and dispersed with an efficiency of 50 percent to an average depth of 5 m, will receive a dilution of 10^5, resulting in an immediate drop in concentration to ~10 ppm.

Dispersion results in a distribution of droplet sizes with the smaller droplets being transported deeper and longer. If Q is the mass of oil entrained per unit area of the slick, and d is a characteristic droplet diameter, it is clear that the goal of chemical dispersants is to increase Q and decrease d. And while it is obvious that use of chemical dispersants increases the *mass* of oil within the water column, it may or may not increase the *concentration* of oil, because the greater dilution achieved by smaller droplets may offset the increase in mass. This question will be revisited at the end of this subsection.

The initial depth of droplet penetration, h_i, is proportional to the wave height, h_w, with many studies showing that $h_i \cong 1.5h_w$ (Nilsen et al., 1985; Delvigne and Sweeney, 1988). [Variables used in this chapter are summarized in Table 4-1.] Subsequent vertical transport depends on a balance between vertical diffusion (characterized by a vertical diffusivity E_z, with dimensions of L^2/T) and buoyant rise (characterized by a terminal velocity w_s). Vertical diffusivity transports droplets deeper into the water column, while buoyancy makes them return to the surface.

Vertical diffusivity generally ranges between 1 and 200 cm^2/s depending on a number of environmental factors. Near the surface, diffusivity is a strong function of wave height, and a number of investigators report $E_z \sim h_w^2$ (Koh and Fan, 1970). Because wave energy decreases with depth, E_z decreases below the surface. For example, Ichiye (1967) suggests that, in the absence of density stratification,

$$E_z = 0.028(h_w^2/T)\exp[-4\pi z/L] \tag{4-1}$$

where L is wave length, T is wave period, and h_w is taken as the significant wave height. Other formulations suggest a stronger cut-off with depth, attributed to the depth of Langmuir circulation (windrows), which is caused by the interaction of wind and waves (Leibovich and Lumley, 1982;

TABLE 4-1 Variables Used in Scaling Arguments in Chapter 4

Variable	Definition	Dimension
C_{diss}	Oil concentration in dissolved phase	ML^{-3}
C_{drop}	Oil concentration in droplet phase	ML^{-3}
d	Droplet diameter	L
E_z	Vertical diffusion coefficient	L^2T^{-1}
E_r	Horizontal (radial) diffusion coefficient	L^2T^{-1}
h_{char}	Characteristic depth of oil droplets	L
h_i	Initial depth of oil droplets	L
h_w	Wave height	L
L	Wave length	L
Q	Mass of oil entrained per unit area of slick	ML^{-2}
T	Wave period	T
w_s	Droplet slip (rise) velocity	LT^{-1}
λ_z	Vertical velocity gradient	T^{-1}
σ_r	Radial standard deviation of spreading patch	L
ν	Kinematic viscosity	L^2T^{-1}
P_1	Water density	ML^{-3}

and references in Champ, 2000). Diffusivity also decreases under the influence of vertical density stratification, and a host of formulations suggest that E_z is inversely proportional to the vertical density gradient (Koh and Fan, 1970; Broecker and Peng, 1982). A thermocline is a region of maximum density gradient suggesting small E_z, and if stratification is strong enough, a "diffusion floor" may be assumed. Some models assume that the depth of this floor is simply proportional to wave height.

Unless there is significant interaction with suspended particulates, most oil droplets will be positively buoyant and will rise toward the surface. Those with a diameter less than about 300 mm will obey Stokes Law and rise with a velocity of:

$$w_s = (\Delta\rho/\rho)gd^2/18\nu \quad (4\text{-}2)$$

where ν is the kinematic viscosity of water, $\Delta\rho/\rho$ is the normalized density difference between seawater and oil, g is gravitational acceleration, and d is droplet diameter. The quadratic dependence of rise velocity on droplet diameter suggests that the smallest droplets will rise very slowly, accentuating dispersion. For example, with $\Delta\rho/\rho = 0.13$ (for an oil with a density of 0.89 mg/mL and seawater at 1.025 mg/mL), $\nu = 10^{-2}$ cm^2/s and $g = 981$ cm/s^2, droplets with a diameter of 300 μm will rise with a velocity of 0.6 cm/s while droplets with a diameter of 30 μm will rise with a velocity of 0.006 cm/s. The former will take less than 8 minutes to rise a height of 3 m, while the latter will take over 12 hours. And, because of vertical diffusion, the smaller droplets will most likely reside deeper in the water column, further prolonging their ascent.

The above discussion can be used to estimate how the concentration of droplet and dissolved phase oil might depend on dispersion efficiency and vertical transport mechanisms. The concentration of oil in the droplet phase is proportional to the mass of oil entrained per unit area, Q, divided by a characteristic depth of droplet penetration, h_{char}, or

$$c_{drop} \sim Q/h_{char} \tag{4-3a}$$

The rate of dissolution of dispersed oil per volume of seawater is proportional to the number of droplets per volume ($\sim Q/h_{char}d^3$) times the surface area of a drop ($\sim d^2$). Hence the concentration of dissolved oil

$$c_{diss} \sim Q/h_{char}d \tag{4-3b}$$

A simple model for the characteristic depth is $h_{char} \sim E_z/w_s$, where $E_z \sim h_w^{\ 2}$ (independent of depth), and $w_s \sim d^2$ (from Eq. 4-2). The wave flume experiments by Delvigne and Sweeney (1988) suggest that $Q \sim h_w^{\ 1.14}$, while d is independent of h_w. Thus, from Eq. (4-3a), $c_{drop} \sim d^2/h_w^{\ 0.86}$, and from Eq. (4-3b), $c_{diss} \sim d/h_w^{\ 0.86}$. With this "model" both droplet and dissolved phase concentrations *decrease* with wave height and increase with droplet diameter. In reality, diffusivity is not likely to be constant with depth so an alternative model assumes a characteristic depth that is proportional to wave height, or $h_{char} \sim h_w$. In this case, equations (4-3a) and (4-3b) give $c_{drop} \sim h_w^{\ 1.14}$ and $c_{diss} \sim h_w^{\ 1.14}/d$. Here both droplet and dissolved phase concentrations *increase* with wave height and either decrease with, or are independent of, droplet diameter, i.e., quite different from the conclusions of the first model.

These arguments are qualitative, and more precise information should come from computer models that integrate multiple mechanisms in a quantitative manner as later discussed. But computer models are no better than our understanding of the individual mechanisms upon which they are based, and the uncertainty in even the *direction* of change noted above suggests we need better understanding of dispersant effectiveness (i.e., the dependence of Q and d on oil properties and environmental parameters), as well as better models of the vertical distribution of E_z, in order to accurately predict the concentrations of dispersed oil.

Horizontal Subsurface Transport

Subsurface advection of dispersed and dissolved phase oil by a uniform current affects the location of the oil, but does not, in itself, cause additional mixing. However, mixing *is* produced when the currents are non-uniform, and this mixing is responsible for the ultimate dilution of the oil. Without horizontal mixing, and under sufficiently calm weather conditions, vertically dispersed oil droplets could all ultimately resurface given enough time.

Horizontal mixing consists of two fundamental processes. The first process is called scale-dependent diffusion and represents the fact that large eddies will advect a patch of marked fluid if the patch is smaller than the scale of the eddies, but mix and dilute the patch if it is larger than the eddies (Csanady, 1973). The second process is termed shear dispersion and results from the combination of velocity gradient(s) in combination with mixing (or other transport mechanism) in the direction of the gradient(s) (Fischer et al., 1979). The latter effect is enhanced with the use of chemical dispersants, because the smaller droplets that are produced are transported deeper, where they experience greater differences in horizontal velocity. Unfortunately, field measurements cannot always distinguish the two processes, and frequently their effects are combined.

Horizontal mixing is determined best using site-specific measurements, but as these are often not available, literature values should be used. Okubo (1971) summarizes a number of coastal tracer studies and shows that

$$\sigma_r^2 = 0.011t^{2.34} \quad (4\text{-}4)$$

where σ_r is a characteristic radius (standard deviation) of an equivalent circular tracer patch (cm) and t is time (sec). Other investigators report similar trends. Okubo's data apply to patch sizes ranging from ~30 m to ~100 km, and more recent data suggest the approximate relationship applies to even larger scales (Ledwell et al., 1998). Simple relationships such as this are useful because dilution resulting from horizontal mixing is proportional to patch variance, σ_r^2, and hence Eq. (4-4) can be used to directly compute changes in concentration due to horizontal mixing. Also, predictive models make use of horizontal diffusion coefficients (E_r, with dimensions of L^2/T) defined by the time rate of change of patch variance. For example, using Eq. (4-4)

$$E_r = d\sigma_r^2/4dt = 0.006t^{1.34} = 0.085\sigma_r^{1.15} \quad (4\text{-}5)$$

For $\sigma_r = 100$ m, $E_r = 0.3$ m^2/s, while for $\sigma_r = 1{,}000$ m, $E_r = 5$ m^2/s. Note that these values of horizontal diffusivity are orders of magnitude larger than the corresponding vertical values (E_z) suggesting that horizontal mixing is much stronger than vertical mixing. However, horizontal mixing is also much less effective, because horizontal plume dimensions are much larger (and hence horizontal concentration gradients are much smaller) than in the vertical.

It should also be recognized that different investigators define horizontal diffusion coefficients differently. For example, as implied above, some data used to determine mixing coefficients include the effects of vertical shear, while others do not. Also some analyses separate E_r into separate components in the longitudinal and lateral direction (i.e., an E_x and E_y), and some analyses define an *apparent* diffusivity based on a cumulative, rather than instantaneous, change in σ_r^2 (i.e., $E_{ra} = \sigma_r^2/4\Delta t$). In order for a predictive model not to over or under account for mixing, care should be taken to define E_r in the same way in the model that it was defined in the analysis of field measurements used to determine its value.

Horizontal mixing can be considered important to the dilution process when it has caused the patch concentration to be diluted by a significant amount. Again using Eq. (4-4), the time required for patch size to increase from σ_r to $\sqrt{2}\,\sigma_r$ (a two-fold increase in dilution) is

$$\Delta t_{\text{double}} = 2.4\,\sigma_r^{0.85} \tag{4-6}$$

where Δt_{double} is in sec, and σ_r is in cm. For example, $\Delta t = 12$ hours for $\sigma_r =$ 1,000 m, and only about 1.7 hours for $\sigma_r = 100$ m. The fact that this time increases with σ_r suggests that horizontal mixing is more important for small spills, and that dispersants can be used more effectively when applied before substantial spreading has occurred (i.e., small σ_r). Of course, other factors affecting dispersant effectiveness are also time dependent. Tank studies, or small-scale field experiments, cannot be used to directly simulate horizontal mixing because the spills in such tests are too small, and there are additional artifacts due to the presence of walls.

While horizontal mixing data such as those compiled by Okubo (1971) usually include the effects of shear dispersion, it is interesting to consider this component separately and evaluate how it varies with sea state and dispersant effectiveness. One type of shear dispersion that was discussed previously involves larger droplets that become vertically entrained into the water column and later rise to the surface. Because the slick generally travels faster than the underlying water, the droplets will re-enter the slick at the "back-of-the-pack," leading to a long tail. This effect can be especially important nearshore, where vertical circulation is more pronounced. Indeed, this effect has been proposed as the reason oil from the *Braer* spill off the Shetland Islands was observed to travel in the opposite direction of the surface current (Proctor et al., 1994; Ritchie and O'Sullivan, 1994; Spaulding et al., 1994).

Smaller droplets that are (nearly) permanently dispersed, and hence behave like water, are also affected by conventional shear dispersion. Consider a parcel of marked seawater occupying a depth h_{char}. If there is a vertical gradient in the near-surface velocity of magnitude λ_z (dimensions of velocity per depth, or T^{-1}), the patch will experience a top-to-bottom velocity difference of $\Delta u = \lambda_z h_{char}$. Following Taylor's analysis of longitudinal dispersion (see Fischer et al., 1979), a shear dispersion coefficient E_{sd} (part of E_r) ~ $(\Delta u)^2 h_{char}^2 / E_z$ ~ $\lambda_z^2 h_{char}^4 / E_z$. Based on the previous discussion of h_{char}, E_{sd} is expected to increase strongly with increasing wave height and decreasing droplet diameter, suggesting an increase in shear-induced mixing, and hence dilution of dispersed oil, as sea state and dispersant effectiveness increase.

The above discussion clearly implies that the vertical dimension needs to be included in modeling the transport of dispersed oil—not just to represent the concentration field, but also to properly represent the velocity field (i.e., a model needs to realistically represent the vertical gradients in velocity). Normally this requires a 3-D model. In shallow water, dispersed oil may become distributed over the entire water depth. However, even in this case, vertical gradients in velocity are important for dispersing the oil and these gradients should be accounted for, either by explicitly simulating the vertical shear in a 3-D model, or by computing horizontal shear dispersion coefficients for use in a 2-D (depth-integrated) model. In deeper locations where the dispersed oil is not uniformly distributed over depth, the oil will tend to be concentrated in a relatively thin horizontal layer near the surface. As with models of thermal or salinity stratification, this horizontal layering can present numerical challenges associated with resolving strong near-surface gradients. Resolution can be enhanced by employing models with stretched coordinates, such as σ-coordinates (that use a constant number of vertical grid cells regardless of water depth) or γ-coordinates (that, in addition, provide unequal grid spacing, allowing greater resolution near the surface). However, care should be taken to minimize or counteract the spurious vertical mixing that may result with such models due to the fact that the "horizontal" grid lines are not parallel with the stratification (Huang and Spaulding, 1995).

FATE AND WEATHERING

In addition to spreading and drift as discussed earlier, there are numerous processes that affect the ultimate fate of spilled oil or petroleum products (Figure 4-1). These include evaporation, dissolution, dispersion of whole oil droplets into the water column (entrainment), interaction of dissolved and dispersed components with suspended particulate material (SPM), photooxidation, biodegradation, uptake by organisms, water-in-oil emulsification (mousse formation), and stranding on shorelines (NRC, 1985, 1989, 2003).

Chapter 3 summarized the changes in rheological properties (viscosity, interfacial tension, density, etc.) that begin to occur immediately after oil is spilled. The changes in physical properties caused by water-in-oil emulsification are particularly important because they affect how spilled oil is physically dispersed (entrained) into the water column (with and without dispersants), the ability of oil spill skimmers to recover oil from the sea surface, the ability of pumps to transfer the collected oil, and the volume of collected material that requires storage and disposal.

In the following sections, the chemical and physical changes to oil on the water surface (generally thought of as weathering) caused by evaporation, photooxidation, and water-in-oil emulsification are discussed, with particular emphasis given to the latter (including identification of the chemical constituents within oil that largely control emulsion behavior) because of its importance in controlling dispersant effectiveness. After that, the fate of physically and chemically entrained oil droplets in the water column is considered. In evaluating the fate of entrained oil droplets, the primary focus is on biodegradation of dispersant-treated oil and the interaction of both physically entrained and dispersant-treated oil droplets with suspended particulate material.

Surface Oil Evaporation Weathering

Evaporation of lower-molecular-weight volatile components from a surface slick is important for dispersant applications because it can indirectly affect the formation of stable water-in-oil emulsions through the precipitation of asphaltenes and resins that help to stabilize the emulsion (Fingas and Fieldhouse, 2003). As the solvent components are evaporated from the slick, these higher-molecular-weight components can precipitate to coat entrained water droplets in the emulsion and inhibit water-water droplet coalescence and phase separation (Sjoblom et al., 2003). In addition, the evaporative loss of mono-aromatic components (benzene, toluene, xylenes, etc.) and two- and three-ring polynuclear aromatic hydrocarbons (PAH) and their alkyl-substituted homologues can significantly reduce the toxicity of the oil and the concomitant water-soluble fractions generated after physical or chemically enhanced entrainment of oil droplets into the water column.

Evaporation is the single most important and rapid of all weathering processes (McAuliffe, 1989), and it can account for the loss of 20–50 percent of many crude oils, 75 percent or more of refined petroleum products, and 10 percent or less of residual fuel oils (Butler, 1975; Butler et al., 1976; NRC, 1985; 2003). Most of the early studies on evaporation focused on the loss of individual hydrocarbon components as a function of their vapor pressures and other factors such as temperature, wind speed, and sea state. Traditionally, loss of lower-molecular-weight hydrocarbons as a function of time and weathering conditions was tracked by gas chromatographic (GC) analyses of the residual components in the oil. For example, Figure 4-2 shows time-series chromatograms from Prudhoe Bay crude oil weathered in subarctic summer conditions in the flow-through wave-tank systems described in Chapter 3. Note that all compounds with volatilities (vapor pressures) greater than n-C_{11} (b.p. $< 400°$ F, roughly 204° C) were lost within the first twelve days. (In these and subsequent chromatographic profiles, individual peaks are identified by relative GC retention time indices (i.e., n-C_{11} = 1100, n-C_{12} = 1200, etc., with a peak eluting midway between n-C_{11} and n-C_{12} = 1150) as defined by Kovats (1958). Between the late spring and the following summer and fall period, there was little additional evaporative loss. There was, however, a significant change in the straight-chain/branched-chain hydrocarbon ratios due to selective bacterial degradation of the n-alkanes in preference to the isoprenoid components (Figure 4-2D).

While compound-specific and gravimetric pan-evaporation studies were useful in characterizing rates of evaporation processes and served as input for early model development (Mackay and Matsugu, 1973; Mackay and Leinonen, 1977; Mackay et al., 1980a,b; 1982; Payne et al., 1983), in the mid 1980s the concept of using True Boiling Point (TBP) distillation data (available for all crude oils and refined products) to generate "pseudo-components" for modeling evaporation behavior was introduced by Payne et al. (1984). The pseudo-component approach allowed a mass balance to be calculated for the oil remaining after evaporation (something that couldn't be done on a compound-specific basis), and it was validated for Alaska North Slope crude oil by National Oceanic and Atmospheric Administration (NOAA) and Minerals Management Service (MMS) sponsored laboratory and outdoor wave-tank studies and ultimately comparison of model predictions with measured oil weathering behavior after the T/V *Exxon Valdez* oil spill (Payne et al., 1991a).

Many oil weathering observations have been reported at accidental spills (Mackay, 1993; Thomas and Lunel, 1993; Harris, 1997; Law et al., 1997; Lunel et al., 1996, 1997b) and various controlled dispersant field trials in Canada and Europe (Green et al., 1982; Bocard et al., 1987; Humphrey et al., 1987; McDonagh and Colcomb-Heiliger, 1992; Lunel and Lewis, 1993a,b; Lunel, 1993, 1994a,b; Lunel et al., 1995b,c; Walker and Lunel, 1995; Lewis et al., 1995a,b; Strom-Kristiansen et al., 1995; Brandvik et al., 1995, 1996; Lunel and Davies, 1996), plus numerous shorter-term laboratory and wave-basin studies on a variety of different oils with and without dispersants (Mackay and Chau, 1986; Brown and Goodman, 1987; Cormack et al., 1987; Daling and Lichtenthaler, 1987; Bobra, 1990; Brandvik and Daling, 1990; Daling et al., 1990b; Brandvik et al., 1991, 1992; Ross and Belore,

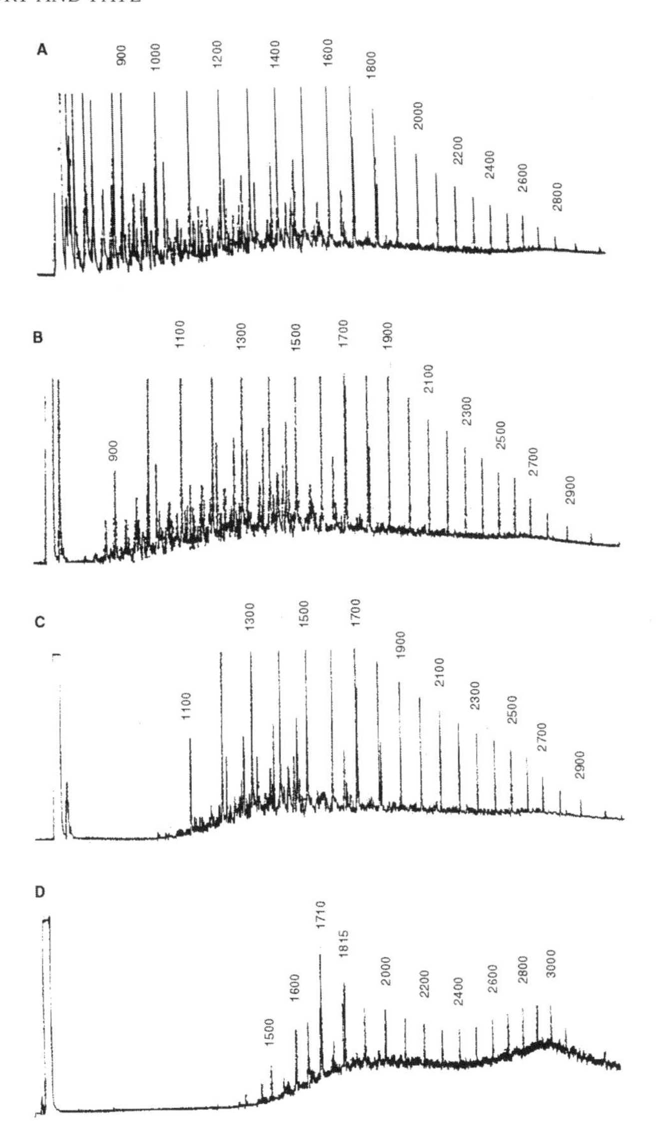

FIGURE 4-2 FID gas chromatograms of oil samples obtained from 2,800 L flow-through open-air summer wave-tank experiments using 16 liters of Prudhoe Bay crude oil. A) fresh oil; B) the oil slick after 48 hours of weathering, showing loss of the most volatile compounds; C) the oil slick after 12 days of weathering, showing the loss of all compounds with molecular weights less than n-C_{11} (Kovats Index 1100); and D) the slick after 12 months of weathering, showing significant biodegradation (Kovats Indices 1710 and 1815 represent the isoprenoids pristane and phytane, respectively).
SOURCE: Modified from Payne et al., 1984.

1993; Knudsen et al., 1994; Major et al., 1994; Strom-Kristiansen et al., 1994; Lunel et al., 1995a; Mackay, 1995; Hokstad et al., 1996; Venosa et al., 1999).

Many of these studies have been reviewed and recently summarized by S.L. Ross (1997) and in the proceedings of a two-day symposium held in Anchorage, Alaska, in March 1998 on oil spill dispersant applications in Alaskan waters (Trudel, 1998). The data from these studies have led to refinements in the overall ability to mathematically model oil spill behavior, and significantly more oil-specific data are now available on the magnitude and rates of change of density, viscosity, and water content for numerous oils and water-in-oil emulsions (Fingas and Fieldhouse, 2003; 2004a,b). No attempt will be made to duplicate or even briefly cover the findings presented in each of these most recent papers. Instead, several of the most significant observations specifically related to predicting evaporation and spilled oil behavior will be highlighted below.

A recent oil-weathering development has been the evaporation modeling approach proposed by Fingas (1996; 1997; 1999a) who has taken issue with the boundary layer regulation model based on earlier work by Mackay (Mackay and Matsugu, 1973; Mackay et al., 1980a,b; Stiver and Mackay, 1984; Berger and Mackay, 1994) that has been the basis for most evaporation algorithms used for predicting oil-weathering behavior. Fingas now proposes that evaporation rates are independent of oil film thickness and surface area and, instead, he has developed a set of empirical equations for estimating oil-specific evaporation rates as a function of exposure time (natural log or square root time-dependence) and the percentage of oil distilled at 180° C (roughly 356° F). Most of these experiments were conducted with a calculated film thickness (based on the cross-sectional area of the experimental evaporation dish and the volume of oil added) ranging from 0.8 to 10 mm. Tasaki and Ogawa (1999) have also reported that evaporation processes are not affected by oil film thicknesses in the range of 1 to 4 mm. A similar representation for evaporation from film thicknesses around 15 mm was also reported earlier by Bobra (1992). Additional discussions of the significance of Fingas' approach and counter arguments based on modifications of more traditional pseudo-component approaches are considered by Jones (1996,1997), who has proposed a simplified pseudo-component (SPC) model relating molar volume, vapor pressure, and molecular weight to the boiling point of the component. Thus, only the boiling points and initial volume fractions of the components need to be specified to implement the model.

Overprediction of evaporation rates can be a problem with oil-weathering models that assume a well-mixed oil phase (which is probably valid for very thin slicks) and also assume that resistance to mass transfer is entirely in the air phase (Berger and Mackay, 1994). Results from several studies suggest that evaporation rates may be controlled in the oil phase, especially at low temperatures and for higher viscosity water-in-oil emulsions (Payne et al., 1984, 1987c; Ross and Buist, 1995) and for waxy oils where a skin may form on the oil surface inhibiting component loss from within the oil phase (Berger and Mackay, 1994). As a result, it may be inappropriate to always model oil as a well-mixed phase, and algorithms for both well-mixed and diffusion-controlled fluids may need to be sequentially utilized as a function of oil weathering-dependent viscosity changes to better approximate spilled oil evaporative behavior. The possibility of oil-phase diffusion-controlled evaporative weathering was discussed at length by Payne et al. (1984). Experimental evidence for the importance of liquid-phase resistance for lower molecular weight compounds (e.g., hexane, cyclohexane, toluene, p-xylene) was presented by Berger and Mackay (1994), and experimental confirmation of the phenomenon for intermediate molecular weight components (decane through tetradecane) was obtained by Payne et al. (1987c; 1991b) during cold-room experiments examining the evaporation behavior of oil spilled onto ice.

In a related study, Ross and Buist (1995) reported that hydrocarbon evaporation was reduced when oil is mixed with water to form a stable water-in-oil emulsion. The degree of evaporation inhibition appears to increase with increasing water content and increasing slick thickness, which again suggests internal resistance to mass transfer within an emulsified slick, in line with the observations of Payne et al. (1987c; 1991b).

Most research indicates that differences in evaporation rates due to different slick thicknesses should be considered in evaporation weathering algorithms. These observations are in direct contrast with the findings by Fingas (1996; 1997; 1999a), who concluded that evaporation was not a function of wind speed, turbulence level, slick area, or thickness. In the modeling approach used by S.L. Ross (1997), smaller slicks are emulsified faster, yielding higher viscosities because of faster evaporation caused by thinner films. This is a subtle effect, but it is worth noting in developing models to predict oil weathering and slick behavior. The S.L. Ross model also predicts that smaller slicks will dissipate faster, which is at variance with the viscosity prediction, because Payne et al. (1984), Lunel et al. (1997b), and others have observed that as viscosity increases, natural dispersion of oil droplets decreases and eventually becomes self-limiting.

Photooxidation

Numerous laboratory studies have been completed on photochemical oxidation of oil, and in general, increases in the water-soluble fraction of most crude oils are readily apparent. As discussed in greater detail below, the photochemical generation of additional polar products (resins, carboxylic acids, ketones, aldehydes, alcohols, and phenols) with low hydrophilic-lipophilic balance (HLB) values that remain in the oil phase can also lead to the formation and stabilization of water-in-oil emulsions with greater water content (Payne and Phillips, 1985b; NRC, 1985; Daling and Brandvik, 1989; Daling et al., 1990b; Lewis et al., 1994). In addition, photooxidation of oil on the water surface can result in higher-molecular-weight products through the condensation of peroxide and other free-radical intermediates to yield intractable tar and gum residues (NRC, 1985; 2003). To the extent that chemical dispersion of surface oil can remove it from exposure to direct sunlight (or prevent it from stranding on shorelines where additional direct photo-transformations and tar/gum formation can occur), the effects of photooxidation as described below may be reduced.

Payne and Phillips (1985a) reviewed the earlier literature on the photooxidation of petroleum, and details on component-specific transformations, photooxidation products, viscosity changes, reaction mechanisms, the role of various sensitizers, reaction rates, etc., are summarized in that paper. An even more comprehensive treatise on organic photochemistry is available in Schwarzenbach et al. (1993) who review the basic principles of photochemistry, the roles of direct and indirect (sensitized) photolysis of numerous organic compounds in aqueous solutions, and the effects of particulates on photolytic transformations. While several PAH components are discussed, their treatment doesn't focus specifically on petroleum, and additional details on oil-related compounds can be found in Kochany and Maguire (1994) who completed a critical review of the chemical and photooxidation of PAH and polynuclear aromatic nitrogen heterocycles (PANHs) in water. More recently, Garrett et al. (1998) studied photooxidation of PAH in a variety of crude oils, and a general overview is presented in NRC (2003).

In general, aliphatic hydrocarbons in oils are more resistant to photochemical oxidation whereas aromatic compounds are particularly sensitive, and alkyl substitution increases the sensitivity of the aromatic compounds. Aliphatic sulfur compounds were more easily oxidized compared to aromatic thiophene compounds, with the sulfur in the aliphatic components being oxidized to sulfoxides, sulfones, sulfonates, and sulfates in approximately equal amounts. PAH degrades to relatively stable quinones via reactions initiated by electron transfer from singlet state PAH to molecular oxygen (Sigman et al., 1998), and natural organic mater (humic and fulvic acids) in seawater may enhance indirect photolysis of PAH through the generation of triplet excited states (NRC, 2003).

The extent of photooxidation of dissolved petroleum constituents is controlled by the spectrum and intensity of incident light, and photooxidation occurs faster with shorter-wavelength light (<300 nm), which is rapidly absorbed by seawater and natural dissolved organic matter (which can both enhance and inhibit photochemical processes).

Another more recent finding with regard to photochemical processes is the apparent increase in toxicity to transparent oil-exposed organisms when they are subsequently exposed to sunlight. Phototoxicity may occur by two processes: photomodification and photosensitization. Photomodification (or photooxidation) is the structural modification of a chemical in the oil or water column to more toxic or reactive oxidation products as described in NRC (1985, 2003), Garrett, et al. (1998), Kochany and Maguire (1994), and Payne and Phillips (1985a). In photosensitization, the bioaccumulated chemical transfers light energy to other molecules within the organism causing tissue damage. Phototoxic components in oil are primarily three- to five-ring PAH and heterocycles. The importance of this phenomenon as it relates to the toxicological effects of dispersed oil is discussed in Chapter 5.

Water-in-Oil Emulsification

Significant progress has been made in the identification of factors affecting water-in-oil emulsification (Bridie et al., 1980a,b; Zagorski and Mackay, 1982; Payne and Phillips, 1985b; Mackay, 1987; Bobra,1990; 1991; Fingas and Fieldhouse, 1994; 2003; 2004a,b; Fingas et al., 1995a,b; 1996b; 2002a,b; 2003a; Walker et al., 1993a,b, 1995; McLean and Kilpatrick, 1997a,b; McLean et al., 1998; Sjoblom et al., 2003). It has long been recognized that the indigenous petroleum emulsifying agents are contained in the higher boiling fractions (boiling points >350–400° C [roughly >662–752° F]), and particularly in the non-distillable residuum (Lawrence and Killner, 1948). These higher boiling fractions contain the higher-molecular-weight asphaltenes and resins that are now recognized as the necessary emulsifying agents for stable water-in-oil emulsion formation. These higher-molecular-weight components are believed to orient within the continuous oil phase at the water-droplet/oil interface where they retard recoalescence of the water droplets to form separate water and oil phases.

It is now known that, to be effective, these emulsifying agents should be in the form of precipitated, finely divided, submicron particles (Bobra, 1990, 1991; McLean and Kilpatrick 1997a,b; McLean et al., 1998; Sjoblom et al., 2003). Secondly, it has been shown that the lower-molecular-weight alkane and aromatic components in fresh crude oils serve as solvents to control the *in-situ* solubility and precipitation behavior of these higher-molecular-weight constituents within the oil phase. The chemical composition of the oil also determines the amount and size of the precipitated asphaltene and resin particles, as well as the "wetting" properties of those particles.

Fingas et al. (2002a, 2003a) and Fingas and Fieldhouse (2003, 2004a,b) classified four "states" that describe how water can exist in combination with oil. These include: stable emulsions, unstable water-in-oil mixtures, meso-stable emulsions, and (simply) entrained water. These states are differentiated by rheological properties as well as by differences in visual appearance, and very few emulsions were reported by these authors to have questionable stability. The viscosity of a stable emulsion can be as much as three orders of magnitude greater that the starting oil, and the product has significant elasticity. Stable emulsions are also usually reddish or red-brown in color. An unstable emulsion usually has a viscosity no more that about 20 times greater than the starting oil, and no elasticity is observed. Unstable or entrained-water mixtures are always the color of the starting oil (brown or black). Meso-stable emulsions are emulsions that have properties between stable and unstable emulsions, and can be either reddish or brown/black in color. Fingas et al. (1999) hypothesized that meso-stable emulsions lack sufficient asphaltenes to render them completely stable or that they still contained too many de-stabilizing materials such as the smaller aromatic solvent components that solubilize the asphaltenes. If the viscosity of the initial oil is high enough, it can stabilize some water droplets for a period of time in a meso-stable state. However, meso-stable emulsions may evolve to form either separate layers of oil and water or stable emulsions. Unstable emulsions are those that rapidly decompose to separate water and oil phases after mixing energy is removed, generally within a few hours. Some water (usually <10 percent) may be retained by the oil, especially if the oil is viscous.

The type of emulsion produced is determined primarily by the properties of the starting oil, and the most important of these are the asphaltene and resin content as noted above, and the initial viscosity of the oil (Fingas and Fieldhouse, 2003). At one time, waxes were thought to be important in the formation of water-in-oil emulsions, but Fingas et al. (2000a) have shown that they are not a factor in the formation of either stable or meso-stable emulsions. They may, however, play a role in certain circumstances by temporarily stabilizing entrained water with highly viscous oils. Stable emulsions have more asphaltenes and fewer resins, and interestingly, a narrow initial viscosity window. Instability results when the parent oil has too high or too low an initial viscosity, and as a result, the formation of stable emulsions may not occur with highly viscous oils. While this may initially appear to be contrary to intuitive reasoning, it has been explained by diffusion-controlled migration of asphaltenes and resins being too slow in the highly viscous oils to permit water-water droplet stabilization. In line with these observations, Daling and Brandvik (1989) previously reported that the maximum water uptake versus initial parent oil viscosity actually decreases as higher initial oil viscosities increase (Figure 4-3).

Turbulent kinetic energy is the most important form of energy related to emulsion formation. Although they were unable to measure turbulent

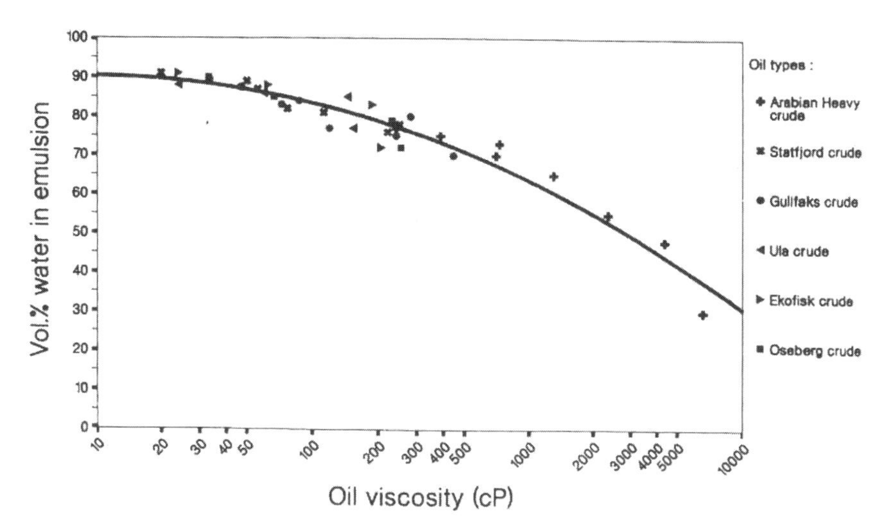

FIGURE 4-3 Maximum water uptake in water-in-oil emulsions versus initial parent oil viscosity. The mixing time is 24 hours at 6 and 13° C. Note that water uptake decreases as initial oil viscosity increases.
SOURCE: Daling and Brandvik, 1989.

energy directly in their laboratory experiments, Fingas et al. (1999) and Fingas and Fieldhouse (2003) correlated emulsion stability with total kinetic energy (proportional to the rotational speed of their rotary agitator and measured in ergs) and work (a measure of the power input to the agitator integrated over time, recorded in joules—same dimensions as energy—and proportional to time for a given energy). Neither of these metrics was normalized by the mass of oil, so results can only be interpreted in a relative sense. Oil that forms an entrained water state required relatively little threshold energy (200–300 ergs), and showed no increase in stability with increasing energy. Oil that forms a mesoscale emulsion required a relatively high level of energy (about 25,000 ergs) but also displayed no increase in stability with additional energy. Meanwhile oils that form stable emulsions showed increasing stability with increasing energy. Figure 4-4 displays the trend in emulsion state as a function of time (work) for oils displaying various final emulsion states. An implication is that formation of a given emulsion type at sea may require both a threshold energy level (corresponding to a given sea state) and a finite period of time. Because of the lack of higher-molecular-weight asphaltenes and resins, most light refined products (such as gasoline, kerosene, heating oil, and diesel fuels) do not easily form a stable water-in-oil emulsion.

In addition to the influence of indigenous asphaltenes and resins in

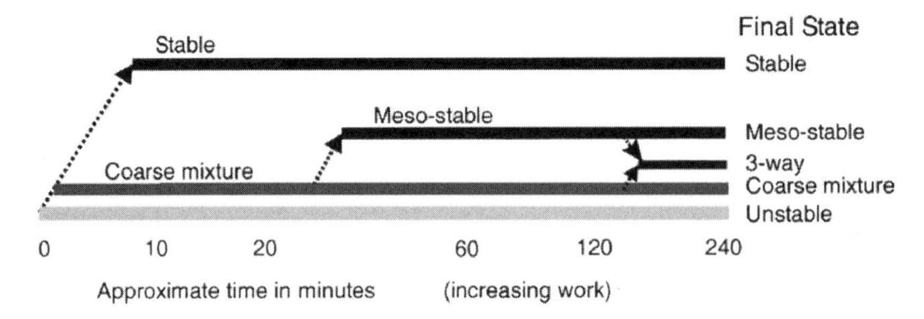

FIGURE 4-4 The overall concept of state and approximate kinetics in emulsion and water-in-oil state formation.
SOURCE: Fingas and Fieldhouse, 2003; courtesy of Elsevier.

the oils themselves, oxidation products from oil photolysis-weathering have also been suspected to play a significant role in water-in-oil emulsification behavior (Payne and Phillips, 1985a,b; NRC 1985). Daling and Brandvik (1989) correlated the increase in resin components in photolyzed oils with increased water contents and smaller-sized water droplets in the resultant water-in-oil emulsions. Specifically they found that photolysis results in:

- the formation of polar compounds (resins)
- significant reductions in oil-water interfacial tension
- slightly increased rates of water-in-oil emulsion formation (not for waxy crudes)
- slightly higher maximum water content in the water-in-oil emulsion
- significantly higher stability in viscosity of the water-in-oil emulsion
- a higher concentration of chemicals needed to break or inhibit water-in-oil emulsion formation, and
- oil-specific changes in chemical dispersibility

In subsequent dispersant tests, Daling and Brandvik (1989) reported that naphthenic crudes were more highly dispersible after photolysis. Waxy crudes showed drastic reduction in dispersibility. Photolyzed components were believed to facilitate the accumulation of waxes at the oil-water interface, which were believed to "block" the access and penetration of the dispersant's surfactant into the oil phase. Without such stabilizing agents the water droplets in a water-in-oil emulsion will tend to coalesce and separate from the oil phase.

In an attempt to generate a database of oil properties that might be useful in predicting water-in-oil emulsification and dispersibility behavior as a function of the degree to which oil weathering has occurred, Daling et al. (1990b) undertook the systematic study of emulsification behavior on three standard test oils that had been artificially weathered in the laboratory. The results demonstrated that the rate of weathering processes (water uptake, viscosity increases, etc.) was critically dependent on the type of oil studied and the different environmental conditions. Even when the variations in the physical-chemical parameters of the starting oils were relatively small, the weathering behavior of the oil and effectiveness of the different dispersants varied significantly. The authors suggested that this approach be considered by other laboratories and used in a similar way to generate a larger database of valuable information, which could be used in oil weathering modeling. Using experimental weathering data available for a wide range of oils, they hoped that correlations could be developed for model predictions of oil-weathering behavior based on only generally available crude oil assay data (e.g., true boiling point curves, density, pour point, wax, resin, and asphaltene content).

Lewis et al. (1994) expanded Daling et al.'s weathering approach and incorporated a mesoscale flume to investigate the chemical dispersion of oil and water-in-oil emulsions after different stages of weathering. In addition to the more common laboratory bench-scale methods used for assessing chemical dispersibility, the flume approach allowed testing of higher viscosity water-in-oil mixtures and more closely approximated conditions that might be encountered at sea.

This approach, or modifications of it, has now been successfully used to investigate the oil weathering properties and dispersibility of a wide variety of heavier and more viscous crude oils and emulsions (Guyomarch et al., 1999a,b,c; Fiocco et al., 1999), and the more recent data have significantly expanded the viscosity-limited range of dispersant effectiveness (and the concomitant time window available for responding to an oil spill at sea). In addition, the flume approach has allowed the rapid and cost-effective evaluation and testing of various emulsion breaking chemicals and the effectiveness of multiple dispersant applications and sequential emulsion breaker/dispersant combinations to disperse particularly recalcitrant slicks.

Without question, water-in-oil emulsification is a critically important process that affects oil droplet entrainment and dispersant effectiveness, and for many years it was believed to be the most difficult process to model or predict on a oil-specific basis (S.L. Ross, 1997). Except for the few oils that had been extensively tested, it was virtually impossible to predict when a particular oil would start to emulsify or how long it would take for the spilled oil to form a "stable" highly viscous emulsion. Likewise, it was not possible to predict the final water content that a water-in-oil emulsion might contain.

Nevertheless, oil spill models have to deal with the problem of emulsification because it is such an important process. Traditionally, many oil spill models used data from an older laboratory test, called the Mackay-Zagorski Test (Zagorski and Mackay, 1982) that was developed to measure: (1) an oil's tendency to form an emulsion, and (2) the stability of the emulsion once formed. The test did not, however, predict the rate of emulsification under field conditions. Likewise, the conventional emulsification equation (Mackay and Matsugu, 1973) that had been generally used in oil-weathering modeling includes some inconsistencies. For example, it does not include the influence of initial oil thickness and the progress of evaporation on emulsification. Emulsified oil takes up water as evaporation progresses, thus maintaining a constant oil-to-water (OWR). Therefore, Tasaki and Ogawa (1999) developed a new equation and governing parameters for emulsification of crude oil. Through the differentiation of a formula defining the water content ratio, a water-in-oil emulsification equation for crude oil was derived to include the effects of evaporation loss and entrained water with time. The equation reveals that the emulsification is governed by two fixed parameters (related to the evaporation process) and four free constant parameters selected to fit the numerical solutions to the measured values obtained in flume tests on six types of Middle East crude oils.

Noting the requirement for the loss of lower-molecular-weight alkane and aromatic solvents to precipitate asphaltenes for the formation of stable mousse with Alaska North Slope (ANS) crude oil, S.L. Ross (1997) and NOAA (in their Automated Data Inquiry for Oil Spills [ADIOS] 2 oil weathering model) do not initiate the formation of a water-in-oil emulsion in their computer models until after a specified percent evaporation for the crude oil has occurred.

To further this research effort and develop an empirical database that can be used in a predictive sense, Fingas and Fieldhouse (2003) have examined the emulsion forming tendencies of over 200 oils. The resulting emulsified products were characterized as a function of time (day of formation and after one week) by viscosity, complex modulus, elasticity modulus, viscosity modulus, water content, and several other parameters, including visual appearance. These empirical data were then used to develop a numerical model that uses the density, viscosity, and the saturate, asphaltene, and resin contents to compute a class index, which in turn yields either an unstable or entrained water-in-oil state or a meso-stable or stable emulsion (Fingas and Fieldhouse, 2004a,b). This approach has been used to develop a prediction scheme to estimate the water content and viscosity of the resulting water-in-oil state and the time to formation with input of wave-height. When compared to the laboratory data upon which the empirical approach was based, this model was reported to provide accurate predictions of stability class about 50 percent of the time, and 90 percent of the predictions were no more that one category off. Predictions could not be compared to field data because there are very few data available for the comparison; however, this empirical approach clearly advances the ability to make *a priori* predictions about how a particular oil might behave in a spill situation.

Fate of Physically and Chemically Entrained Oil Droplets in the Water Column

Physical Entrainment of Untreated Oil

Notwithstanding the fact that most oils will not readily sink (NRC, 1999), fresh oil can be temporarily entrained/driven into the water column by wind and wave turbulence as described previously. In such instances, however, droplets above a certain size range (generally greater than 60–80 µm or 0.06–0.08 mm) would be expected to quickly resurface after the turbulence regime subsides. Then, as the viscosity of the surface slick increases due to evaporation and water-in-oil emulsification (mousse formation), it becomes increasingly difficult for wind-driven waves to plunge discrete oil droplets into the water column.

In the wave-tank systems used by Payne et al. (1983; 1984; 1991a) for subarctic oil weathering studies with Prudhoe Bay crude oil, the total entrained oil droplet concentrations in the water column were over 9,000 µg/L immediately after the oil release, 2–3 µg/L after 4 hours, 0.5 µg/L after 2 days, and less than 0.1 µg/L after 12 days. Initial chromatograms of the entrained oil droplets appeared essentially identical to those from the simultaneously collected surface oil samples (Figure 4-2), but between 8 and 48 hours, there was evidence for slightly enhanced evaporation and/or dissolution loss of lighter molecular weight components in the physically entrained oil droplets compared to the surface oil slick. Presumably, this loss resulted from the increased surface-area-to-volume ratio of the smaller physically entrained oil droplets compared to the more continuous surface oil slick. After 12 days with continued constant turbulent mixing in the wave tank there was little evidence of physically dispersed oil droplets in the water column.

Energy dissipation rates and oil droplet-size distributions were not measured in the wave-tank studies by Payne et al., but these parameters have been measured in other studies (Delvigne and Sweeney, 1988; Sterling et al., 2004a). Delvigne and Sweeney (1988) measured droplet sizes in their turbulent grid and breaking-wave experiments with both Ekofisk and Prudhoe Bay crude oils. These studies showed that droplet sizes between 60 and 200 μm were obtained with the non-dispersant-treated oils in their turbulent-grid column experiments at the highest turbulence regimes tested, and that droplet size diameters increased significantly as the oil viscosity increased and the turbulence regime decreased. In their breaking wave experiments, Delvigne and Sweeney (1988) measured droplet sizes from 6 to >800 μm with the highest number concentrations in the 6–50 μm size range with greater numbers of the smaller droplets driven deeper into the water column and greater numbers of larger droplets near the surface. Clearly, as oil viscosity increased, droplet dispersion was inhibited with a concomitant increase in the proportions of larger droplets with faster rising velocities being generated. Similar findings of inhibited oil droplet dispersion/entrainment from higher-viscosity water-in-oil emulsions have also been reported by Lewis et al. (1994).

Enhanced Entrainment of Smaller Droplets with Dispersants

Franklin and Lloyd (1986) presented various size distributions for oil:dispersant mixtures studied in the laboratory using a toxicity test developed by the United Kingdom's Ministry of Agriculture, Fisheries and Food. All droplet sizes were reported as volume median diameter, which is the droplet diameter that divides the sample distribution into two equal parts by volume. The hydrocarbon solvent based dispersants yielded droplet histogram plots that peaked at 20 μm (the mode diameter). Water-dilutable concentrates yielded drops in the 25–65 μm range. Dispersants that were concentrates applied to the oil undiluted generated a flat bimodal distribution with a large proportion of droplets <5 μm.

More recently, Lunel (1993a,b; 1995b) reported the first successful field measurements of oil droplet-size distribution below experimental dispersant-treated oil slicks at sea from a premixed oil-dispersant combination (medium fuel oil and Slickgone NS) measured at sea using a Phase Doppler Particle Analyzer (see Figure 3-3 in Chapter 3). For a variety of test oils and dispersants, the range of mean diameters was between 15 and 25 μm (volume distribution 35 to 50 μm). Smaller droplet sizes (or increased number densities of smaller droplets) were observed in both instances.

Compound-Specific Dissolution Behavior

True dissolution of individual components from an oil slick is not generally significant in terms of the overall mass balance of an oil spill (NRC, 1985; 2003). As a result, many oil-weathering models generally do not include dissolution in their mass-balance calculations. Dissolution of individual components is important, however, when considering the potential for biological impacts.

The dissolved concentrations of individual components from an oil slick are controlled by partition coefficients, rather than the solubilities of individual oil components. Payne et al. (1984) and Payne and McNabb (1984) presented data for Prudhoe Bay crude oil:seawater partitioning reporting that the truly dissolved components were almost exclusively alkyl-substituted lower-molecular-weight mono-aromatic hydrocarbons (MAH) and two-ring PAH. The water-soluble fraction contained no appreciable n-alkanes. Published octanol:water partition coefficients, K_{ow}, for many parent (and fewer alkyl-substituted) PAH can be used to predict dissolution behavior (Nirmalakhandan and Speece, 1988; Hodson et al., 1988; Blum and Speece,1990; McCarty, 1986; McCarty et al., 1992; Mackay et al., 1992; McCarty and Mackay, 1993; Varhaar et al., 1992; Swartz et al., 1995; French-McCay et al., 1996; French-McCay, 1998; 2001; 2004). For modeling dissolution behavior, partition coefficients, not pure component solubility data, should be used.

In the case of an oil spill, true dissolution of individual components is controlled by the mole fraction of each component in the slick, the oil/water partition coefficient, the oil-water interfacial surface area (which significantly increases with successful dispersant application), and the interphase mass transfer coefficient. During a spill, however, a static equilibrium can never be established because the dissolved components are removed (diffused and advected) away from the surface oil source (and dispersed oil droplets), mixed with fresh uncontaminated water, and subjected to evaporation loss from the water column itself.

As a result of these processes, dissolution should not be modeled as an equilibrium process, but instead as a kinetics-controlled process where the driving force is determined by the distance of the system from equilibrium. The oil-water interfacial area should be a term in the rate equation, and the aqueous-phase concentration could be modeled as being arbitrarily low (e.g., zero) or through a mass balance on the water-soluble components. In a study related to this conceptual approach, Page et al. (2000b) reported on the importance of kinetics, thermodynamics, and colloidal phenomena in controlling the partitioning of naphthalenes from West Texas and Arabian medium crude oils into seawater. Under conditions of light turbulence, the alkyl-substituted naphthalenes showed an inverse correlation between both the dissolution rate coefficients and saturation concentrations and the degree of alkyl-substitution. At higher turbulence levels and variable oil loadings, there was a direct correlation between the measured total petroleum hydrocarbon concentrations in the water and the nominal oil loading; however, there was no such correlation between the naphthalenes and oil loading. It was concluded that the first experiment was controlled by dissolution kinetics across the oil/water boundary and solubility phenomenon, while the second experiment also included a colloidal oil-phase contribution to the measured constituent concentrations.

Sterling et al. (2003) examined the partitioning of PAH components from oil droplets into the water column and the influence of the shear stresses used to generate those droplets. Their approach was based on Raoult's Law using pure component solubilities and individual-component mole fractions in the oil phase. At low shear stresses generating relatively low dispersed oil droplet concentrations, the influence of PAH solubility dominated the observed PAH concentrations. At higher shear stresses and dispersed oil droplet loadings, the PAH concentrations in the water column were influenced primarily by the mole fraction of the individual PAH in the entrained oil droplets.

To assess the time-series water-column concentrations in the subarctic flow-through wave-tank studies discussed above, Payne et al. (1984) filtered 20-liter subsurface seawater samples through 0.7 μm pore-size 293 mm diameter glass-fiber filters for separate analyses of the dissolved components and dispersed- and/or particulate-bound oil droplets (Gordon et al., 1973). Chromatograms of dissolved components measured in time-series filtered seawater samples from the wave-tank systems are shown in Figure 4-5. These chromatograms are characterized only by the individual aromatic components, not the evenly repeating series of n-alkanes that predominate in the dispersed oil droplets and surface oil samples (Figure 4-2). As shown in Figure 4-5A, dissolution of lower-molecular-weight aromatic components began immediately, as little as five minutes after the spill. The major dissolved components measured over time include benzene, toluene, xylene(s), ethylbenzenes, C_3-substituted benzenes, naphthalene, methylnaphthalenes, C_2- and C_3-substituted naphthalenes, phenanthrene, and C_1- and C_2-substituted phenanthrenes.

As shown in Figures 4-6 and 4-7, maximum dissolved-component concentrations were observed very early in the spill. Dissolved-phase total benzene, toluene, ethylbenzene, and xylenes (BTEX) concentrations reached a maximum of 250 μg/L during the first 2 hours of the spill and were less than 25 μg/L after 2 days. The maximum total PAH concentration was somewhat delayed, occurring between 4 and 12 hours (Figure 4-7). Subsequent water-column concentrations decreased from a combination of evaporative losses from the water surface and advective/diffusion processes (simulated by one tank-volume water exchange every 4 hours). Concomitant (and orders of magnitude greater) evaporative losses of these same components also occurred from the surface slick over the same time frame. However, the higher-molecular-weight PAH were persistent in the

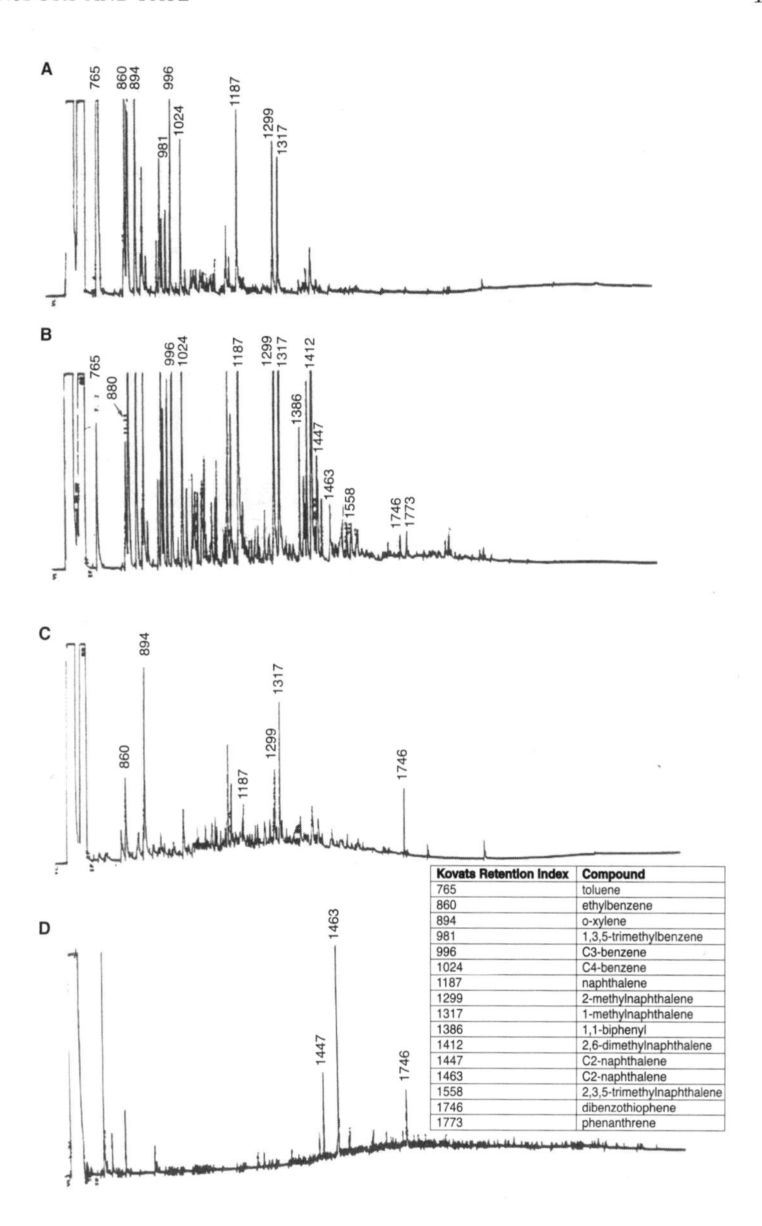

Kovats Retention Index	Compound
765	toluene
860	ethylbenzene
894	o-xylene
981	1,3,5-trimethylbenzene
996	C3-benzene
1024	C4-benzene
1187	naphthalene
1299	2-methylnaphthalene
1317	1-methylnaphthalene
1386	1,1-biphenyl
1412	2,6-dimethylnaphthalene
1447	C2-naphthalene
1463	C2-naphthalene
1558	2,3,5-trimethylnaphthalene
1746	dibenzothiophene
1773	phenanthrene

FIGURE 4-5 FID gas chromatograms of filtered water samples showing dissolved-phase components obtained from 2,800-liter flow-through open-air summer wave-tank experiments using 16 liters of Prudhoe Bay crude oil after: A) 5 minutes; B) 48 hours; C) 12 days and D) 12 months of weathering. Selected components are identified by Kovats Retention Index (see text and inset).
SOURCE: Modified from Payne et al., 1984.

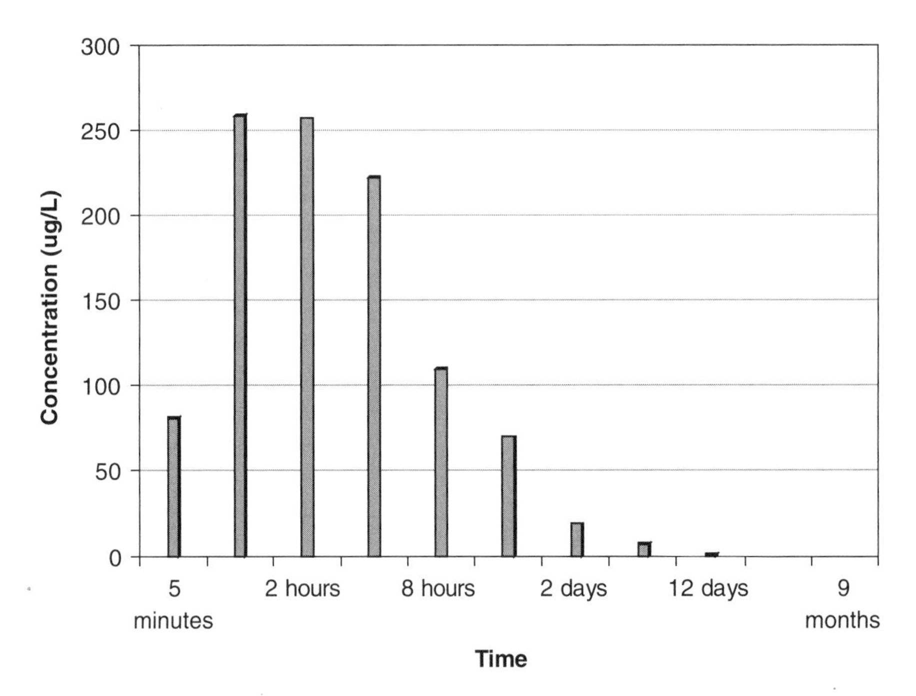

FIGURE 4-6 Total dissolved benzene, toluene, ethylbenzene, and xylenes (BTEX) concentrations in ppb over time from flow-through open wave-tank experiments using Prudhoe Bay crude oil. BTEX concentrations peaked at 1–2 hours.
SOURCE: Data from Payne et al., 1984.

oil, and continued dissolution of C_2-substituted naphthalenes and alkyl-substituted phenanthrenes from the surface oil occurred for periods of up to 4 to 7 months (Figure 4-8) even though there was little or no physical dispersion of oil droplets occurring at that time. There was no evidence of significant aromatic hydrocarbon dissolution into the water column after 13 months in the wave-tank systems.

Clearly, the absolute concentrations of oil droplet- and dissolved-phase components presented in Figures 4-5 through 4-8 were influenced by the size of the wave tank systems (2,800 L), the water-column turnover rate or residence time (one tank volume every 4 hrs), and the volume of oil (16 L) used for the experiments. Slick spreading was also inhibited by the walls of the tanks, so extrapolating these data to actual open-ocean conditions must be done with caution. Nevertheless, it is worthy of note that excellent agreement was obtained between the wave-tank data and observed changes in oil rheology and oil chemistry after the T/V *Exxon Valdez* oil spill (Payne et al., 1991a). Both the wave-tank studies and the

EVOS observations are representative of ice-free Alaskan subarctic conditions, and the rates of evaporation, emulsification, (and possibly to a lesser extent, dissolution) would be different in warmer environments, but the processes themselves would still occur. Computer models can help define the anticipated changes in these rates for other environmental conditions and oil types. The data are included here to provide insight on the dynamic nature of the dissolved- and oil-phase component concentrations in the water column that ultimately drive toxicity considerations.

Although the use of chemical dispersants will clearly increase the upper water column concentration of entrained oil droplets, and theoretically should lead to enhanced dissolution of water-soluble PAH components (French-McCay and Payne, 2001), no field measurements of this phenomenon have been successfully completed to date. It is known that enhanced dissolved-phase concentrations occur with subsurface blowouts (Brooks et al., 1980; Payne et al., 1980a,b; Boehm and Fiest, 1980, 1982; Fiest and Boehm, 1980) and after extensive surface entrainment of fresh

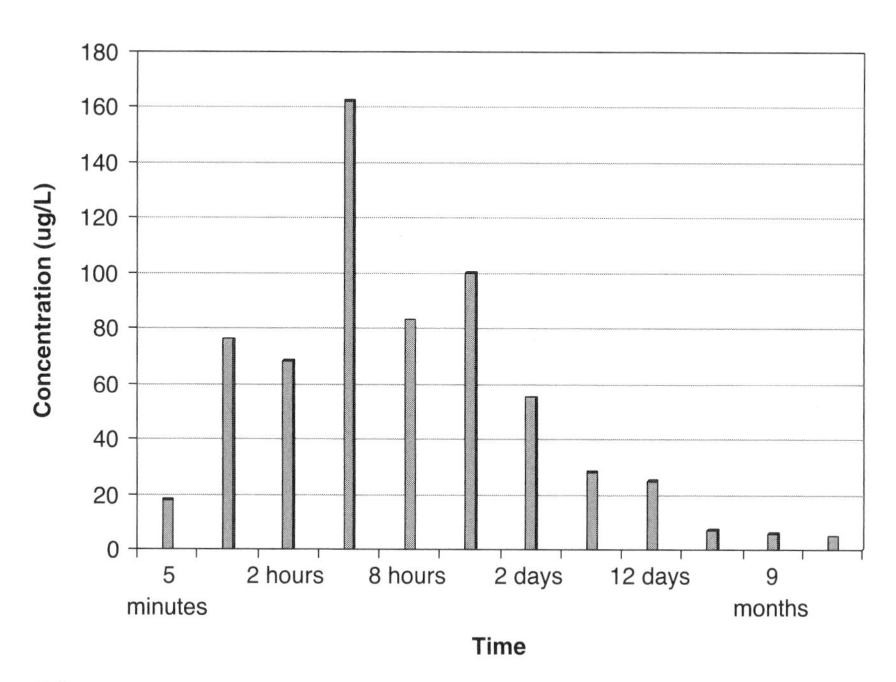

FIGURE 4-7 Total dissolved polynuclear aromatic hydrocarbons (PAH) concentrations in ppb over time from flow-through open wave-tank experiments using Prudhoe Bay crude oil. PAH concentrations peaked at 4 hours but continued to dissolve from the surface slicks for one year.
SOURCE: Data from Payne et al., 1984.

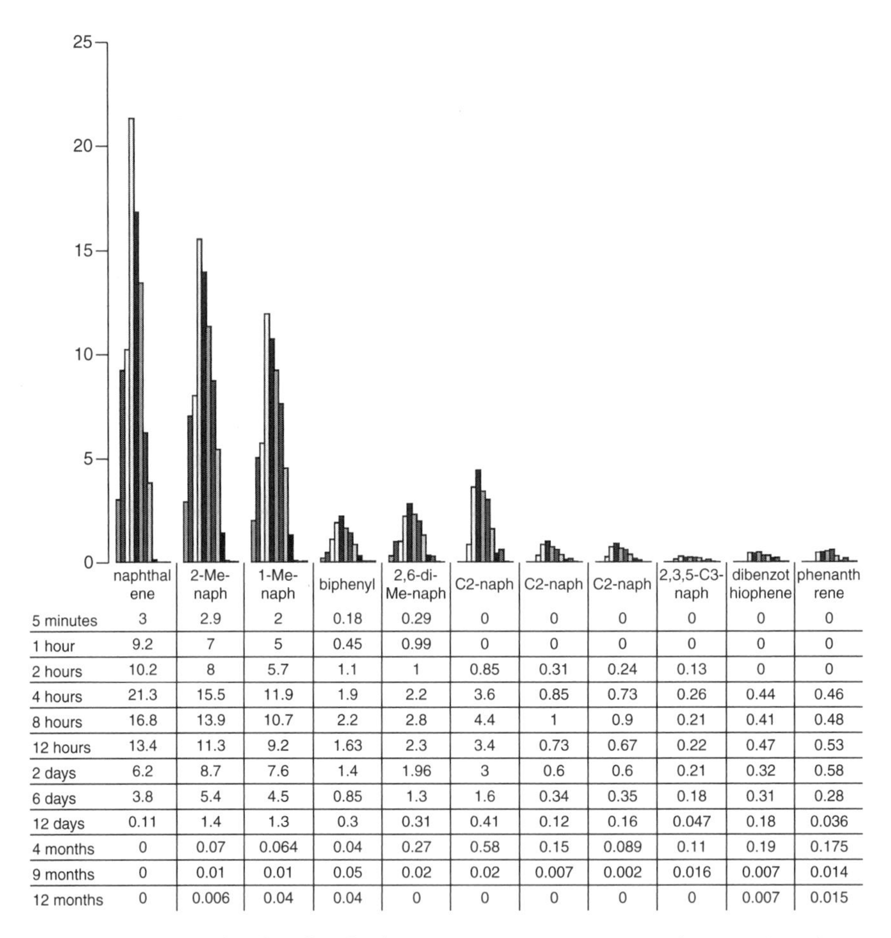

	naphthalene	2-Me-naph	1-Me-naph	biphenyl	2,6-di-Me-naph	C2-naph	C2-naph	C2-naph	2,3,5-C3-naph	dibenzothiophene	phenanthrene
5 minutes	3	2.9	2	0.18	0.29	0	0	0	0	0	0
1 hour	9.2	7	5	0.45	0.99	0	0	0	0	0	0
2 hours	10.2	8	5.7	1.1	1	0.85	0.31	0.24	0.13	0	0
4 hours	21.3	15.5	11.9	1.9	2.2	3.6	0.85	0.73	0.26	0.44	0.46
8 hours	16.8	13.9	10.7	2.2	2.8	4.4	1	0.9	0.21	0.41	0.48
12 hours	13.4	11.3	9.2	1.63	2.3	3.4	0.73	0.67	0.22	0.47	0.53
2 days	6.2	8.7	7.6	1.4	1.96	3	0.6	0.6	0.21	0.32	0.58
6 days	3.8	5.4	4.5	0.85	1.3	1.6	0.34	0.35	0.18	0.31	0.28
12 days	0.11	1.4	1.3	0.3	0.31	0.41	0.12	0.16	0.047	0.18	0.036
4 months	0	0.07	0.064	0.04	0.27	0.58	0.15	0.089	0.11	0.19	0.175
9 months	0	0.01	0.01	0.05	0.02	0.02	0.007	0.002	0.016	0.007	0.014
12 months	0	0.006	0.04	0.04	0	0	0	0	0	0.007	0.015

FIGURE 4-8 Dissolved individual PAH concentrations in ppb over time from flow-through open wave-tank experiments using Prudhoe Bay crude oil.
SOURCE: Data from Payne et al., 1984.

petroleum products, such as occurred with the *North Cape* (French-McCay, 2003). However, attempts to measure this during dispersant applications to accidental spills have not documented similar behavior.

Dispersants were used during the response to the *Mega Borg* spill of 3.9 million gallons (roughly 13,000 tonnes) of Angola Planca crude oil off Galveston, Texas, in 1990, and Payne et al. (1993) examined target and control areas of the treated slick to see if evidence of enhanced PAH dissolution could be observed. The measured concentrations of dispersed oil droplets were very heterogeneous and reflected the patchy distribution of oil on the water surface before dispersant application. Nevertheless, the ratio data for aliphatic:aromatic concentrations showed no evidence of significantly enhanced dissolution of lower- and intermediate-molecular-weight aromatics as a result of the dispersant treatment.

Payne et al. (1991d) demonstrated that dissolved aromatic compounds from oil introduced into refreezing leads can be advected as conservative components in the brine generated during frazil ice formation to the benthic bottom boundary layer in field experiments completed in the Chukchi Sea. If dispersants were applied to oil released in open water during freezing conditions or to oil contained in open leads/broken sea ice during a refreezing event, it is conceivable that the enhanced dissolution process predicted by French-McCay and Payne (2001) could lead to transport of dissolved aromatic components to the benthos before significant evaporative weathering could otherwise occur. Clearly, any such enhanced transport would be spill or location specific, as it assumes only minor horizontal transfer.

Biodegradation

The effects of surfactants and commercial oil dispersant mixtures on the rate and extent of biodegradation of crude oil, petroleum products, and individual hydrocarbons have been intensively investigated for over thirty years with mixed results. In some studies, biodegradation is stimulated, others find evidence of inhibition, and others observe no effects attributable to the presence of surfactants or commercial dispersants. Experimental systems have used a wide variety of substrates (e.g., crude oil, individual hydrocarbons), surfactants (e.g., commercial dispersant mixtures, pure surfactants), and microbial communities (e.g., natural seawater microbiota, microbial communities enriched by growth on crude oil, pure cultures). None of these factors appear to systematically affect the outcome. Instead, the effects of surfactants or commercial dispersant mixtures on the biodegradation rates of crude oil and defined hydrocarbons appear to depend on the chemical characteristics of the surfactants, the hydrocarbons, and the composition of the microbial community. Other factors, such as nutrient concentrations, oil-water ratios, and mixing energy, can also be expected to affect the observed biodegradation rate of dispersed oil.

One source of confusion in the literature on dispersant effects on oil biodegradation is that conclusions are often based on indirect evidence. For example, Corexit 9527 (DOR = 1:10) was shown to increase the rate of oxygen uptake in suspensions of South Louisiana and Kuwait crude oils relative to suspensions of physically dispersed oil (Traxler and Bhattacharya, 1978), and several dispersants increased the rate or extent of microbial growth on Arabian (Mulkins-Phillips and Stewart, 1974) or Forties (Swannell and Daniel, 1999) crude oil. Unfortunately, this stimulation cannot be unambiguously attributed to growth on dispersed oil as opposed to growth on the dispersants themselves, which are usually readily biodegradable and support microbial growth (Mulkins-Phillips and Stewart, 1974; Bhosle and Row, 1983; Bhosle and Mavinkurve, 1984; Lindstrom and Braddock, 2002). Corexit 9500 (DOR = 1:10), however, had no effect on the growth rate of hydrocarbon degraders on Alaska North Slope crude oil, and it slightly decreased the gross rate of oil mineralization (Davies et al., 2001; Lindstrom and Braddock, 2002). Total extractable material (TEM), which is a gravimetric measurement of the concentration of hydrophobic (e.g., oil and grease) compounds, is a somewhat more direct measure of oil biodegradation than is oxygen consumption or microbial growth rate, but it is still subject to interference from surfactants or other dispersant components. Based on this metric, Dispolene 34 S (DOR = 1:5) decreased the rate of biodegradation of Kuwait crude oil by about two-thirds relative to physically dispersed oil when both were present at realistically low concentrations in natural seawater (Literathy et al., 1989).

More direct evidence for the effects of dispersants on oil biodegradation rates involve measurement of oil composition, usually by gas chromatographic-flame ionization detector (GC-FID) or gas chromatographic-mass spectroscopy (GC-MS), or measurement of the rate of biotransformation of specific hydrocarbons to carbon dioxide (e.g., mineralization). Based on changes in oil composition, Corexit 9527 was found to inhibit the biodegradation of normal and branched alkanes and sulfur heterocycles in Prudhoe Bay crude oil, but biodegradation of PAH was not affected (Foght and Westlake, 1982; Foght et al., 1983). Corexit 9500 had no effect on the rate of n-alkane biodegradation in Alaskan North Slope crude oil when compared to physically dispersed oil (Davies et al., 2001). The rate of biodegradation of naphthalene and phenanthrene in the water-accommodated fraction (WAF) of a heavy residual fuel oil was not affected by the presence of a dispersant (Taiho Self-Mixing S-7), but the biodegradation rates of the 4-ring PAH pyrene and chrysene were faster (Yamada et al., 2003). It should be noted, however, that the initial concentrations of pyrene and chrysene in the WAF that was prepared with chemically dispersed oil were sufficiently high that their final aqueous-phase concentrations were higher after biodegradation than the initial concentrations in the water-only WAF.

Although disappearance of target compounds from samples provides a very detailed and sensitive view of changes in oil composition that occur during incubation under specified experimental conditions, the mechanism that causes the observed compositional changes and the fate of the compounds that disappear often must be inferred. It is, for example, often difficult to determine whether specific compounds were lost from the sample by physical, chemical, or biological processes, and even when biological transformation is indicated, the fate of the biotransformation products is not known (i.e., a single metabolic reaction, such as hydroxylation by a nonspecific oxygenase enzyme, is sufficient to cause a compound to disappear from a gas chromatogram, but it might not reduce the toxicity of the oil). Isotopically labeled substrates can provide more information on the ultimate products of biotransformation reactions, but these data are much more restricted in that they provide information only on the fate of specific compounds, which may or may not be representative of larger classes of petroleum hydrocarbons. Like other measures of dispersant effects on oil biodegradation, the effects on mineralization of specific hydrocarbons vary with target compound, dispersant, oil, and microbial community composition. For example, when n-[1-^{14}C]-hexadecane, [1-^{14}C]-pristane, or [9-^{14}C]-phenanthrene were added to Norman Wells crude oil (Foght et al., 1987), mineralization of pristane was inhibited by all dispersants that were tested, whereas hexadecane mineralization was stimulated by Corexit 7664 and Corexit 9600 and inhibited by Dispersol SD and W-1911. Although phenanthrene mineralization by the oil-degrading enrichment culture was stimulated by Corexit 9600, this dispersant inhibited its mineralization by indigenous bacteria in natural river water. Note that the position of the radiolabel is important in these studies, because it determines the extent of metabolism that must occur before the labeled carbon atom can be released as $^{14}CO_2$. In general, end-labeled substrates, such as n-[1-^{14}C]-hexadecane and [1-^{14}C]-pristane, will release $^{14}CO_2$ following much more limited metabolic transformation than is required for a molecule that is internally labeled, like [9-^{14}C]-phenanthrene. This distinction is probably relatively unimportant for n-alkanes, which are usually completely degraded following the initial oxygen-insertion reaction, but it can be somewhat more important for molecules for which partial metabolic transformations are more common (e.g., branched and cyclic alkanes, PAH).

The effect of dispersants on oil biodegradation rate is very sensitive to the chemical characteristics of the dispersant, even when all other factors (e.g., oil, microbial community) are kept constant. A survey of thirteen dispersants and two surfactants showed that none significantly stimulated the biodegradation rate of Norman Wells crude oil by an oil-degrading enrichment culture in an artificial freshwater medium (Foght et al., 1987). About half of the dispersants—and both of the surfactants—that were tested inhibited biodegradation of one or more classes of petroleum hydrocarbons (e.g., aliphatics, aromatics, sulfur heterocycles) as measured by GC-MS. The class of compounds whose degradation was inhibited and the degree of inhibition that was observed varied from dispersant to dispersant, even though the oil and the microbial culture were the same. A systematic investigation of the relationship between surfactant chemical structure and oil biodegradation involved the use of a homologous series of nonylphenol ethoxylate surfactants that varied in the length of the polyethoxylate chain, which confers the hydrophilic character to the molecule. The hydrophilic-lipophilic balances (HLBs) of this homologous series ranged from 4.6 to 18.2, but only compounds with HLBs between about 12 and 14 stimulated the biodegradation rate of Bow River crude oil by a culture of oil-degrading bacteria enriched from refinery sludge (Van Hamme and Ward, 1999). The most effective surfactant in this series, Igepal CO-630, stimulated the biodegradation of both aliphatic and aromatic hydrocarbons. The degree of stimulation was strongly dependent on surfactant concentration; the optimal surfactant-to-oil ratio (SOR) was 1:32. Notably, HLB alone was not an adequate predictor of the ability to stimulate oil biodegradation: of seven surfactants with similar HLBs but which varied in chemical structure, two stimulated oil biodegradation by this enrichment culture, two inhibited it, and three had no effect. A similarly systematic, but conceptually distinct, approach was used in another investigation of the relationships between dispersant chemical composition and effect on oil biodegradation rate (Varadaraj et al., 1995). A series of model dispersant mixtures that differed in HLB was created by varying the relative proportions of two surfactants that are components of Corexit 9527 and 9500: Span 80 (sorbitan monooleate, HLB = 4.3) and Tween 80 (eicosethoxy sorbitan monooleate, HLB = 15). In this study, the maximum rate of biodegradation of Alaska North Slope crude oil by sludge from a refinery wastewater treatment system was obtained at a dispersant HLB of about 8, which is significantly lower than the HLB that was most efficient for the homologous series of Igepals (Van Hamme and Ward, 1999). The extent to which the effects of HLB and surfactant structure on biodegradation efficiency depend on the dispersion efficacy (i.e., do surfactants that stimulate biodegradation entrain higher concentrations of oil as very small droplets than those that do not?) is not known.

Further evidence that the effects of surfactants and dispersant mixtures on oil biodegradation involve complex interactions between the oil, the surfactants, and the composition of the microbial community was obtained through studies with pure cultures. The effect of surfactants on the biodegradation of Statfjord crude oil (topped at 210° C [roughly 410° F]) by *Rhodococcus* sp. O94, an alkane-degrading bacterium, varied with the physiological state of the cells (Bruheim et al., 1997). Oil biodegradation and hexadecane mineralization by exponentially growing cells, which were highly hydrophobic and adhered strongly at oil-water interfaces, were inhibited by all of the dispersants and surfactants that were tested, including Corexit 9527, Finasol OSR-5, Inipol IPF, and Tween 85 (a component of Corexit 9527). Stationary-phase cells, which were relatively hydrophilic and degraded the oil slowly relative to exponential-phase cells, were generally stimulated by the pure surfactants that were tested, but the dispersants either inhibited oil biodegradation and hexadecane mineralization or had no effect on these processes (Bruheim et al., 1997). Further research showed that the difference between the effects of Tween 85 and Corexit 9527 are probably due to synergistic interactions between the nonionic and anionic surfactants that are present in the dispersant mixture (Bruheim et al., 1999). The anionic surfactant in Corexit 9527 and 9500, sodium dioctyl sulfosuccinate, was highly inhibitory to *Rhodococcus* sp. O94 and *Acinetobacter calcoaceticus* ATCC 31012, another alkane-degrading bacterium. Corexit 9527 inhibited the rate of crude oil biodegradation by stationary-phase cells of *A. calcoaceticus* ATCC 31012 by about 40 percent relative to physically dispersed oil, but the nonionic surfactants stimulated the oil biodegradation rate under the same conditions when they were tested alone. These surfactants apparently acted directly at the cell surfaces, possibly interacting with proteins in the cytoplasmic membranes, because they also affected transport and/or oxidation of acetate, a completely water-miscible substrate (Bruheim et al., 1999). For Gram-negative bacteria, surfactants with chemical characteristics that stimulated the rate of n-alkane oxidation also stimulated the rate of penetration of fluorescein diacetate (FDA) through the outer membrane (Bruheim and Eimhjellen, 2000). Thus, these surfactants appeared to affect the permeability of the outer membrane. Although these surfactants also stimulated the rate of alkane oxidation by some Gram-positive bacteria (e.g., *Rhodococcus* sp. 094 and 015), they did not affect the rate of FDA hydrolysis, implying that surfactants interacted more strongly with the outer membrane than with the cytoplasmic membrane (Bruheim and Eimhjellen, 2000).

The variable effects of dispersants and surfactants on oil biodegradation rate are probably due to their effect on microbial uptake of hydrocarbons. Three main mechanisms are recognized by which microorganisms take up hydrocarbons as a prelude to metabolic transformation: transport of aqueous-phase substrates through a variety of well-characterized membrane transport mechanisms (e.g., passive diffusion, active transport); direct contact with nonaqueous-phase liquids or solids followed by poorly understood incorporation into cell membranes or intracellular vesicles; and uptake of water-accommodated hydrocarbons present in surfactant micelles (Singer and Finnerty, 1984; Watkinson and Morgan, 1990). There is good evidence for each of these mechanisms in specific cases. For example, microbial uptake of low-molecular-weight PAH, such as naphthalene and phenanthrene, has been shown to occur by transport of aqueous-phase (i.e., truly dissolved) molecules (Wodzinski and Bertolini, 1972; Wodzinski and Coyle, 1974), whereas direct attachment to nonaqueous-phase liquid droplets is essential for growth of certain alkane degraders (Miura et al., 1977; Rosenberg and Rosenberg, 1981). Some hydrocarbon degraders produce biological emulsifying agents that appear to function in substrate transport (Rosenberg and Rosenberg, 1981; Reddy et al., 1983).

These different transport mechanisms for hydrocarbons will almost certainly be affected differently by dispersants; so, it is perhaps not surprising that the effects of dispersants on oil biodegradation rate vary from system to system. The mechanism that would be expected to be affected least, and possibly enhanced, by surfactant-mediated dispersion of oil is uptake of dissolved hydrocarbons. Dispersion of oil into small droplets will increase the oil-water interfacial area and, therefore, increase the transport rate of hydrocarbons from the nonaqueous to the aqueous phase. It may be significant that dispersants are often reported to have no effect on the degradation rate of low-molecular-weight PAH (Foght and Westlake, 1982; Lindstrom and Braddock, 2002; Yamada et al., 2003), which may reflect the fact that these PAH are more water soluble than other petroleum components. Conversely, direct attachment is the mechanism that should be affected most by chemical dispersion, because accumulation of surfactants at the oil-water interface will change its chemical characteristics and, perhaps, interfere with normal attachment mechanisms. For example, low concentrations of a rhamnolipid biosurfactant reduced the apparent hydrophobicity, measured by the efficiency of attachment to hexadecane, of four strains of *Pseudomonas aeruginosa* (Zhang and Miller, 1994), and a hydrophobic *Rhodococcus* sp. that adhered strongly to oil droplets suspended in water could be released by addition of a nonionic surfactant, Igepal CO-630 (Van Hamme and Ward, 2001). Interference with bacterial attachment to oil droplets may have been the mechanism by which several dispersants and surfactants inhibited oil biodegradation by the highly hydrophobic exponential-phase cells of *Rhodococcus* sp. O94 (Bruheim et al., 1997), and the biosurfactant emulsan strongly inhibited mineralization of n-[1-^{14}C]-hexadecane and [1-^{14}C]-pristane in crude oil by six pure cultures of alkane-degrading bacteria (Foght et al., 1989). Notably, emulsan had no effect on mineralization of [9-^{14}C]-phenanthrene in oil by three pure cultures of PAH-degrading bacteria, which presumably accumulated the phenanthrene directly from the aqueous phase, in the same study.

Although it is clear that surfactants can interfere with attachment of hydrophobic bacteria to oil droplets, the overall effects of chemical dispersion of crude oil on its biodegradation rate are likely to be very complex. For example, although low concentrations of a rhamnolipid biosurfactant interfered with microbial attachment to hexadecane, higher concentrations (above the critical micelle concentration [CMC]) promoted attachment of two hydrophilic strains of *P. aeruginosa* but did not affect the attachment efficiency of two hydrophobic strains (Zhang and Miller, 1994). Also, treatment of ANS crude oil with Corexit 9500 only marginally affected bacterial colonization of oil droplets, with 40 percent of chemically dispersed oil droplets being colonized by at least one bacterium after four days compared to colonization of 60 percent of physically dispersed oil droplets after about one week (Davies et al., 2001). Note, however, that chemical dispersion of ANS crude oil had no significant effect on the oil biodegradation rate in this study. Alternatively, although Igepal CO-630 decreased the biodegradation of the aliphatic fraction of Bow River crude oil by the hydrophobic bacterium *Rhodococcus* sp. F9-D79, which attaches to oil droplets, it did not affect the rate of biodegradation of the aromatic fraction (Van Hamme and Ward, 2001). In addition, the presence of the surfactant had no effect on the biodegradation rate of either hydrocarbon fraction by *Pseudomonas* sp. JA5-B45, which did not attach to oil droplets, and it significantly enhanced the oil biodegradation rate by a coculture of these two organisms (Van Hamme and Ward, 2001). In the long term, the effect of dispersants on bacterial attachment to oil droplets may be less important than is indicated by these studies, because the surfactants will partition out of the droplets into the aqueous phase as the dispersed-oil plume dilutes into a large volume of seawater. The ultimate result of this dilution and partitioning will be small isolated oil droplets lacking a surfactant coating that can interfere with microbial attachment. Thus far, no studies have specifically investigated the biodegradability of these surfactant-depleted droplets of dispersed oil.

The third mechanism of hydrocarbon uptake, transport of micelle-accommodated hydrocarbons, is probably relatively unimportant within the context of the fate of chemically dispersed oil. It is clear that this is an important transport mechanism in some cases (Miller and Bartha, 1989; Bury and Miller, 1993; Garcia et al., 2001), and it may be particularly important for hydrocarbon-degrading bacteria with relatively hydrophilic surfaces (Churchill and Churchill, 1997; Van Hamme and Ward, 2001). Nonetheless, surfactants must be present in the aqueous phase at concentrations greater than their CMC before micelles can form, and the micelle-accommodated hydrocarbon concentration is almost always less than the surfactant concentration (Miller and Bartha, 1989; Zhang and Miller, 1992; Bury and Miller, 1993; Churchill and Churchill, 1997; Schippers et al., 2000; Garcia et al., 2001). Given the large dilution potential of the surface mixed layer in the ocean and the relatively low treatment rate of oil slicks with dispersants (DOR usually less than 1:10), accommodation of a significant fraction of the oil slick in micelles is extremely unlikely.

Implications for the Fate of Dispersed Crude Oil No systematic and reproducible effects of chemical dispersion on the biodegradation rate of crude oil have been demonstrated. In most cases, however, the experimental systems used to investigate these effects may have been inappropriate for extrapolation to behavior in the environment, because they generally applied high mixing energy in an enclosed, usually nutrient-sufficient, environment and allowed sufficient time for microbial growth to result in a substantial enhancement of the extent of physical dispersion of the oil. Microbial growth on open-ocean oil slicks is likely to be nutrient limited and may be slow relative to processes that lead to formation of water-in-oil emulsions, which tend to be extremely resistant to biodegradation. The only way to predict the contribution of biodegradation to the fate of dispersed crude oil is to incorporate this process as a term in a comprehensive fate and transport model. At a minimum, this term would be a function of dispersed oil-water interfacial area (L^{-1} or $L^2 L^{-3}$), a heterogeneous rate coefficient (LT^{-1}), and the oil density ($M_{oil} L^{-3}$) or the concentration of specific components or pseudocomponents in the oil (M_i). If specific components or pseudocomponents are considered, the rate coefficient should be specific for the target component or pseudocomponent. For example, the rate of change of the concentration of phenanthrene ($M_{phe}L^{-3}T^{-1}$) in a dispersed oil plume would be given by:

$$r_{phe} = kaC_{phe,oil} \tag{4-7}$$

The heterogeneous rate coefficient, k, is probably a function of the nutrient concentration and the concentration of hydrocarbon-degrading bacteria. A second equation that links oil degradation to microbial growth could be used to account for changes in the size of the microbial population over time.

Unfortunately, existing studies do not provide the type of data that is needed to estimate kinetic parameters for this simple model. The size of hydrocarbon-degrading microbial populations is often measured in oil-biodegradation studies (Mulkins-Phillips and Stewart, 1974; Bhosle and Row, 1983; Bhosle and Mavinkurve, 1984; Lindstrom et al., 1999; Swannell and Daniel, 1999; Van Hamme and Ward, 1999; Davies et al., 2001; Lindstrom and Braddock, 2002; MacNaughton et al., 2003; Yamada et al., 2003), but few studies have measured droplet size distributions in conjunction with biodegradation experiments (Varadaraj et al., 1995; Swannell et al., 1997; Swannell and Daniel, 1999; MacNaughton et al., 2003) and none attempted to estimate interfacial-area normalized biodegradation rates. Good quantitative degradation rates for specific components or pseudocomponents are also missing in many studies, especially if one is interested in components other than normal alkanes and pristane or phytane. In some cases, when information on changes in the oil composition is presented, it is qualitative (i.e., gas chromatograms; Foght and Westlake, 1982;

Foght et al., 1983, 1987, 1989) or given for only one incubation time, usually the endpoint (Swannell et al., 1997; Burns et al., 1999; Swannell and Daniel, 1999; Davies et al., 2001; MacNaughton et al., 2003). Other studies use radiolabeled tracers to monitor the biological mineralization of specific hydrocarbons (Foght et al., 1987, 1989; Bruheim et al., 1999; Davies et al., 2001; Lindstrom and Braddock, 2002), but the selected substrates were usually easily degradable hydrocarbons, such as hexadecane and phenanthrene, that do not provide any information regarding the biodegradation rates of compounds that may be of concern with regard to the long-term effects of dispersed oil (e.g., high-molecular-weight PAH). The use of tracer compounds to monitor the mineralization of high-molecular-weight PAH is important, because relatively minor structural modifications (e.g., hydroxylation by a nonspecific monooxygenase) will result in removal of the target compound when GC-MS is the primary analytical tool, but the biotransformation products may not be significantly less toxic than the parent substrate, and in fact could be much more toxic. For example, whereas some studies have demonstrated relatively rapid biodegradation of 4-ring PAH, such as pyrene and chrysene, in chemically dispersed crude oil when the process was monitored by GC-MS (Yamada et al., 2003), others have been unable to detect any mineralization when [4,5,9,10-^{14}C]-pyrene was used to monitor the process (Lindstrom and Braddock, 2002). Only PAH mineralization can be confidently equated with toxicity reduction.

Interactions with Suspended Particulate Material

In its most recent review of oil in the sea, the NRC (2003) stated, "Understanding the distribution of petroleum hydrocarbons between the dissolved phase and the variety of aquatic particles is important for determining the fate of hydrocarbons in the sea and the bioavailability of these chemicals to marine biota." For chemically dispersed oil, the formation and fate of whole oil/suspended particulate material (SPM) aggregations are of particular importance.

Several of the earlier and fundamental studies in this area included examination of sediments following oil spills (Hoffman and Quinn, 1978; 1979), controlled experimental ecosystems (Gearing et al., 1980; Gearing and Gearing, 1982a,b; Wade and Quinn, 1980), estimates of sedimentation rates in regions of petroleum activity (Malinky and Shaw, 1979), and more fundamental studies of adsorption (Bassin and Ichiye, 1977; Herbes 1977; Rogers et al., 1980; Karickhoff, 1981). Studies examining the direct interaction of physically (and chemically) dispersed oil droplets with suspended particulate material include those by Mackay and Hossain (1982), Boehm (1987), Payne et al. (1984, 1987a,b, 1989, 2003), Wood et al. (1998), Guyomarch et al. (1999a,b,c; 2002), Hill et al. (2003), Le Floch et al. (2002), Muschenheim and Lee (2002), and Sterling et al. (2004a,b). Also, related studies have now demonstrated that removal and biodegradation of stranded oil in the intertidal zone can be enhanced by augmenting oil/mineral-particle interactions through berm relocation and so-called surf-washing (Bragg and Owens, 1994, 1995; Lee et al., 1997a,b, 2001; Hill et al., 2003; Owens and Lee, 2003). These studies are not considered in detail in this section except to the extent of how specific measured parameters such as salinity, mineral type, and oil type have been shown to affect oil/SPM binding, aggregation, and sedimentation as reviewed by Lee (2002) and Muschenheim and Lee (2002). Lee (2002) stated that the principal environmental parameters affecting oil/SPM interactions include: (1) quantity, type, and surface properties of associated minerals; (2) quantity, viscosity, and composition of the oil; (3) physical energy of the system; and (4) salinity.

Wood et al. (1998) reported that oil/SPM interactions drive the reversible process of oil droplet entrainment and surface slick recoalescence in favor of the dispersed oil phase. Specifically, as entrained oil droplets interacted with SPM and their density increased, the agglomerates were removed from the upper water column and were no longer able to recoalesce and rejoin the surface slick. Their studies also showed that oil/SPM aggregate formation occurs primarily with dispersed oil droplets in the water column interacting with SPM rather than SPM scavenging oil from the surface. The minerals used in their studies did not increase the rate of oil droplet formation from the surface slick, but they did prevent recoalescence and the effect was greatest with minerals with greater cation exchange capacities. The more readily entrained oils (which formed oil/SPM aggregates more easily) were characterized by relatively lower percentages of resins and asphaltenes relative to higher percentages of alkanes, which presumably limited the formation of water-in-oil emulsions.

Schlautman and Morgan (1993) examined the effects of aqueous chemistry on the binding of PAH by dissolved humic materials. Henrichs et al. (1997) and Braddock and Richter (1998) examined the partitioning of naphthalene and phenanthrene onto representative Alaskan sediments and SPM, and a recent special edition of *Environmental Toxicology and Chemistry* (Volume 18, No. 8, 1999) was devoted to causes and effects of resistant sorption and desorption of hydrophobic organic contaminants (including several PAH) onto natural particulate material.

Adsorption of oil droplets onto suspended particulates may provide a relatively efficient mechanism for sedimenting significant fractions of the oil mass. For example, following the *Tsesis* oil spill in the Baltic Sea, approximately 10–15 percent of the 300 tons of spilled oil were removed by sedimentation of the SPM-adsorbed oil (Johansson et al., 1980). The high oil flux was due to the large SPM concentrations resulting from turbulent resuspension of bottom sediments. Likewise, as a result of the hurricane-force winds during the *Braer* oil spill in the Shetland Islands (as well as the influence of a local gyre) it was estimated that as much as 30 percent (30,000 tonnes) of the oil was deposited in the subtidal sediments around the Shetlands Islands, with hot spot concentrations of 2,000 to greater than 10,000 ppm total hydrocarbons (Ecological Steering Group on the Oil Spill in Shetland, 1994).

Payne et al. (1989, 2003) investigated the interaction of physically dispersed (entrained) oil droplets and individual dissolved constituents from fresh and weathered Alaska North Slope (ANS) crude oil and commercially available No. 1 fuel oil with nine SPM/sediment types collected from a variety of Alaskan coastal regions at high sediment loadings (200–1,000 mg/L). In these studies, entrainment of oil droplet-SPM interactions overwhelmed dissolved constituent-SPM adsorption by many orders of magnitude. Results of statistical analyses indicated that particle number density per unit mass showed the highest correlation ($r = 0.902$) with the values for the oil/SPM reaction rate. A slightly lower degree of correlation ($r = 0.798$) existed with the values for sediment fractions comprising the 0–2 μm particle-size range. The remaining three variables (total organic carbon [TOC], specific density, and background total GC resolved hydrocarbon content) showed no significant correlation with the oil/SPM reaction rate.

More recent studies on the fractal dimensions of oil/SPM agglomerates under differing turbulence conditions have been reported by Sterling et al. (2004a). Smaller and more compact aggregates were observed with increasing velocity gradients, and the authors concluded that colloidal oil and mixing shear were the more dominant factors (compared to salinity and mineral type) affecting aggregate morphology in nearshore waters. In a related study Sterling et al. (2004b) described a modeling approach to simulate changes in particle-size distribution and density as a result of aggregation with oil. Aggregation studies were reported for clay, colloidal silica, crude oil, clay-oil, and silica-crude oil systems. Clay and crude oil by themselves were characterized as cohesive particles while silica was classified as noncohesive. The introduction of crude oil increased the aggregation of the noncohesive silica. Apparent first-order aggregation rates for oil, clay, and silica and apparent second-order aggregation rates for oil and clay in clay-oil systems and oil and silica in silica-oil systems were obtained. For oil and clay systems alone, droplet coalescence and clay aggregation were observed to occur on the same time scales as oil resurfacing and clay settling, respectively. For the mixed oil-clay studies, the relative time scales for clay settling and clay-oil aggregation were within an order of magnitude, and it was concluded that oil-clay aggregation should be considered when modeling crude oil transport in nearshore waters. In this regard, excellent agreement was obtained between observed and model-predicted behavior for the oil-clay systems. Conversely, the data for the silica and silica-oil systems suggested that the silica aggregation and oil-silica aggregation both occurred more slowly than aggregate settling. Because of the greater volume mean diameters and densities of the silica particles compared to the clay, it was concluded that the removal of oil by silica was less efficient than that by clay.

Additional modeling of the effects of sediment size on the size of subsequently generated oil/SPM agglomerates has been reported by Khelifa et al. (2004). Both model simulation and laboratory results showed negative effects of sediment size on oil/mineral/aggregate formation. Variations of the concentration of stabilized oil with sediment size showed a maximum when the ratio between the sediment and oil droplet sizes varied between 0.1 and 0.4. The highest concentration of stabilized oil was observed when sediment size varied between 0.3 and 1.2 μm. The model results showed that the sticking efficiency between oil droplets and sediment particles is a significant factor in oil/SPM aggregate formation.

Not as much is known about the longer-term fate of oil-SPM agglomerates while still in suspension in the water column; however, Wood et al. (1998) implied that an association of mineral particles and bacteria may be more efficient at biodegrading dispersed oil compared to bacteria alone. Based on what is known about weathering of free oil droplets, it can be inferred that oil-SPM agglomerates would still be subject to the same oil-phase diffusion-controlled weathering behavior (although possibly at an altered rate due to changes in the surface area-to-volume ratios for the agglomerates and the presence of bacteria associated with the SPM). It is known that bacteria are more likely to be associated with particulate surfaces in the water column (Subba-Rao and Alexander, 1982; van Loosdrecht et al., 1990), and introduction of clay-sized particles into oil trapped in sandy intertidal sedimentary regimes to form oil-SPM agglomerates has been shown to enhance bacterial utilization of the hydrocarbons by generating increased surface area (Lee et al., 1997a; Jezequel et al., 1998, 1999; Weise et al., 1999). Likewise, Jahns et al. (1991) demonstrated the enhanced removal and biodegradation of previously buried oil in low-energy cobble beaches due to the natural incorporation of clay-sized SPM over time following the T/V *Exxon Valdez* oil spill, and these mechanisms may also apply to oil SPM agglomerates while still suspended in the water column.

Once formed, oil-mineral aggregates appear to be very stable structures, and the buoyancy will ultimately depend on the ratio of oil to mineral in each individual aggregate (Stoffyn-Egli and Lee, 2002). Because oil-mineral aggregates are typically less dense than mineral-only aggregates and in many cases buoyant (Stoffyn-Egli and Lee, 2002), associated residual oil is believed to be kept in suspension long enough to be dispersed over a wider area by physical processes.

Dispersant-Treated Oil Droplet/SPM Interactions

Mackay and Hossain (1982) completed one of the earliest exploratory studies of naturally and chemically dispersed oil in which several crude oils were dispersed with varying amounts of chemical dispersants in seawater in the presence of differing quantities of sedimenting mineral and organic matter. Mineral concentrations ranged from 40 to 160 mg/L and DOR ratios ranged from 1:5 to 1:20 with no dispersant added in some cases. Oil and dispersant were premixed and shaken with water to form the dispersion in the test apparatus.

Settling velocities in the test chamber were quite low, and it was concluded that little actual settling would occur in a turbulent ocean surface layer. Corexit 9527 consistently gave lower amounts of oil settled than BP 1100 WD. In generally all cases, more oil was settled to the bottom in the absence of any chemical dispersant, than when dispersants were used. Compared to kaolinite, higher settling values were obtained with higher organic content sediments such as humic acid and dead algae. Higher concentrations of oil and sediment tended to give higher fractions settled.

Dispersant dosage had a major effect in Mackay and Hossain's study, as illustrated by the observation that at zero dosage, 30 percent of the oil settled, but at higher dispersant dosage this dropped to 10–15 percent at 1:10 (dispersant:oil) and to 6 percent at 1:5 (dispersant:oil). It appears that either smaller oil droplets are less able to associate with sedimenting particles than are larger droplets, or the association results in particles that do not settle. The higher the dispersant to oil ratio, the more the oil/SPM aggregate tends to return to the surface, and the less the oil/SPM aggregate tends to settle to the bottom.

Surprisingly, there have only been a few additional studies published over the last 22 years since Mackay and Hossain's seminal work. Guyomarch et al. (1999c) used laboratory-scale experiments to demonstrate that very high suspended mineral loads (from 1,300 to 3,600 mg/L) were required to form aggregates with dispersant-treated oil, and the threshold mineral concentration value depended on the oil and clay type, their relative concentrations, and the water salinity (Guyomarch, et al., 1999c). Not surprisingly, the maximum amount of oil trapped on the mineral particles also depended on the dispersant (Inipol IP 90) efficiency, and Guyomarch et al. (1999c) observed that lower clay concentrations were required to sediment 40 percent of the oil as the polar fraction in the oil phase increased. Conversely, as the salinity increased, higher clay concentrations were required to sediment 40 percent of the oil. In their larger-scale Polludrome tests, aggregates were formed at slightly lower SPM loads (750 mg/L). In all experiments, the oil concentrations were quite high (e.g., 120 ppm in the Polludrome tests), and the authors stated that additional tests should be conducted at lower oil and SPM concentrations such as those that might be encountered in outer estuaries or other coastal systems. It should be noted that keeping a steady-state SPM load in a wave tank is very difficult and depends on the turbulence regime utilized in the experiment and the SPM size range being studied.

In a subsequent laboratory study, Guyomarch et al. (2002) focused on the size distribution of oil-mineral aggregates when the oil was chemically dispersed with Inipol IP 90. Four oil types or blends were examined to provide a wide variety of initial oil viscosities (25 to 20,000 mPa-s) and asphaltene content (1 to 16 percent). Mixtures of oil, dispersant, and clay were stirred at 75 rpm for 2.5 min in a 250 mL beaker and allowed to settle for 1 hour. Clay concentrations were varied from 200 to 2,000 mg/L at a constant oil loading of 35 ppm.

Photo-microscopic observations (Figure 4-9) demonstrated that at low clay concentrations (<200 mg/L), small aggregates were mainly composed of clay with an occasional oil droplet (Figure 4-9A). At intermediate clay concentrations, much larger aggregates containing up to 15 oil droplets in a single cluster were observed (Figure 4-9B), and these higher oil-droplet density aggregates were reported to be much more common than in a previous study completed without dispersants (Lee et al., 1998). At higher clay concentrations (Figure 4-9C), the smaller agglomerates were explained by the hypothesis that all the oil droplets were completely covered by clay particles, which prevented interactions with each other and the concomitant formation of larger (multiple droplet) structures.

The effects of salinity examined by Guyomarch et al. (2002) confirmed their early findings that increased clay concentrations were needed to form aggregates as the salinities increased. Likewise, the influence of oil type (presence of asphaltenes) supported the hypothesis that interactions between polar compounds in the oil and the negatively charged clay particles by the intermediary action of a cation, as proposed by Bragg and Yang (1995), was important for stable aggregate formation. In a competitive sense, however, it is also known that as asphaltene content in the crude oil increases, its propensity to form higher viscosity water-in-oil emulsions also increases, and this inhibits dispersant effectiveness and oil droplet formation. Guyomarch et al. (2002) did not observe any significant difference in the maximum size of the oil-mineral aggregates for the different oils studied.

Taken in total, these studies might lead to the conclusion that because

A

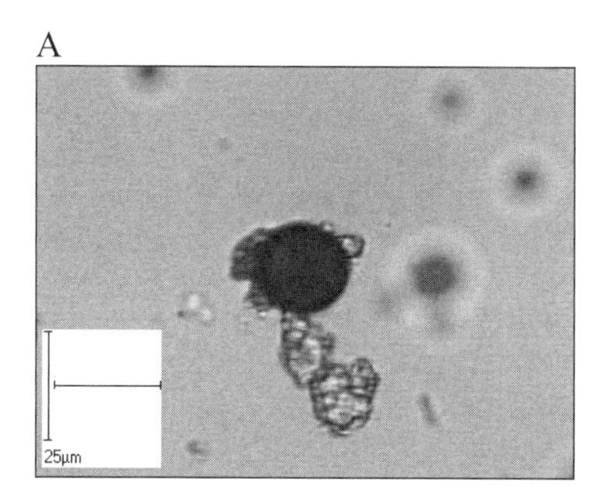

B

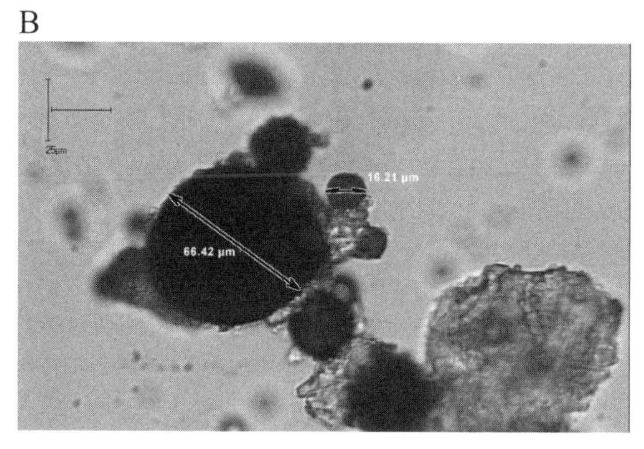

C

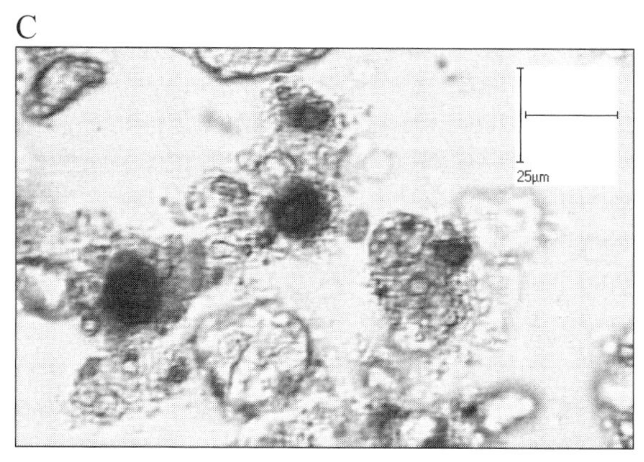

FIGURE 4-9 Photomicrographs of: A) oil droplet aggregate formed at low clay concentrations. Oil appears as the darkest areas; B) multiple oil droplet aggregate formed at intermediate clay concentration; and C) oil droplet aggregate at high clay concentration.
SOURCE: Guyomarch et al., 2002; courtesy of Elsevier.

dispersant treatment results in the formation of greater numbers of oil droplets, the potential for their interaction with SPM will increase (i.e., because of higher oil droplet number densities as described and modeled by Payne et al., 2003), but the ultimate agglomerate size distribution will be controlled by the SPM loading in the water column as described by Guyomarch et al. (2002).

The size and composition of an oil/SPM aggregate will control the buoyancy of the agglomerate affecting its ultimate fate (transport vs. sedimentation), which in turn will vary with the local salinity and hydrodynamic regime. That is, the behavior of dispersant-enhanced oil droplets in the presence of SPM will be stochastic, depending on a number of variables that can complement or compete with one another to determine the ultimate disposition of the oil. As such, the transport of oil-mineral aggregates should be studied further to determine if it is beneficial to apply dispersants to oil on seawater that is loaded with SPM. This will be particularly important in estuaries and coastal zones where elevated SPM levels are often encountered. Likewise, additional studies on the biodegradation of dispersed oil/SPM agglomerates (both in the water column and after sedimentation) are warranted.

MODELS

Models of oil transport and fate integrate into one system the major physical, chemical, and biological processes discussed previously in this chapter (and some models also include the toxicological processes discussed in the following chapter). As such, models can provide decision-makers with a more complete picture of what happens to spilled oil—and what effects it has on the environment—both with and without the use of chemical dispersants. Models of the trajectory of floating oil are regularly used, along with field measurements, in real time during a spill. To date in the United States, models of the transport and fate of dispersed oil have primarily been used in pre-planning exercises to simulate hypothetical spills in order that response measures, including use of chemical dispersants, can be evaluated *before the fact*, or to assist with natural resource damage assessment of real spills *after the fact*. So far models have not been used to evaluate the use of dispersants in real time, but we argue that this should be possible. The remainder of this chapter presents a brief overview of available models and their attributes (focusing on transport and fate only), summarizes a sensitivity study that is described in more detail in Appendix E, and provides recommendations on how models could be used more effectively.

The literature contains many reviews of oil transport and fate models (Yapa and Shen, 1994; ASCE, 1996; Reed et al., 1999; NRC, 1989, 1999,

2003). Computer codes are available for different water bodies—offshore open sea, nearshore water, semi-confined coastal water, estuaries, rivers, lakes, and reservoirs—and include ADIOS2 (Lehr et al., 2002), Spill Impact Model Application Package (SIMAP; French-McCay and Payne, 2001; French-McCay, 2003, 2004), Natural Resource Damage Assessment Model (NRDAM; Reed et al. 1991; French-McCay et al., 1996), Oil Spill Information System (OSIS; Leech et al. 1993), Oil Spill Contingency and Response (OSCAR; Aamo et al. 1997), OILMAP (Howlett et al. 1993), IMMSP (Institute of Mathematical Machines and Systems Problems; Brovchenko et al., 2003), Zhang (Zhang et al., 1997), General NOAA Oil Modeling Environment (GNOME; Simecek-Beatty et al., 2002), and River Oil Spill Simulation (ROSS) Model (Yapa et al., 1994). A few models, such as ROSS (Yapa et al., 1994), Water Planning and Management Branch (WPMB) model (Fingas and Sydor, 1980), and RiverSpill (Tsahalis, 1979) are specifically developed for rivers because major transport mechanisms and concerns in rivers are significantly different from those in open seas. Some models (e.g., ADIOS2) predict oil transport and fate on the water surface only; others (e.g., 3-D GNOME, IMMSP) simulate oil movements on the water surface, water column, and shorelines; some (e.g., SIMAP) predict oil on the water surface, in the water column and shorelines, and also quantify the biological impacts.

Most models simulate oil as a single substance, but a few treat oil as a composite of multiple hydrocarbons, usually sorted by distillation cut. Because various constituents weather at different rates, respond differently to chemical dispersants, and have different impacts on biota, the latter approach is superior (though often limited by toxicological data). It is recommended that models be formulated by constituent wherever this can be supported by available data.

Among their capabilities, one would like models to predict the effectiveness of chemical dispersants (i.e., the amount of oil entrained below the surface as droplets, and the resulting droplet size distribution) as a function of environmental conditions, type of oil and the extent of oil weathering, type and quantity of dispersant, etc. However, all models that purport to simulate the effect of chemical dispersants include such effectiveness measures as *model inputs*, rather than model outputs. Developing the ability to predict dispersant effectiveness, and integration of this predictive ability into models, is strongly recommended.

Sensitivity Study

In order to understand the effects of various processes on the transport and fate of spilled oil, a series of sensitivity tests was performed with the NOAA surface oil fate model ADIOS2 (Lehr et al., 2002), and the

NOAA oil transport code 3-D GNOME (Simecek-Beatty et al., 2002). Hypothetical spills were simulated off the coast of Florida using a range of oils, wind speeds (that in turn affect surface transport, wave height, and horizontal and vertical mixing), percentage effectiveness of dispersant application, and oil droplet-size distribution. These models were selected because they are three-dimensional, they treat oil as a composite of pseudocomponents, and they have limited computing requirements to operate the models. They are also used in real time during oil spills by NOAA to provide scientific support to the FOSC. This review was not intended as an endorsement of these particular models and indeed similar sensitivity analysis should be conducted using other models. Results of some sensitivity runs are presented below while the entire set is contained in Appendix E.

It is very difficult to predict where, when, and how much spilled oil moves. For example, the effect of wind on the movement of Alaskan North Slope crude oil is illustrated in Figure 4-10. This figure shows predicted changes in oil distributions 24 hours after oil is spilled on the water surface in the south Florida nearshore area under 2, 10 and 25m/s wind. As the wind becomes stronger, more oil is naturally entrained into the water column—from 0 volume percent at 2 m/s wind to 3 volume percent at 10

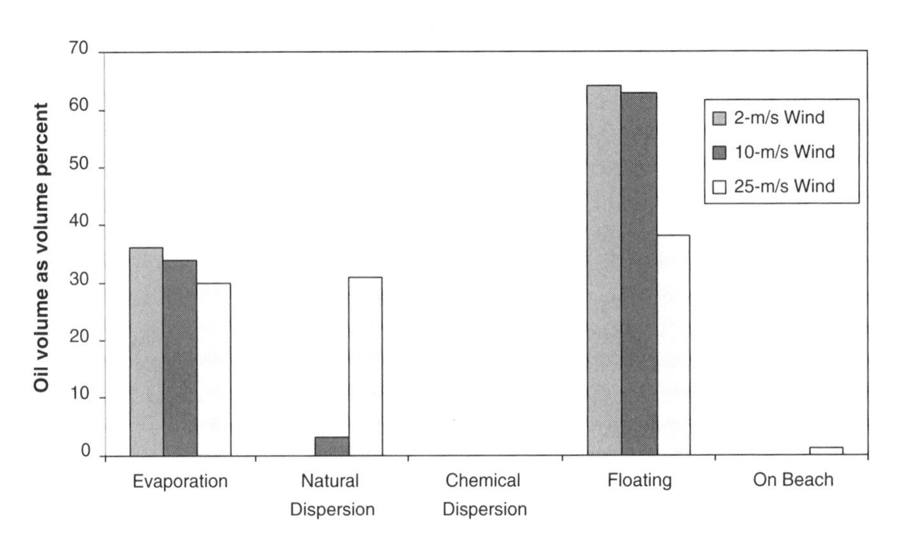

FIGURE 4-10 Predicted oil distributions 24 hours after the release of Alaskan North Slope crude oil (no dispersant applied) under 2-, 10-, and 25-m/s wind in nearshore off Florida Keys. There is no oil dispersed by a chemical dispersant for these three cases.

m/s, to 31 volume percent at 25 m/s. The greater entrainment means that less oil floats on the water surface, and hence less is available for evaporation, a result that might seem counterintuitive. Oil concentrations in the water column vary depending on the amount of oil naturally dispersed (entrained), but they also reflect the diffusivity (which increases at higher wind speed) in the water column. This example indicates some of the complexity involved with the way that currents and wind, and in turn waves and diffusion, affect the horizontal and vertical movement of oil.

When a dispersant is applied, more oil is entrained into the water column, and the droplets changes size distribution from the original size, further complicating oil transport and fate processes (see Figure 3-1). For example, assuming 50 percent effectiveness for a dispersant applied between 6 and 12 hours after the oil spill, 40 percent of the oil discussed previously was predicted to end up in the water column (37 percent by chemical entrainment and 3 percent by natural entrainment) under 10-m/s wind. Figure 4-11 shows the location of the predicted plume 24 hours after the spill. The oil spill location is marked by "+." Black spots represent oil floating on the water surface, and the shaded areas show different ranges of oil concentrations in the top 1 m of the water column. The oil plume in the top 1 m of the water column is following a different trajectory at a different speed than the oil on the surface that is moved by the wind and the current. This figure also indicates that the area of the top 1 m of water column containing oil is about 64 km^2, 2.5 times more than the contaminated top 1 m water area without dispersant application. Clearly such quantitative estimates would not be possible without a model.

Further complexity comes from the fact, mentioned previously, that oil consists of a wide ranges of hydrocarbons. Although oil toxicity comes from the cumulative impacts of multiple hydrocarbon components, low- and intermediate-molecular-weight components such as BTEX and PAH tend to cause more acute risks to aquatic biota, as is discussed in Chapter 5. These components usually evaporate faster and to a greater extent than large-molecular-weight components such as wax, resins, and asphaltenes. The latter are contributing components in the formation of mousse, which makes it more difficult for a dispersant to work effectively (see Chapter 3). Table 4-2 presents Alaska North Slope crude oil's chemical components (Environment Canada, 2005), as indicated in their distillation cuts (built into the ADIOS2 code), together with those of intermediate fuel oil (IFO) 300 and marine diesel oil used in the sensitivity analysis. As shown in this table, the Alaska North Slope crude oil has more low molecular-weight components than the two refined oils. Oil composition changes and emulsification occurring during the transport of spilled oil significantly alter the physical properties of oil, especially viscosity and dispersant effectiveness, as previously discussed. Thus, it is important to

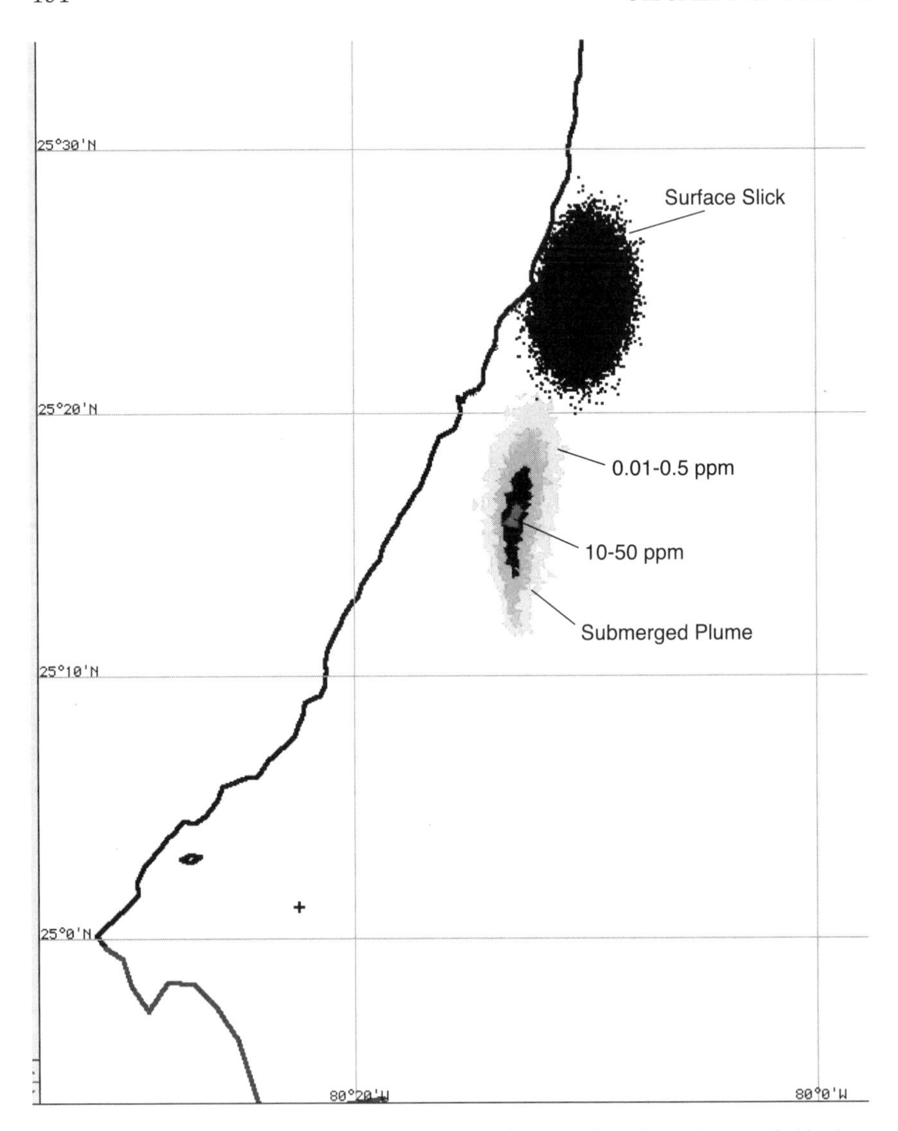

FIGURE 4-11 Predicted oil movement at 24 hours after the release of Alaskan North Slope crude at point + under 10-m/s wind with a dispersant application (additional details contained in the text of Chapter 4).

simulate the behavior and transport of the components of the various hydrocarbons rather than treating oil as one substance.

Figure 4-12 presents the predicted composition (a relative volume fraction of each distillation cut) of these three oils floating on the water surface 0 and 6 hours after the spill. Because of evaporation, the composi-

TABLE 4-2 Distillation Cuts of the Three Oils Used in the Modeling Sensitivity Analysis

	Alaska North Slope Crude Oil		Intermediate Fuel Oil 300		Diesel Fuel Oil	
Oil Cut Number	Weight Fraction, wt percent	Temperature, °C	Weight Fraction, wt percent	Temperature, °C	Weight Fraction, wt percent	Temperature, °C
1	1.0	42	1.1	180	1.1	120
2	4.0	98	1.1	200	1.1	140
3	5.0	127	6.4	250	1.1	160
4	5.0	147	9.4	300	3.2	180
5	5.0	172	7.2	350	5.2	200
6	10.0	216	8.1	400	20.4	250
7	10.0	238	6.0	450	31.9	300
8	5.0	247	3.0	500	25.5	350
9	5.0	258	4.9	550	9.7	400
10	5.0	265	9.8	600	1.0	450
11	5.0	272	14.7	650	—	—
12	10.0	282	10.7	700	—	—
13	30.0	>282	17.4	>700	—	—

SOURCE: Data from Environment Canada, 2005.

tion of each oil changed significantly over time. Most of the cuts that distill at about 200° C [roughly 392° F] or lower, including alkanes with <10 carbons plus the monocyclic aromatics, benzene and toluene, ethylbenzene, o-, m-, and p-xylene, and most of the C2- and C3-substituted benzenes), evaporated within six hours. Thus, if a dispersant is applied six hours after the oil spill, it would not be expected to introduce these compounds into the water column. The IFO 300 does not naturally disperse into water due to its high viscosity (~15,000 cP), according to the modeling. On the other hand, diesel, with very low viscosity (~4 cP), disperses naturally (73 percent) without adding a chemical dispersant, and after 16 simulation hours, no diesel would be floating on the water surface. The combination of natural and chemical dispersal would disperse 78 percent of the diesel into the water, so there is no merit to applying a dispersant in this particular case. Because these refined oils have a low percentage of low-temperature distillation cuts, the IFO 300 and diesel evaporated only 10 and 18 volume percent, respectively, over 24 and 14 hours (much less than Alaska North Slope crude oil). This example clearly shows contrasting behavior of these three oils having different hydrocarbon composition.

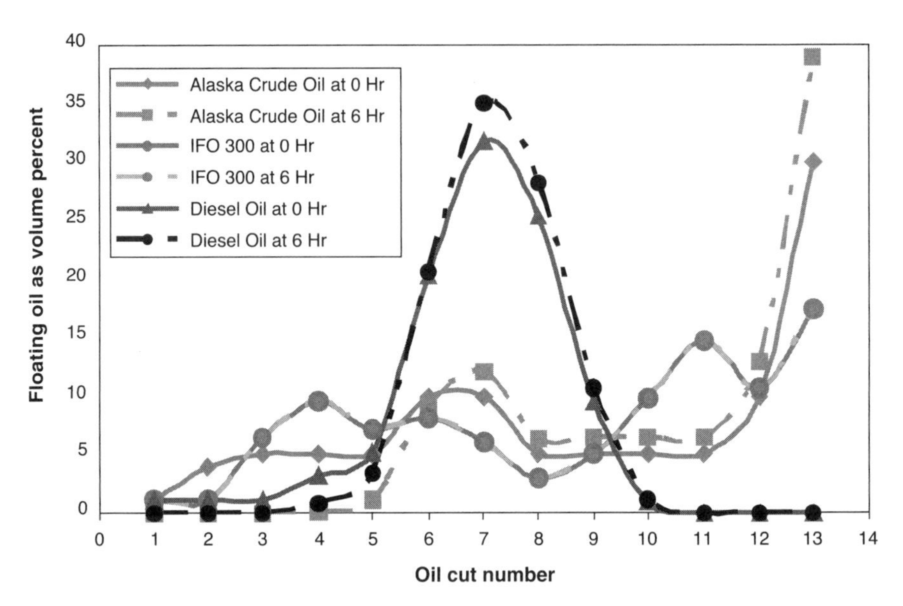

FIGURE 4-12 Predicted compositions of floating oils initially and 6 hours after the releases of Alaskan North Slope crude oil, Intermediate Fuel Oil (IFO) 300, and Marine Diesel Oil (additional details contained in the text of Chapter 4).

These examples of the sensitivity analysis results illustrate very complex and sometimes competing interactions among oil types, environmental conditions, and dispersant use. Quantitative estimates of oil concentration distributions clearly require the use of computer models, especially those with oil pseudo-component modeling capabilities.

A final motivation for the sensitivity study was to assess whether models could be used in real time to help decide whether or not to use chemical dispersants during an actual spill. These questions are particularly important in nearshore areas where the impacts of using—and not using—dispersants are likely to be most significant. Unfortunately nearshore areas are also the most complicated hydrodynamically. Although 3-D GNOME can accept a three-dimensional flow field, it presently uses two-dimensional flows that are calculated based on a simplified force balance involving pressure, Coriolis, bottom friction, and variation in water density adjusted by tide and wind. This simplified approach is justified because of the need to make simulations very quickly for real-time model predictions, and because field observations can be used to update model output. Because chemical dispersants help transport oil into the water column, realistic simulation of subsurface transport becomes more important when evaluating the use of chemical dispersants, and the same formulation may not be sufficient. It is recommended that a range of 3-D hydrodynamic formulations be evaluated with the goal of identifying approaches that are sufficiently accurate, yet still efficient, for real-time use.

APPLYING KNOWLEDGE ABOUT THE TRANSPORT AND FATE OF DISPERSED OIL TO SUPPORT DECISIONMAKING

As discussed in Chapters 2 and 3, many ultimate conclusions about the wise and effective use of dispersants in nearshore settings will need to be based on an accurate and adequate understanding of many processes controlling the transport and fate of dispersed oil. These processes may play a significant role from the instant the oil enters the environment, and they constrain a number of operational decisions and play a significant role in evaluating potential impacts of whole and dispersed oil on sensitive species or habitats.

Fate and Weathering of Oil

Oil on the Surface

Better information is still needed to determine the window of opportunity and percent effectiveness of dispersant application for different oil types and environmental conditions. **Coordinated research should be undertaken at bench and wave-tank scales to define those parameters that control oil dispersability as the oil is allowed to weather under carefully controlled but realistic environmental conditions.**

Overprediction of evaporation rates can be a problem with oil-weathering models that assume a well-mixed oil phase (which is probably only valid for very thin and relatively unweathered oil slicks) and also assume that resistance to mass transfer is entirely in the air phase. As a result, it may be inappropriate to always model oil as a well-mixed phase. Algorithms for both well-mixed and diffusion-controlled fluids may need to be sequentially utilized as a function of oil weathering-dependent viscosity changes to better approximate spilled oil evaporative behavior. **Additional work is recommended to reconcile the differences between the empirical evaporation approach utilized by Fingas (1996, 1997, 1999a) and more traditional pseudo-component approaches as considered by Jones (1996, 1997), who has proposed a simplified pseudo-component (SPC) model relating molar volume, vapor pressure, and molecular weight to the boiling point of the components.**

Sediment Particle Interactions

The ultimate fate of dispersed oil is poorly understood. Of particular concern is the fate of dispersed oil in areas with high suspended solids and areas of low flushing rates. Although this has been an area of recent research, there is still insufficient information on which to determine how chemically dispersed oil interacts with a wide variety of suspended sediment types, both short- and long-term, compared to physically dispersed oil. In this regard, there appears to be more information on short-term comparisons versus the longer-term fate of oil/SPM agglomerates generated with and without dispersant addition. In particular, the longer-term biodegradation of oil/SPM agglomerates in the water column has not been adequately studied. Likewise, there are uncertainties in how dispersed oil might be consumed by plankton and deposited on the seafloor with fecal matter or passed through the food chain. **Relevant state and federal agencies, industry, and appropriate international partners should develop and implement a focused series of studies to quantify the weathering rates and final fate of chemically dispersed oil droplets in high SPM-concentration regimes compared with non-dispersed oil.**

Biodegradation

Past research on the effects of dispersants on the biodegradation of petroleum hydrocarbons cannot be used to predict the fate of chemically dispersed crude oil at sea. The results of many of these studies may be confounded by metabolism of the dispersant or short-term effects of dispersants on bacterial attachment to oil droplets. When dispersed oil plumes become diluted by the transport processes that act in the surface layer of the ocean, however, the surfactants present in the dispersant will partition out of the oil into the surrounding seawater. If this partitioning is fast relative to the kinetics of bacterial attachment to oil droplets, the dispersant may not interfere with microbial uptake of the petroleum hydrocarbons (i.e., the dispersed oil droplets will behave like physically dispersed oil except the oil-water interfacial area will be larger due to entrainment of a larger number of small droplets in the water column). **Therefore, future research on the kinetics of dispersed oil biodegradation should be conducted at low oil-water ratios to simulate conditions that represent those that follow significant dilution of the dispersed oil plume.** In addition, the experimental designs of laboratory studies that have been used are probably inappropriate for estimating the *in-situ* biodegradation rate of oil that is floating on the sea surface, because the mixing energies that are typically applied are usually sufficient to result in substantial physical dispersion (i.e., oil droplets continuously break away from the floating slick and are entrained into the aqueous phase due to vigorous mixing) and there is little opportunity for formation of water-in-oil emulsions, which can dramatically reduce *in-situ* biodegradation rates. Therefore, the biodegradation rates for chemically dispersed and undispersed oil that have been compared in most laboratory studies are probably skewed in opposite directions relative to their *in-situ* rates: the biodegradation rates that have been measured in the laboratory for chemically dispersed oil are probably lower than what would prevail in a dispersed oil plume and those measured for undispersed oil are probably higher than could be realized in a floating oil slick that is not subject to a high degree of natural dispersion. **Due to the difficulty of designing laboratory-scale experimental systems that adequately simulate the *in-situ* processes that are expected to affect the biodegradation rate of chemically dispersed oil, future biodegradation studies should be designed to support dispersed oil fate and transport modeling. Ideally, droplet-scale models of biodegradation kinetics should be developed and the appropriate kinetic parameters should be estimated.** In general, existing oil biodegradation kinetics data cannot be used to support modeling of biodegradation in dispersed-oil fate and transport models, because one or more important variables (e.g., oil-water interfacial area, microbial population size, hydrocarbon concentrations as a function of time) were not monitored.

Another major limitation for predicting the fate (and effects) of chemically dispersed oil based on available laboratory studies is that few studies have quantitatively investigated the biodegradation rates and products of compounds that are of most long-term concern. These include the high-molecular-weight PAH (e.g., 4- and 5-ring compounds), which are degraded slowly if at all by microorganisms, have the potential to bioaccumulate, and can exert chronic toxic, mutagenic, or developmental effects. Most studies have focused on bulk measurements of oil degradation (e.g., carbon dioxide production or reductions in TPH) or degradation of major components, such as n-alkanes. Although these are important metrics, because they measure the extent of reduction in the total oil mass, they may not be the most important drivers of long-term effects, because normal and branched alkanes are well known to be easily biodegradable by bacteria that are ubiquitous. So, while the rate of degradation of these compounds is of interest from a model mass-balance perspective, their ultimate fate is not in doubt. High-molecular-weight PAH, on the other hand, are likely to persist in the residual oil droplets, which may be ingested by animals in the water column or benthos where they can exert chronic effects. **Therefore, the biodegradation kinetics and ultimate biotransformation products of high-molecular-weight PAH should be investigated using indigenous microbial communities from seawater.** The ecological impact of these persistent compounds will be determined by their transport characteristics, which can best be predicted by accurate fate and transport models that include all relevant processes (including biodegradation) and robust estimates of the model parameters.

Present and Possible Role of Models

As discussed previously, various processes constrain a number of operational decisions and play a significant role in evaluating potential impacts of whole and dispersed oil on sensitive species or habitats. Models are, therefore, powerful and necessary tools to support decisionmakers during all phases of oil spill planning, response, and assessment. Currently trajectory analysis is a key component of contingency planning, real-time prediction of slick trajectory, size, and thickness, and in natural resource damage assessment. These models are not currently used in real time to support decisionmaking for dispersant use, but in principle they could be. The required sophistication of the models for these purposes varies, but their performance could be improved for all purposes. Specifically, they are incomplete in terms of their representation of the natural physical process involved, verification of the codes, and validation of the output from these models in an experimental setting or during an actual spill. Thus, their ability to predict the concentrations of dispersed oil and dissolved aromatic hydrocarbons in the water column with sufficient accuracy to aid in spill decisionmaking has yet to be fully determined.

The sensitivity analysis identified that dispersant effectiveness and oil droplet size change are the most important parameters for dispersant application modeling. Unfortunately, oil spill models currently available do not simulate physical mechanisms and chemical reactions in order to predict these parameters. Emulsification is also an important process that greatly influences dispersant effectiveness. Predicting emulsification requires accurate oil properties, as well as conducting a detailed mechanistic investigation on emulsification processes and their influence on dispersant effectiveness. It is also important to evaluate turbulence in the open sea and reflect it more accurately in the transport and fate modeling.

Models show significant progress for supporting real-time spill-response decisions regarding dispersants use, especially in complex nearshore regions; however, any improved models should be evaluated for their ability to satisfy this need. **Oil trajectory and fate models used by relevant state and federal agencies to predict the behavior of dispersed oil should be improved, verified, and then validated in an appropriately designed experimental setting or during an actual spill.** Specific steps that should be taken to improve the value of models for dispersant use decisionmaking include:

- Improve the ability to model physical components of dispersed oil behavior (e.g., shear in vertical dimension, distribution of horizontal velocities as a function of depth, variations in the vertical diffusivity as a function of depth, sea-surface turbulence, etc.)
- Improve the ability for models to predict concentrations of dissolved and dispersed oil, expressed as specific components or pseudo-components, that can be used to support toxicity analysis
- Validate how advective transport of entrained oil droplets is modeled through specifically designed flume/tank studies and open-ocean (spill of opportunity) tests.
- Develop an ability to predict the formation of water-in-oil emulsions under a variety of conditions
- Conduct a sensitivity analysis based on three-dimensional, oil-component, transport and fate models, and develop necessary databases (evaporation, dissolution, toxicity, etc.) for the oil-component based assessment approach

Once the models are improved, they will be valuable tools for transport and fate modeling and associated biological assessments with and without dispersants. They should be used as part of the overall effort to define operational guidelines for dispersant use, including what oils are dispersible and for how long, the predicted effectiveness of dispersant application (which will be a key input into predicting the dispersed oil concentrations in the water column), likely extent and duration of different oil concentrations of concern, and guidelines for buffers around sensitive resources.

Because this study did not explicitly evaluate the pseudo-components and their dissolved chemical components of the oil in the water column with and without dispersant application, additional sensitivity analyses should be conducted with three-dimensional oil-component transport and fate models. It is also important to develop the necessary database (evaporation, dissolution, toxicity, etc.) for the pseudo-component-based assessment approach. This evaluation focused more on nearshore water, and it is recommended to also conduct sensitivity modeling for offshore, semi-confined waters and rivers. A consensus regarding "how good is good enough" needs to be developed among decisionmakers and model developers, and used to guide the future development of models and to optimize their use in real time.

In discussions with NOAA modelers, it was noted that predicting the three-dimensional flow distribution as a part of the oil transport and fate modeling within several hours after an oil spill is difficult. A real-time model application uses actual environmental conditions and oil properties, but, because of time limitations, uses simple approaches for approximating hydrodynamic data. To reflect three-dimensional flow and mixing, NOAA is implementing simple schemes to handle vertically varying diffusion and horizontal velocity fields. There have been some attempts to incorporate surface flow measurements into real-time oil transport models (Hodgins et al., 1993; Ojo and Bonner, 2002). However, these require pre-installation of data acquisition (e.g., high frequency radar) and transmission systems, and are currently applicable only to horizontal surface current and diffusion with relatively coarse grid resolution—not for the three-dimensional distributions needed for the three-dimensional modeling (Ojo and Bonner, 2002). The growing availability of ocean observing systems in coastal waters will likely improve the availability of real-time data useful for improved modeling of physical processes. Unlike real-time model applications, a pre-planning assessment uses hypothetical environmental conditions and oil properties, but can use detailed models, including complex three-dimensional flow fields. Thus, real-time and pre-planning modeling efforts should complement each other to provide better information to a decisionmaker.

One of the greatest weaknesses in correlating laboratory-scale and mesoscale experiments with conditions in the open ocean derives from a lack of understanding the turbulence regime in all three systems. Likewise, one of the biggest uncertainties in computer modeling of oil spill behavior (with and without dispersant addition) comes from obtaining appropriate horizontal and vertical diffusivities. It is difficult to integrate all interacting transport and fate processes and oil properties to predict how much oil will be found in specific areas during an actual oil spill without the use of models. **Relevant state and federal agencies, industry, and appropriate international partners should develop a coordinated program to obtain needed information about turbulence regimes at a variety of interrelated scales.** This effort should include a field program to measure the upper sea-surface turbulence, under a variety of conditions with particular emphasis on quantifying horizontal and vertical diffusivities and the rate of energy dissipation, which can be compared to similar turbulent regimes in mesocosm systems.

5

Toxicological Effects of Dispersants and Dispersed Oil

One of the most difficult decisions that oil spill responders and natural resources managers face during a spill is evaluating the environmental trade-offs associated with dispersant use. The objective of dispersant use is to transfer oil from the water surface into the water column. When applied before spills reach the coastline, dispersants will potentially decrease exposure for surface dwelling organisms (e.g., seabirds) and intertidal species (e.g., mangroves, salt marshes), while increasing it for water-column (e.g., fish) and benthic species (e.g., corals, oysters). Decisions should be made regarding the impact to the ecosystem as a whole, and this often represents a trade-off among different habitats and species that will be dictated by a full range of ecological, social, and economic values associated with the potentially affected resources. Comparing the possible ecological consequences and toxicological impacts of these trade-offs is difficult. First, each oil spill represents a unique situation and second, it is often difficult to extrapolate from published research data into field predictions, especially regarding the possibility of long-term, sublethal toxicological impacts to resident species (Box 5-1 provides definitions for most the common terms used in discussions of toxicological effects).

Historically, the use of dispersants in the United States has been restricted primarily to deepwater (>10 m), offshore spills. In addition, the focus and the recommendations of the 1989 NRC report on oil dispersants were based on expected impacts of dispersants and dispersed oil during open ocean spills (NRC, 1989). As the potential use of dispersants is expanded into nearshore, estuarine, and perhaps even freshwater systems,

BOX 5-1
Common Toxicological Terms Related to Dispersant Toxicity Testing

Exposure—Contact with a chemical by swallowing, breathing, or direct contact (such as through the skin or eyes). Exposure may be either acute or chronic.

Acute—An intense event occurring over a short time, usually a few minutes or hours. An acute exposure can result in short-term or long-term health effects. An acute effect happens within a short time after exposure. Acute toxicity to aquatic organisms can be estimated from relatively short exposures (i.e., 24, 48, or 96 hr) with death as the typical endpoint.

Chronic—Occurring over a long period of time, generally several weeks, months or years. Chronic exposures occur over an extended period of time or over a significant fraction of a lifetime. Chronic toxicity to aquatic organisms can be estimated from partial life-cycle tests of relatively short duration (i.e., 7 days).

Sublethal—Below the concentration that directly causes death. Exposure to sublethal concentrations of a material may produce less obvious effects on behavior, biochemical and/or physiological function (i.e., growth and reproduction), and histology of organisms.

Delayed Effects—Effects or responses that occur some extended time after exposure.

Static Exposures—Exposures for aquatic toxicity tests in which the test organisms are exposed to the same test solution for the duration of the test (static non-renewal) or to a fresh solution of the same concentration or sample at prescribed intervals such as every 24 hr (static renewal). The concentration of the test material may change during the test due to bio-

the trade-offs become even more complex. For example, the protection of sensitive habitats, such as tropical coral reefs and mangroves, is a priority in oil spill response decisions. Many studies have shown that oil, floating above subtidal reefs, has no adverse effects on the coral; however, if allowed to reach the shoreline, the oil may have long-term impacts to a nearby mangrove system. In addition, oil may persist in the mangrove system creating a chronic source of oil pollution in the adjacent coral reefs. The trade-off would be to consider the use of dispersants. Application of

logical uptake, volatilization, adherence to the test vessel, chemical degradation, etc.

Flow-Through Exposures—Sample to be tested is pumped continuously into a dilutor system and then to the test vessels. This method is used to control sample concentration throughout the duration of the test.

Spiked Exposures—Spiked Declining (SD) Exposures: Concentration of dispersant sample is highest at start and then declines to non-detectable levels after 6–8 hr using a flow-through exposures protocol developed by Chemical Response to Oil Spills Environmental Research Forum (CROSERF) participants.

LC_p—Lethal Concentration: The toxicant concentration that would cause death in a given percent (p) of the test population. For example, the LC_{50} is the concentration that would cause death in 50 percent of the population. The lower the LC, the greater the toxicity.

EC_p—Effective Concentration: A point estimate of the toxicant concentration that would cause an observable adverse effect on a quantal ("all or nothing") response in a given percent (p) of the population.

IC_p—Inhibition Concentration: A point estimate of the toxicant concentration that would cause a given percent (p) reduction in a non-quantal biological measurement such as reproduction or growth.

NOEC—No-Observed-Effect-Concentration: The highest concentration of toxicant to which organisms are exposed in a full or partial (short-term) life-cycle test that causes no observable adverse effects on the test organisms (i.e., the highest concentration of toxicant at which the values for the observed responses are not statistically different from the control).

SOURCES: Singer et al., 1991; Rand, 1995; Grothe et al., 1996; EPA, 2002a,b, 2005; New York Department of Health, 2005.

dispersant would result in dispersion of the oil in the water column and so provide some degree of protection to the mangroves; however, the reef system would now have to endure the consequences of an increase in dispersed oil in the water column (see section on coral reefs later in this chapter). Therefore, for oil spill responders to decide upon appropriate response strategies, it is important that decisions are based on sound scientific data. Ecological factors that go into this decision include: expected sensitivity of exposed resources, proportion of the resource that would be affected, and recovery rates (Pond et al., 2000). There is a tremendous need to reduce the uncertainty associated with each of these decision criteria.

This chapter reviews recent laboratory, mesocosm, and field studies on the toxicological effects of dispersants and dispersed oil, particularly those published since the 1989 NRC report on oil dispersants (NRC, 1989). The intention is first to summarize the current state of understanding regarding the biological effects of dispersants and dispersed oil, and second to make recommendations for additional studies that will help fill critical data gaps in the knowledge and understanding of the behavior and interaction of dispersed oil and the biotic components of ecosystems. The following discussion is limited primarily to studies of the toxicological effects on individual organisms, as opposed to populations or communities. This narrower scope reflects the current state of science in ecotoxicology (see Box 5-2). Although the research and management communities recognize the importance of considering higher order ecological effects, not enough is known to extrapolate from toxicity tests to population or community-level impacts—an issue that concerns all applications of ecotoxicology. Consequently, the explicit consideration of these impacts, and formulation of research to address them, is beyond the scope of this report on the application of ecotoxicological principles to oil spill research.

Due to implementation of several of the recommendations made in 1989 (NRC, 1989), particularly the standardization of toxicity testing methods and information garnered from long-term monitoring of field studies, some general conclusions about the toxicity of dispersants and dispersed oil can be reached. However, there are still areas of uncertainty that will take on greater importance as the use of dispersants is considered in shallow water systems. Specifically, there is insufficient understanding of the fate of dispersed oil in aquatic systems, particularly interactions with sediment particles and subsequent effects on biotic components of exposed ecosystems. In addition, the relative importance of different routes of exposure, that is, the uptake and associated toxicity of oil in the dissolved phase versus dispersed oil droplets versus particulate-associated phase, is poorly understood and not explicitly considered in exposure models. Photoenhanced toxicity has the potential to increase the impact "footprint" of dispersed oil in aquatic organisms, but has only recently received consideration in the assessment of risk associated with spilled oil. One of the widely held assumptions is that chemical dispersion of oil will dramatically reduce the impact to seabirds and aquatic mammals. However, few studies have been conducted since 1989 to validate this assumption. Finally, more work is needed to assess the long-term environmental effects of dispersed oil through monitoring and analysis of spills on which

BOX 5-2
Assessing Population and Community-Level Impacts: A Central Issue in Ecotoxicology

The decision of whether or not to use chemical dispersants in aquatic systems involves evaluation of the trade-offs between potential impacts on various natural resources. Toxicity tests are one of the primary tools that are used to predict these impacts. Much of the toxicological literature focuses on the effects of dispersed oil on individual organisms, because this is the level of biological organization that is most readily studied. Of far greater significance—and of far greater complexity as well—are the effects of dispersed oil on populations and communities of organisms. How to make meaningful predictions about toxicological effects on populations or communities is a problem that is not unique to the assessment of the impacts of an oil spill, but rather is a central question in the field of ecotoxicology. How does the loss or impairment of one or more individual organisms impact a population? How does damage to single or multiple populations impact a community? In the case of dispersed oil, numerous ecological factors may affect the impacts to, and recovery of, these higher levels of biological organization, including the proportion of the resource affected (which in turn involves an understanding of the toxicological sensitivity of organisms as well as the behavior, habits, and habitats that will affect the probability of a species being exposed to oil), birth and death rates of the affected species, the current status of the population (e.g., endangered or common species), life stages that are present, and time of year (e.g., nesting or spawning season, seasonal migration).

Population and community models are tools that show promise in enhancing our understanding of the toxicological impacts to these higher levels of biological organization. Despite recent efforts to advance these approaches (SETAC, 2003), there is no scientific consensus on this issue. Consequently, the majority of ecological risk assessments of environmental chemicals are still based on species-specific tests of toxicological effects on individual organisms. Until population and community-level approaches are more widely accepted and utilized in ecotoxicology, evaluations regarding the impacts of oil spills will remain largely based on qualitative assessments and best professional judgment. However, progress has been made in our understanding of the long-term effects of oil spills on biological communities. The NRC (2003) report on *Oil in the Sea III: Inputs, Fates and Effects* provides a good summary of some of the long-term studies that have been conducted after oil spills, especially those assessing effects on benthic communities and seabirds. For the moment, these types of studies represent the best chance of improving our understanding of the effects of spilled and dispersed oil on biological populations and communities.

SOURCE: SETAC, 2003.

dispersants have been used. Interestingly, several of these data gaps were also identified in 1989 (NRC, 1989).

TESTING PROCEDURES FOR DISPERSANT AND DISPERSED OIL TOXICITY

Toxicity Tests

Much that is currently known about the toxicity and biological effects of dispersants and dispersed oil has been derived from bench-scale acute toxicity tests. These tests typically consist of exposing a single species to varying dilutions of dispersant or dispersed oil preparations under carefully controlled laboratory conditions. Factors that influence such tests include:

- choice of test organism and life stage
- condition of oil (fresh versus weathered)
- method of preparing test solutions
- exposure conditions
- choice of response parameters

Commonly used test organisms include fish, mollusks, arthropods, annelids, and algae. The choice of test organism is dictated by a combination of factors including potential risk, comparative sensitivity, suitability of the species to the testing conditions, and relative ecological and economic significance. An additional consideration is the specific life stage to be tested, because larvae and adults may respond to exposure in significantly different ways.

The method of preparing test solutions is particularly important in the case of dispersed oil testing. Water and oil are not easily miscible, so factors such as mixing energy and loading method can readily affect the relative concentrations of oil components to which test organisms are exposed. Dispersants can also separate and form films on water unless test solutions are properly prepared and mixed.

Exposure conditions in toxicity tests for dispersants and dispersed oil vary with the choice of test chamber (e.g., open or closed), the exposure model (e.g., static or flow-through, spiked or continuous), route of exposure (e.g., water or food), test duration, and other factors such as temperature, salinity, and buffering capacity. The choice of test duration alone can significantly overestimate or underestimate toxicity depending on the actual oil spill situation being simulated.

The choice of response parameters measured in a test can be significant as well. Current generation dispersants appear to cause toxicity through disruptive effects on membrane integrity and a generalized narcosis mechanism (NRC, 1989). Dispersed oil, on the other hand, exerts a toxic effect through multiple pathways including narcosis, more specific receptor-mediated pathways associated with elevated dissolved phase exposures, and possibly by additional pathways such as smothering by dispersed oil droplets. The presence of receptor-mediated pathways suggests that relatively short-term toxicity tests with death as the primary or sole endpoint may not be sufficient to adequately assess the potential risks of dispersed oil. Short-term tests are also incapable of addressing potential delayed effects due to metabolism of oil constituents, bioaccumulation, or possible photoenhanced toxicity.

Although much of the literature on the toxicity of dispersants and dispersed oil is based on typical static exposures of 48–96 hr duration, such tests have been criticized as potentially overestimating the toxicity of oil and dispersed oil in actual spill scenarios (NRC, 1989; George-Ares, et al., 1999). In response to these concerns, a university-industry-government working group, the Chemical Response to Oil Spills Environmental Research Forum (CROSERF), was organized to coordinate and disseminate research on oil spill dispersant use. CROSERF developed toxicity test protocols involving spiked exposures of shorter durations and standardized preparations of water accommodated fractions (WAF) of oil and chemically enhanced water accommodated fractions of dispersed oil (CEWAF) (Singer et al., 1991, 1993, 1994a,b, 1995, 2000, 2001a,b; Clark et al., 2001; Rhoton et al., 2001). For clarity, the term "CEWAF" will only be used in this chapter when referring to a dispersed oil water accommodated fraction that is prepared using the CROSERF protocols. "Chemically dispersed oil" will be used to describe non-CROSERF preparation methods. The CROSERF test methods are summarized in Table 5-1.

The main focus of CROSERF was to standardize methods (i.e., preparation and quantification of fractions and exposure protocols) to allow for greater comparability of toxicological data. In this regard, CROSERF was quite successful. Significant toxicological information was generated using these protocols that successfully addressed the relative toxicity of different dispersants and oil, as well as the relative sensitivity of test organisms.

Refinements to the CROSERF protocols may be warranted for future toxicity testing of dispersants and dispersed oil, either to address specific concerns with the current test procedures (as highlighted below) or to provide greater site-specificity for risk assessment purposes (e.g., dispersant use in nearshore areas). For example, several refinements to the CROSERF procedures have been proposed to adapt the test to subarctic conditions, including changes in WAF preparation, exposure and light regimes, analytical chemistry, and use of subarctic test organisms (Barron

TABLE 5-1 CROSERF Toxicity Test Specifications[a]

Parameter	CROSERF Procedure
WAF and CEWAF Preparation	
Water	Local seawater recommended; minimal 0.45 μm filtration
Oil	Fresh and artificially weathered[b]
Oil loading	Variable loading (0.01–25 g of oil per liter of water); serial dilution not recommended
Vessel	1–20 L carboys or aspirator bottles as appropriate for amount of solution required
Head space	20–25% by volume
Mixing energy/duration[c]	Original: 18–24 h at low mixing energy (approximately 200 rpm with no vortex) and no settling time for WAF, and moderate mixing energy (20–25% vortex) with 3–6 h settling time for CEWAF; Modified[d]: WAF and CEWAF both prepared with moderate mixing energy and settling
Mixing conditions	Sealed in dark at test temperatures
Analytical chemistry[e]	TPH and $<C_{10}$ volatile hydrocarbons required, other analyses optional; TPH, alkanes measured by GC/FID; VOCs and PAHs measured by GC/MS
Dispersant (dispersant:oil)	Primarily Corexit 9500 and/or 9527 (1:10); occasionally Corexit 9554 and others
Dispersant concentration verification	UV–spectroscopy
Test Procedures	
Test design	Five treatments plus control, each with three replicates
Test concentrations	Exposure concentrations derived from a series of geometrically progressing oil loading rates; for toxicity comparisons, total hydrocarbon content (THC: TPH plus $<C_{10}$ volatile hydrocarbons) recommended as concentration endpoint
Exposure regime	48 or 96 h tests in sealed vessels; static-renewal exposures for duration of test, aeration discouraged; flow-through "spiked exposures" with concentrations decreasing to non-detectable levels in <8 h
Test maintenance	Renew solutions at unspecified intervals for static renewal tests, removing dead organisms; dead organisms not removed in flow-through exposures; feeding as specified for test species, with food amount adjusted for loss of test organisms

TABLE 5-1 Continued

Parameter	CROSERF Procedure
Species/life stage	Temperate aquatic species/early life stages
Temperature; salinity	Temperatures appropriate to species; salinity full-strength seawater
Light regime	Laboratory lighting (fluorescent)
Toxicity endpoint	Lethality assessed daily for length of test; sublethal endpoints assessed as appropriate for test organism
Bioaccumulation	Not measured

[a]SOURCE: Singer et al. (1991); Singer et al. (2000); Clark et al. (2001), Rhoton et al. (2001), Singer et al. (2001a).

[b]Modified ASTM Method D-86 (1990 modification); oil "topped" by distillation to 200 °C roughly simulating 1 day at sea (Daling et al. 1990; Singer et al., 2001b).

[c]WAF=Water accommodated fraction; CEWAF=Chemically enhaced WAF, or chemically dispersed oil; stir bar size 1–2 in as appropriate.

[d]Clark et al. (2001) modification of standard CROSERF mixing energy protocol for physically dispersed oil (WAF) using 20–25% vortex, followed by 6 h settling time.

[e]TPH: total petroleum hydrocarbons; alkanes: >10 carbon alkanes; VOC: volatile organic compounds (<10 carbon alkanes and MAHs); PAHs: polycyclic aromatic hydrocarbons; GC: gas chromatography; FID: flame ionization detection; MS: mass spectrometry

and Ka'aihue, 2003). However, the potential benefits of altering test protocols from the CROSERF procedures should be carefully weighed against the implications for potential loss of data comparability and reproducibility.

Some factors to consider in possible refinements to the current CROSERF test protocols for future testing efforts include:

- procedures for making dilutions to be tested
- exposure regimes, including test chambers
- methods for quantifying petroleum exposure
- chemical measurements
- response parameters
- potential photoenhanced toxicity

Two alternate methods for preparing WAF and CEWAF fractions have been suggested, discussed at great length, and remain the subject of scientific debate (see Singer et al., 2000; 2001a; Barron and Ka'aihue, 2003) The CROSERF protocols recommend preparation of toxicity test solutions by variable loading using a series of decreasing concentrations of applied oil and dispersant (Figure 5-1). Other researchers (for example see Barron and Ka'aihue, 2003) have proposed the use of a single oil:water loading rate and the preparation of test solutions using various dilutions of the stock preparation. The decision of which method to use may depend ultimately on the specific scientific question being addressed. Singer et al. (2001a) argue for the variable loading method because they believe it is more "field relevant" since spilled oil slicks tend to be dynamic, continu-

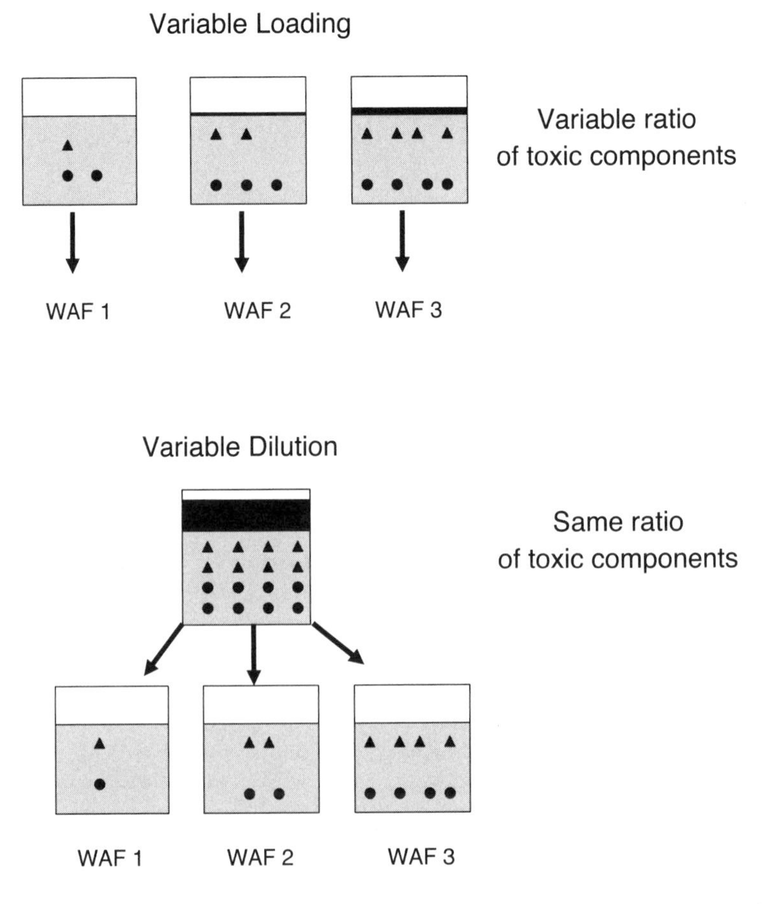

FIGURE 5-1 Comparison of variable loading and variable dilution methods of preparing toxicity test solutions.
SOURCE: Barron and Ka'aihue, 2003; courtesy of Elsevier.

ally changing in size, shape, and thickness. Consequently, these tests address the question: "At what oil to water loading ratio is WAF (CEWAF) toxic?" Barron and Ka'aihue (2003) advocate a variable dilution method for preparing a WAF for testing dispersant that standardizes the oil:water ratio and provides a consistent chemical concentration in a test-series for each oil-dispersant combination (Figure 5-1). This approach answers the question: "At what dilution is a given oil:water ratio of WAF (CEWAF) toxic?" Because it has not been conclusively demonstrated that either method more accurately simulates the temporal dilution of dispersed oil under actual spill conditions, we do not endorse one method over the other. As noted below, there are drawbacks to both approaches.

In the variable loading method, the dispersant:oil ratios do not change and, therefore, each test preparation will have different amounts of oil and dispersant relative to the volume of water in the test chamber. As a result mixing energies change as loading rate (Singer et al., 2000), potentially affecting droplet size or coalescence. The drawback of the variable dilution method has been described as the production of the equal ratio of each specific PAH across the dilution range (Barron and Ka'aihue, 2003). WAF and CEWAF produce significant proportions of oil in the droplet phase, such that increasing dilution may differentially affect the partitioning of the PAH into the aqueous phase. In addition, Barron and Ka'aihue (2003) have argued that the variable dilution approach provides economies in analytical costs by reducing the need to analyze the composition of every tested concentration. However, if chemical analyses were limited to stock solutions, inaccuracies may occur due to differential partitioning in the test dilutions, adsorption of compounds onto test chambers, or loss to the gaseous phase.

The interpretation of the results of toxicity tests can be significantly affected by the method of WAF and CEWAF preparation because of the variable solubilities of the many components in oil. For example, the variable loading method yields different mixtures of petroleum hydrocarbons at different loading rates (see Figure 5-1). The problems that arise between the two methods are due to the fact that often both methods report their data in the same form (i.e., in ppm of some overall metric, such as TPH or tPAH). Therefore, the elimination of any fractional characteristics can lead to a misunderstanding of what that concentration actually represents. For example, LC_{50} data derived from tPAH or TPH alone may result in under- or overestimation of toxicity depending on test preparation method used. Hence, more complete characterizations of chemical analytes are needed.

Another issue with the CROSERF protocols concerns the mixing energies involved in the process of preparing test solutions. The various CROSERF protocols employ equal mixing energies for the production of CEWAF, but differ in the approaches for the production of WAF. For example, initial CROSERF protocols (e.g., Singer et al., 2000) used slow mixing (200 rpm) with no vortex for WAF and a vortex of 20–25 percent for CEWAF preparations. Additional modifications of the method were made (e.g., Clark et al., 2001) so that CEWAF and WAF were prepared using equal mixing energies and a 20–25 percent vortex. Unless a clear rationale can be provided for doing otherwise, it is recommended that equal mixing energies for both WAF and CEWAF be considered for standardization purposes.

A potential issue with the exposure regimes of the CROSERF test is the use of airtight test chambers for flow-through tests. Volatiles, although highly toxic, tend to evaporate very rapidly from spilled oil (NRC, 2003) but are retained in the CROSERF test with unweathered oil because of the sealed nature of the test chamber. The advantage of this approach is that it attempts to standardize the exposure regime, but the drawback is that it may result in an overestimation of toxicity. In most instances, the application of dispersant during an oil spill will happen at least several hours after the initiation of the spill, such that substantial weathering of spilled oil will have occurred (see modeling results in Appendix E). In order to better reflect actual exposure scenarios, open chambers could be considered for use with unweathered oil. Alternatively, tests with closed chambers could be conducted with weathered oil. The choice of experimental protocol will depend on the purpose of the experiment (e.g., standardization or site-specific assessment). Similarly, the temporal exposure regimes of the CROSERF test may not provide an appropriate simulation for some spill situations. For instance, spiked, flow-through exposures in the recommended CROSERF test protocols have oil concentrations decreasing by half about every 2 hr with nondetectable concentrations being reached at about 8 hr. This exposure regime may be a relatively accurate approximation of the exposure situation for the majority of offshore spills in temperate climes. However other temperate zone oil spills (French-McCay, 1998), especially subarctic spills (Neff and Burns, 1996; Short and Harris, 1996), may cause much longer periods of elevated PAH, compounds that contribute significantly to the toxicity of chemically and physically dispersed oil. Furthermore, future potential uses of dispersants in either semi-enclosed inshore waters or freshwater situations could conceivably result in much longer exposure durations than originally envisioned by the CROSERF working group. Thus, the CROSERF spiked protocol may reflect the typical offshore, open-water spill conditions relatively accurately, but longer test durations may yield exposure scenarios that more realistically recreate certain spill conditions. Spiked exposure data yield significantly lower toxicity values than standard constant exposure tests of longer duration (Figure 5-2; also, Clark et al., 2001; Fuller and Bonner, 2001). Consequently, the use of CROSERF spiked exposure data in risk

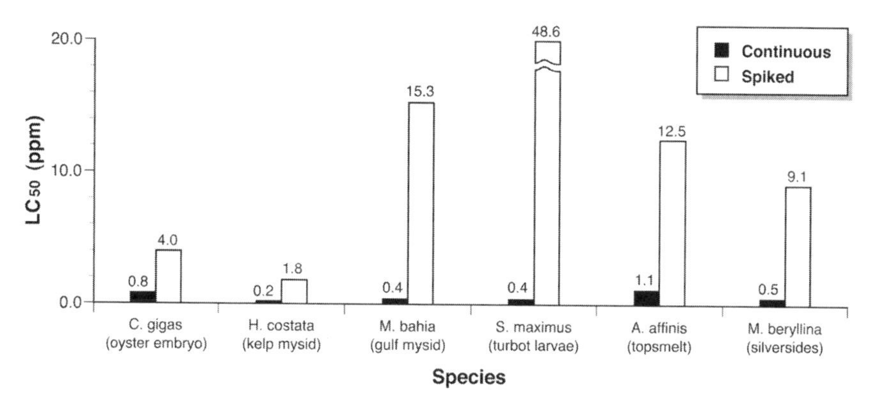

FIGURE 5-2 Comparison of the LC_{50}s for continuous versus spiked exposure regimes using chemically enhanced water accommodated fraction (CEWAF) of different oils. Continuous exposures were 96 hours in duration, except for tests with oyster larvae that were 48 hours. Spiked tests represented an 8-hour declining exposure. Species were exposed to fresh Forties crude oil and Corexit 9500, except for topsmelt, which were exposed to fresh Prudhoe Bay crude oil, and kelp mysid, which were exposed to fresh Kuwait crude oil and Corexit 9527. LC_{50}s for spiked exposures were based on the initial total petroleum hydrocarbon concentration of the CEWAF.
SOURCE: Data are from Clark et al. (2001) and Singer et al. (2001b).

assessment should be evaluated in the context of the specific spill scenarios under consideration.

Additionally, the literature calls for better exposure quantification in testing protocols, moving away from nominal doses and simple estimates of total petroleum hydrocarbon (TPH) to the measurement of specific toxicants in the exposure media, both dissolved and suspended (NRC, 1989; Singer et al., 2000; Shigenaka, 2001; Barron and Ka'aihue, 2003). The CROSERF protocol recommends the measurement of TPH and volatile organic compound (VOC) concentrations in test mixtures, as well as analysis of each PAH in some instances. In comparison with many of the previous studies that reported only nominal concentrations of petroleum products in the test mixtures, the CROSERF protocols were a major improvement. However, future studies should clearly specify at what point during the toxicity test chemical analyses were performed and explain how these measurements were used to calculate the toxicological endpoints. In addition, other methods of quantifying exposure deserve further consideration, including the potential use of toxic units to summarize the toxicity of the various active components of dispersed oil preparations (see discussion under Mode of Action). The primary impediment to applying the toxic unit approach is that not all of the toxic components of petroleum are well-characterized. However, when this issue is better resolved, the toxic unit approach holds considerable promise for more accurately relating exposure and toxicity.

Photoenhanced toxicity is another factor that has not been adequately considered in dispersant and dispersed oil toxicity testing under either CROSERF or non-CROSERF protocols. The toxicity of oil dispersed in water has been shown in some studies to be many times higher in the presence of the ultraviolet radiation from sunlight, yet to date only a single study has examined the photoenhanced toxicity of chemically dispersed oil (Barron et al., 2004). Photoenhanced toxicity as it relates to the effects of dispersed oil is discussed later in the chapter.

Mesocosms

Laboratory experimentation, field trials, and monitoring of spills of opportunity have supplied much of what is currently known of the potential toxicological consequences of oil spills and oil spill response measures. Laboratory experiments cannot adequately address the scale or complexity of actual spills. Field studies to better simulate actual oil spill conditions are restricted by high costs, difficulties in replicating experiments, and regulatory restrictions. Mesocosm-scale tests have been proposed as a way to bridge the gap between laboratory and field studies for testing purposes (Coelho et al., 1999). However, mesocosms have been employed in only a limited number of such studies to date.

The Shoreline Environmental Research Facility (SERF; formerly Coastal Oil Spill Simulation System) in Corpus Christi, Texas discussed in Chapter 3 was used in a series of oil spill experiments to examine bioaccumulation (Coelho et al., 1999) and *in-situ* toxicological responses of various coastal organisms, including fish and various invertebrate species (Lessard et al., 1999; Bragin et al., 1999). Also, laboratory tests were used to evaluate the toxicity of test sediments from these experiments (Fuller et al., 1999). More recently, Ohwada et al. (2003) employed a small-scale mesocosm facility in Japan to examine the fate of soluble fractions of oil and measure their effect on several marine coastal microorganisms, including bacteria, viruses, and heterotrophic nano-flagellates.

The SERF tests indicate both the potential and the limitations of mesocosms in helping explain and predict the ecological effects of oil spill response measures. However, such studies are not as readily controlled as laboratory experiments nor are they as realistic as spill-of-opportunity studies. Additional mesoscale investigations of toxicological responses to oil spill response measures are therefore considered a lower priority for future funding compared to targeted laboratory experimentation and spill-of-opportunity studies. However, if mesocosm studies are conducted for other dispersant-related purposes, consideration should be given to the addition of carefully designed studies that examine the effects of dispersants or dispersed oil on organisms or groups of organisms that cannot be readily studied in laboratory-scale tests.

DISPERSANT TOXICITY

Early dispersant formulations (prior to 1970) were essentially solvent-based degreasing agents adapted from other uses. These early dispersants proved to be highly toxic to aquatic organisms, as seen following treatment of the *Torrey Canyon* spill, resulting in an unfavorable public impression of dispersant use that persists today. Concerns about dispersant use after the *Torrey Canyon* spill were summarized in the previous NRC dispersant review as toxicity of the products themselves, and concern that effective dispersant use would make oil constituents more bioavailable enhancing their toxicity (NRC, 1989). However, the previous NRC report concluded that the acute lethal toxicity of chemically dispersed oil is primarily associated not with the current generation of dispersants but with the dispersed oil and dissolved oil constituents following dispersion (NRC, 1989). There has been little evidence in the intervening years to support a different conclusion.

Dispersants in use today are much less toxic than early generation dispersants, with acute toxicity values (measured in standard 96 h LC_{50} tests) typically in the range of approximately 190–500 mg/L (Fingas, 2002a) as compared with dispersed oil values in the typical range of 20–50 mg/L. An abundant literature exists on the toxicity of the Corexit dispersants currently approved for use in the United States (Tables 5-2 and 5-3; George-Ares and Clark, 2000). Numerous studies have found current dispersants to be significantly less toxic than oil or dispersed oil in direct comparisons (Figure 5-3; also Adams et al., 1999; Mitchell and Holdaway, 2000; Clark et al., 2001; Fingas, 2002a), although a few studies have reported greater dispersant toxicity compared with oil or dispersed oil toxicity (Gulec et al., 1997). Sensitivity to dispersants and dispersed oil can vary significantly by species and life stage. Embryonic and larval stages appear to be more sensitive than adults to both dispersants and dispersed oil (Clark et al., 2001), with LC_{50}s for both oyster and fish larvae reported to be as low as 3 mg/L for dispersant alone and about 1 mg/L for dispersed oil. However, some studies report higher larval toxicity values (i.e., lower sensitivity) for both dispersant and dispersed oil that are closer to the adult values (Coutou et al., 2001). Variable sensitivity of early life stages to dispersants could be related to species-dependent variability in egg permeability (Georges-Ares and Clark, 2000).

TABLE 5-2 Aquatic Toxicity of Corexit® 9527 (Adapted from George-Ares and Clark, 2000)

Common Name[a]	Species	Exposure[b] (h)	Endpoint[c]
Cnidarians			
Green Hydra	*Hydra viridissima*	96	LC_{50}
Green Hydra	*Hydra viridissima*	168	NOEC
Crustaceans			
Brine shrimp	*Artemia sp.*	48	LC_{50}
Brine shrimp	*Artemia salina*	48	LC_{50}
Isopod, F	*Gnorimospaeroma oregonensis*	96	LC_{50}
Amphipod, F	*Anonyx laticoxae*	96	LC_{50}
Amphipod, F	*Anonyx nugax*	96	LC_{50}
Amphipod, F	*Boeckosimus sp.*	96	LC_{50}
Amphipod, F	*Boeckosimus edwardsi*	96	LC_{50}
Amphipod, F	*Onisimus litoralis*	96	LC_{50}
Amphipod, (juvenile), F	*Gammarus oceanicus*	96	LC_{50}
Amphipod, F	*Allorchestes compressa*	96	LC_{50}
Copepod, F	*Pseudocalanus minutus*	48	LC_{50}
Copepod, F	*Pseudocalanus minutus*	96	LC_{50}
Grass shrimp, F	*Palaemonetes pugio*	96	LC_{50}
Grass shrimp, F	*Palaemonetes pugio*	96	LC_{50}
Ghost shrimp	*Palaemon serenus*	96	LC_{50}
Giant freshwater prawn (embryo-larval)	*Macrobrachium rosenbergii*	288	EC_{50} Hatching
Prawn	*Penaeus monodon*	96	LC_{50}
Shrimp	*Penaeus vannemai*	96	LC_{50}
White shrimp (postlarvae), F	*Penaeus setiferus*	96	LC_{50}
Gulf mysid	*Mysidopsis bahia*	96	LC_{50}
Gulf mysid	*Mysidopsis bahia*	48	LC_{50}
Gulf mysid	*Mysidopsis bahia*	SD	LC_{50}
Kelp forest mysid, F	*Holmesimysis costata*	96	LC_{50}
Kelp forest mysid, F	*Holmesimysis costata*	SD	LC_{50}
Kelp forest mysid, F	*Holmesimysis costata*	96	LC_{50}
Kelp forest mysid, F	*Holmesimysis costata*	SD	LC_{50}
Kelp forest mysid	*Holmesimysis costata*	96	LC_{50}
Blue crab (larvae), F	*Callinectes sapidus*	96	LC_{50}
Molluscs			
Scallop, F	*Argopecten irradians*	6	LC_{50}
Scallop, F	*Argopecten irradians*	6	LC_{50}
Scallop, F	*Argopecten irradians*	6	LC_{50}
Red abalone (embryos)	*Haliotis rufescens*	48	EC_{50}
Red abalone (embryos)	*Haliotis rufescens*	SD	EC_{50}

Effect Concentration (ppm)	References
230[f]	Mitchell and Holdaway (2000)[f]
<15[f]	Mitchell and Holdaway (2000)[f]
52–104	Wells et al. (1982)
53–84	Briceno et al. (1992)
>1000	Duval et al. (1982)
>140	Foy (1982)
97–111	Foy (1982)
>175	Foy (1982)
>80	Foy (1982)
80–160	Foy (1982)
>80	Foy (1982)
3.0	Gulec et al. (1997)[e]
8–12	Wells et al. (1982)
5–25	Wells et al. (1982)
640 (27°C)	National Research Council (1989)
840 (17°C)	National Research Council (1989)
49.4[f]	Gulec and Holdaway (2000)[f]
80.4	Law (1995)
35–45	Fucik et al. (1995)
35–45	Fucik et al. (1995)
11.9	Fucik et al. (1995)
29.2,[d] 19–34	Briceno et al. (1992); George-Ares et al. (1999); Exxon Biomedical Sciences (1993a); Pace and Clark (1993)
24.1–29.2[d,f]	Inchcape Testing Services (1995); Clark et al. 2001[f]
>1014[d]	Pace et al. (1995); Clark et al. (2001)[f]
2.4[d]–10.1[d]	Pace and Clark (1993); Exxon Biomedical Sciences (1993b,c); Clark et al. 2001[f]
195[d]	George-Ares and Clark (2000); Clark et al. (2001)[f]
4.3[d]–7.3[d]	Singer et al. (1990, 1991)
120[d]–163[d]	Singer et al. (1991)
15.3[d]	Coelho and Aurand (1996)
77.9–81.2	Fucik et al. (1995)
200 (20°C)	Ordsie and Garofalo (1981)
1800 (10°C)	Ordsie and Garofalo (1981)
2500 (2°C)	Ordsie and Garofalo (1981)
1.6[d]–2.2[d]	Singer et al. (1990, 1991)
13.6[d]–18.1[d]	Singer et al. (1991)

continues

TABLE 5-2 Continued

Common Name[a]	Species	Exposure[b] (h)	Endpoint[c]
Clam, F	*Protothaca stamiea*	96	LC_{50}
Pacific oyster (embryos)	*Crassostrea gigas*	48	LC_{50}
Pacific oyster (embryos)	*Crassostrea gigas*	SD	LC_{50}
Marine sand snail, F	*Polinices conicus*	24	EC_{50}
Fish			
Medaka	*Oryzias latipes*	24	LC_{50}
Rainbow trout	*Oncorhynchus mykiss*	96	LC_{50}
Spot (embryos)	*Leiostomus xanthurus*	48	LC_{50}
Spot (embryo-larval), F	*Leiostomus xanthurus*	48	LC_{50}
Top smelt (larvae)	*Atherinops affinis*	96	LC_{50}
Top smelt (larvae)	*Atherinops affinis*	SD	LC_{50}
Fourhorn sculpin, F	*Myoxocephalus quadricornis*	96	LC_{50}
Mummichog	*Fundulus heteroclitus*	96	LC_{50}
Inland silverside (larvae)	*Menidia beryllina*	96	LC_{50}
Inland silverside (larvae)	*Menidia beryllina*	SD	LC_{50}
Inland silverside (embryos)	*Menidia beryllina*	96	LC_{50}
Red drum (embryo-larval), F	*Sciaenops ocellatus*	48	LC_{50}
Sheepshead minnow	*Cyprinodon variegatus*	96	LC_{50}
Atlantic menhaden (embryo-larval), F	*Brevoortia tyrannus*	48	LC_{50}
Australian bass (larvae)	*Macquaria novemaculeata*	96	LC_{50}
Seagrass			
Turtlegrass, F	*Thalassia tesudimum*	96	LC_{50}
Macroalgae			
Giant kelp (zoospores), F	*Macrocystis pyrifera*	48	NOEC
Giant kelp (zoospores), F	*Macrocystis pyrifera*	SD	NOEC
Giant kelp (zoospores), F	*Macrocystis pyrifera*	SD	IC_{50}
Brown alga	*Phyllospora comosa*	48	EC_{50}
Bacteria			
Microtox™	*Vibrio fisheri*	0.25	EC_{50}

[a]F: field collected.

[b]SD: spiked, declining exposure (107 min half-life).

[c]EC_{50}: concentrations causing effect in 50% of organisms; LC_{50}: concentration causing mortality in 50% of organisms; IC_{50}: concentration causing inhibition in 50% of organisms; NOEC: no effect concentration.

[d]Measured values.

[e]Listed as Gulec et al., 1994 in George-Ares and Clark (2000).

[f]Updated entries not provided in George-Ares and Clark (2000).

Effect Concentration (ppm)	References
ca. 100	Hartwick et al. (1982)
3.1[d]	George-Ares and Clark (2000); Clark et al. (2001)[f]
13.9[d]	George-Ares and Clark (2000); Clark et al. (2001)[f]
33.8	Gulec et al. (1997)[e]
130–150 seawater 400 freshwater	George-Ares and Clark (2000)
96–293	Wells and Doe (1976)
61.2–62.3	Slade (1982)
27.4	Fucik et al. (1995)
25.5[d]–40.6[d]	Singer et al (1990, 1991)
59.2[d]–104[d]	Singer et al. (1991)
<40	Foy (1982)
99–124	Briceno et al. (1992)
52.3,[d] 14.6–57	Briceno et al. (1992); Fucik et al. (1995); Pace and Clark (1993); Inchcape Testing Services (1995); Exxon Biomedical Sciences (1993d); Clark et al. (2001)[f]
58.3[d]	George-Ares and Clark (2000); Clark et al. (2001)[f]
>100	Fucik et al. (1995)
52.6	Fucik et al. (1995)
74–152	Briceno et al. (1992)
42.4	Fucik et al. (1995)
14.3	Gulec and Holdaway (2000)[f]
200	Baca and Getter (1984)
1.3[d]–2.1[d]	Singer et al. (1990, 1991)
12.2[d]–16.4[d]	Singer et al. (1991)
86.6[d]–102[d]	Singer et al. (1991)
30	Burridge and Shir (1995)
4.9–12.8	George-Ares et al. (1999); Exxon Biomedical Sciences (1992)

TABLE 5-3 Aquatic Toxicity of Corexit® 9500 (adapted from George-Ares and Clark, 2000)

Common Name[a]	Species	Exposure[b] (h)	Endpoint[c]
Cnidarians			
Green Hydra	*Hydra viridissima*	96	LC_{50}
Green Hydra	*Hydra viridissima*	168	NOEC
Crustaceans			
Amphipod, F	*Allorchestes compressa*	96	LC_{50}
Brine shrimp	*Artemia salina*	48	LC_{50}
White shrimp, F	*Palaemonetes varians*	6	LC_{50}
Ghost shrimp	*Palaemon serenus*	96	LC_{50}
Gulf mysid	*Mysidopsis bahia*	48	LC_{50}
Gulf mysid	*Mysidopsis bahia*	96	LC_{50}
Gulf mysid	*Mysidopsis bahia*	SD	LC_{50}
Copepod (adult)	*Eurytemora affinis*	96	LC_{50}
Kelp forest mysid, F	*Holmesimysis costata*	SD	LC_{50}
Kelp forest mysid, F	*Holmesimysis costata*	SD	NOEC
Prawn (larval), F	*Penaeus monodon*	96	LC_{50}
Tanner crab (larvae), F	*Chionoecetes bairdi*	96	EC_{50}
Tanner crab (larvae), F	*Chionoecetes bairdi*	SD	EC_{50}
Molluscs			
Marine sand snail, F	*Polinices conicus*	24	EC_{50}
Red abalone (embryos)	*Haliotis rufescens*	48	NOEC
Red abalone (embryos)	*Haliotis rufescens*	SD	NOEC
Red abalone (embryos)	*Haliotis rufescens*	SD	LC_{50}
Fish			
Barramundi (juvenile)	*Lates calcarifer*	96	LC_{50}
Turbot (yolk-sac larvae)	*Scophthalmus maximus*	48	LC_{50}
Turbot (yolk-sac larvae)	*Scophthalmus maximus*	SD	LC_{50}
Rainbow trout	*Oncorhynchus mykiss*	96	LC_{50}
Mummichog	*Fundulus heteroclitus*	96	LC_{50}
Sheepshead minnow (larvae)	*Cyprinodon variegatus*	96	LC_{50}
Sheepshead minnow (larvae)	*Cyprinodon variegatus*	SD	LC_{50}
Mozambique tilapia	*Sarotherodon mozambicus*	96	LC_{50}
Zebra danio	*Brachydanio rerio*	24	LC_{50}
Inland silverside (larvae)	*Menidia beryllina*	96	LC_{50}
Inland silverside (larvae)	*Menidia beryllina*	SD	LC_{50}

Effect Concentration (ppm)	References
160[f]	Mitchell and Holdaway (2000)[f]
13[f]	Mitchell and Holdaway (2000)[f]
3.5	Gulec et al. (1997)[e]
21	George-Ares and Clark (2000)
8103	Beaupoil and Nedelec (1994)
83.1[f]	Gulec and Holdaway (2000)[f]
32.2	Inchcape Testing Services (1995)
31.4[d,f]–35.9[d]	George-Ares and Clark (2000); Fuller and Bonner (2001)[f]; Clark et al. (2001)[f]; Rhoton et al. (2001)[f]
500[d,f]–1305,[d,f] >789[d,f]	Coehlo and Aurand (1997); Fuller and Bonner (2001)[f]; Clark et al. (2001)[f]; Rhoton et al. (2001)[f]
5.2[d]	Wright and Coehlo (1996)
158[d]–245[d]	Singer et al (1996)
41.4[d]–142[d]	Singer et al. (1996)
48	Marine and Freshwater Resources Institute (1998)
5.6[d,f]	Rhoton et al. (2001)[f]
355[d,f]	Rhoton et al. (2001)[f]
42.3	Gulec et al. (1997)[e]
0.7[d]	Aquatic Testing Laboratories (1994)
5.7[d]–9.7[d]	Singer et al. (1996)
12.8[d]–19.7[d]	Singer et al. (1996)
143	Marine and Freshwater Resources Institute (1998)
74.7[d]	George-Ares and Clark (2000); Clark et al. (2001)[f]
>1055[d]	George-Ares and Clark (2000); Clark et al. (2001)[f]
354	George-Ares and Clark (2000)
140	George-Ares and Clark (2000)
170–193[d,f]	Fuller and Bonner (2001)[f]
593–750[d,f]	Fuller and Bonner (2001)[f]
150	George-Ares and Clark (2000)
>400	George-Ares and Clark (2000)
25.2–85.4[d,f]	Inchcape Testing Services (1995); Fuller and Bonner (2001)[f]; Rhoton et al., 2001[f]
40.7[d,f]–116.6,[d,f] 205[d,f]	Fuller and Bonner (2001)[f]; Rhoton et al. (2001)[f]

continues

TABLE 5-3 Continued

Common Name[a]	Species	Exposure[b] (h)	Endpoint[c]
Hardy heads (juvenile), F	*Atherinosoma microstoma*	96	LC_{50}
Australian bass (larvae)	*Macquaria novemaculeata*	96	LC_{50}
Algae			
Diatom	*Skeletonema costatum*	72	EC_{50}
Brown alga (zygotes), F	*Phyllospora comosa*	48	EC_{50}
Bacteria			
Microtox™	*Vibrio fisheri*	0.25	EC_{50}

[a]F: field collected.
[b]SD: spiked, declining exposure (107 min half-life).
[c]EC_{50}: concentrations causing effect in 50% of test organisms; LC_{50}: concentration causing mortality in 50% of test organisms; NOEC: no effect concentration.
[d]Measured values.
[e]Listed as Gulec et al 1994 in George-Ares and Clark (2000).
[f]Updated entries not provided in George-Ares and Clark (2000).

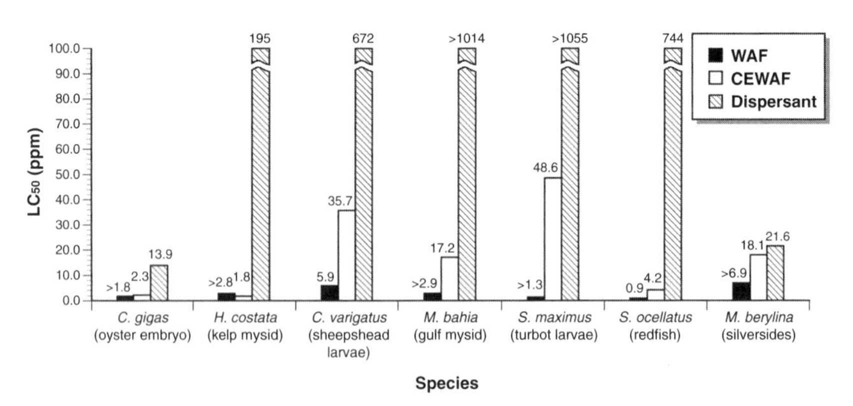

FIGURE 5-3 Comparison of the LC_{50}s derived from spiked exposures of water accommodated fractions (WAF), chemically enhanced water accommodated fraction (CEWAF), and dispersants using either fresh crude oil (Kuwait, Forties, Prudhoe Bay, and Venezuela), weathered-crude oil (Arabian medium) or fresh Medium Fuel Oil, and Corexit 9500 or Corexit 9527. LC_{50}s were based on initial concentrations of total petroleum hydrocarbons.
SOURCE: Data are from Clark et al. (2001); Fuller and Bonner (2001); and Wetzel and Van Fleet (2001).

Effect Concentration (ppm)	References
50	Marine and Freshwater Resources Institute (1998)
19.8	Gulec and Holdaway (2000)[f]
20	Norwegian Institute for Water Research (1994)
0.7	Burridge and Shir (1995)
104[d,f]–242[d,f]	Fuller and Bonner (2001)[f]

In addition to acute toxicity, dispersants may have more subtle effects that influence organism health. Dispersant has been reported to significantly affect the uptake, but not necessarily bioaccumulation, of oil constituents (Wolfe et al., 1998a,b,c; 1999a,b; 2001). In addition, dispersants have been reported to have toxic effects on microbial processes that could potentially interfere with oil decomposition (Varadaraj et al., 1995), but this effect may be offset by other factors that appear to promote oil biodegradation (Swannell and Daniel, 1999). For further discussion on the effect of dispersants and dispersed oil on microbial processes, see section on Microbial Communities (found later in this chapter) and Chapter 4.

TOXICITY OF DISPERSED OIL

Oils are a complex mixture of literally thousands of compounds of varying volatility, water solubility, and toxicity. The purpose of chemical dispersants is to facilitate the movement of oil into the water column. The result is a complex, multi-phase mixture composed of dissolved dispersant, dissolved petroleum hydrocarbons, oil/dispersant droplets, and bulk, undispersed oil. Consequently, aquatic organisms are potentially exposed to many toxicants with different modes of action and through different routes of exposure. Toxicity of dispersed oil in the environment will depend on many factors, including the effectiveness of the dispersion, mixing energy, type of oil, the degree of weathering, type of dispersant, temperature, salinity, duration of exposure, and degree of light penetration into the water column. There is a wealth of information on the biological effects, particularly acute toxicity, associated with exposure to different types of oil (summarized in NRC, 2003). Rather than review these findings, the purpose here is to focus on the issues that are pertinent to understanding the bioavailability and toxicity of chemically dispersed oil.

Route of Exposure

Acute toxicity of oil is the result of a number of interacting chemical, physical, and physiological factors. Thus, toxicity is highly dependent on the conditions of constantly changing exposure. Adverse effects resulting from dispersed oil can be a result of: (1) dissolved materials (e.g., aromatic petroleum hydrocarbons, or dispersant), (2) physical effects due to contact with oil droplets, (3) enhanced uptake of petroleum hydrocarbons through oil/organism interactions, or (4) a combination of these factors (Singer et al., 1998). In general, bioavailability and toxicity of individual hydrocarbons are related to their solubility in water because dissolved hydrocarbons diffuse across the gills, skin, and other exposed membranes of aquatic organisms. The compounds of most concern are the low-molecular-weight alkanes and monocyclic, polycyclic, and heterocyclic aromatic hydrocarbons (Lewis and Aurand, 1997). The monocyclic aromatic hydrocarbons (e.g., benzene, toluene, ethylbenzenes, and xylenes) and low-molecular-weight alkanes are soluble and toxic to aquatic organisms, but these compounds are also very volatile, typically vaporizing rapidly (see Figures 4-2, 4-5, and 4-6 in Chapter 4). As the oil weathers, the concentrations of PAH in the oil plume (including the parent compounds and alkyl substituted homologues) will become relatively enriched compared to the low-molecular-weight alkanes and monocyclic aromatics contributing more to the longer-term toxicity of oil. Because substantial weathering of oil may occur before dispersant is applied (typically at least several hours after the spill), the consequent enrichment of PAH may be particularly important for evaluating the potential toxicity of dispersed oil. Although PAH may drive the toxicity of oil in many instances, some studies have found stronger relationships between TPH concentrations and toxicity than between PAH and toxicity. For example, Barron et al. (1999) conducted studies on the effects of WAF from three different weathered oils on the mysid shrimp, *Mysidopsis bahia*. The median lethal concentrations for the three oils were within a factor of two when expressed as TPH (range from 0.88 to 1.5 mg/L TPH), but differed by nearly a factor of five when expressed as total PAH (range from 2.2 to 9.2 μ/L). Similarly, Clark et al. (2001) found a significant association with TPH, but not PAH or volatiles, in experiments comparing the toxicity of dispersed and untreated oil to early life stages of several marine organisms. McGrath et al. (2003) evaluated the toxicity of various types of gasoline in WAF preparations using an alga, a fish, and a daphnid, and found that both aromatic and aliphatic hydrocarbons contributed to toxicity, with the relative importance of the fractions dependent on the type of gasoline. Furthermore, other components of oil, for example the heterocyclic aromatics, also may be contributing to toxicity (Barron et al., 1999). Some of these fractions are not typically measured in laboratory or field studies, but may be toxicologically important depending on the type of oil and amount of weathering. Another confounding factor in determining the cause of toxicity is that chemical analyses typically measure concentrations in whole samples that include hydrocarbons in the dissolved, colloidal, and particulate phases while the bioavailability of these phases may differ (Fuller et al., 1999). As highlighted below, distinguishing among these phases is important for understanding the fate and effects of dispersed oil.

Oil droplets can physically affect exposed organisms, for example by smothering through the physical coating of gills and other body surfaces. For some organisms, dispersed oil droplets may also be an important route of exposure to petroleum hydrocarbons, through either oil droplet/gill interactions or ingestion of oil droplets. Ramachandran et al. (2004) exposed juvenile rainbow trout to chemically dispersed oil and WAF using Corexit 9500 and Mesa crude oil and then used epifluorescence[1] to microscopically observe PAH uptake in the fish gills. Uptake of PAH from WAF was manifested as an even background of fluorescence on the fish gill with occasional bright spots. Gills of fish exposed to chemically dispersed oil showed localized focal fluorescence (i.e., bright spots), suggesting oil droplets on the gill surface. The authors hypothesized that oil droplets on the fish gill could facilitate uptake of dissolved hydrocarbons.

If dispersion is effective, oil droplets generally range in size from <3 to 80 μm (Franklin and Lloyd, 1986; Lunel, 1993, 1995b). The particle-size distribution of dispersed oil overlaps with the preferred size range of food ingested by many suspension-feeding organisms. For example, common zooplankton, such as copepods, feed on particles in the range of 5 to 60 μm, often switching their preferred particle size depending on the size distribution of available particles (Valiela, 1984). Similarly, benthic and

[1]Method of fluorescence microscopy in which the excitatory light is transmitted through the objective onto the specimen rather than through the specimen; only reflected excitatory light needs to be filtered out rather than transmitted light, which would be of much higher intensity.

epibenthic suspension feeders such as oysters, amphipods, and polychaetes are also known to select particles in size ranges that overlap with dispersed oil droplets, similar to the sizes of some common phytoplankton cells such as *Isochrysis galbana* (4–8 μm), *Chaetocerus spp.* (15–17 μm), and *Skeletonema spp.* (20–25 μm).

The importance of PAH uptake via ingestion of particulate-bound PAH is well known (e.g., Menon and Menon, 1999; Lee, 1992). For example, during the *New Carissa* oil spill near Coos Bay, Oregon, Payne and Driskell (2003) collected dissolved and oil droplet/suspended particulate material (SPM) phase water samples of physically dispersed oil and compared the PAH profiles with those of tissue samples from mussels (a suspension feeder) and Dungeness crabs (an omnivore). The results suggested that mussels accumulated PAH from both the dissolved and the oil droplet/SPM phases, with the latter predominating, while crabs accumulated PAH primarily from the dissolved phase (Figure 5-4). In addition, body burdens of mussels were approximately 500 times greater than those of crabs, indicating the relative importance of these routes of exposure.

Estimating the relative contribution of oil droplets versus particulate-bound oil to total oil exposure is problematic due to the difficulty in distinguishing uptake of these two phases. For physically dispersed oil, interactions with SPM can be very important in the ultimate fate and transport of bulk oil through the formation of oil/SPM agglomerates (see discussion in Chapter 4). Although a limited amount of work has been conducted on the interactions between chemically dispersed oil and SPM, more data are clearly needed to better understand and model the fate and effects of dispersed oil, particularly in shallow water systems with high suspended solids. The limited information available suggests that fairly high oil and SPM concentrations are required before chemically dispersed oil interacts with SPM, and that chemically dispersed oil has a much lower tendency to form SPM agglomerates compared to physically dispersed oil.

Aquatic organisms may also be exposed to oil due to contamination of their food. Wolfe et al. (1998a) evaluated the bioavailability and trophic transfer of PAH from dispersed (Corexit 9527) and untreated Prudhoe Bay crude oil in a simple marine food chain: from phytoplankton, *Isochrysis galbana*, to a rotifer, *Branhionus plicatilis*. Using [^{14}C] naphthalene as a model PAH, direct aqueous exposure was compared to dietary exposure by allowing the rotifers to feed on algae that had been pre-exposed to either WAF or chemically dispersed oil. Results indicated that approximately 20 to 45 percent of uptake was due to dietary exposure, but there was no difference in uptake via the diet between WAF and chemically dispersed oil. Information related to trophic transfer of contaminants is relevant to evaluating the risk of oil exposure, because models based solely on dissolved concentrations may substantially underestimate exposure.

In general, there is insufficient understanding of the fate of dispersed oil in aquatic systems, including interactions with sediment particles and biotic components of ecosystems. In order to better understand the fate and effects of dispersed oil, studies should be conducted to estimate the relative contribution to toxicity of dissolved-, colloidal-, and particulate-phase oil (including an evaluation of oil droplets versus oil/SPM agglomerates) in representative species. Chemical characterization should accompany these tests, including analysis of dissolved and particulate oil concentrations and bioaccumulation. The ability of decisionmakers to estimate the impacts of dispersants on aquatic organisms would be enhanced through greater understanding of these variables used in decision-making tools such as fate and effects models and risk rankings.

Mode of Action

Many oil constituents, most notably the PAH and monoaromatics, are Type I narcotics (DiToro et al., 2000). Narcosis is defined as a reversible state of arrested activity of protoplasmic structures (Bradbury et al., 1989) and is thought to be the primary mechanism of acute toxicity of oil. Often the terms "narcotic" and "anesthetic" are used interchangeably. Type I narcotics are non-polar organic chemicals with a similar mode of action, i.e., narcosis, such that toxicological effects are additive. On the other hand, Type II narcotics, also called polar narcotics, have a different mode of action than the Type I narcotics, and tend to be more toxic. Examples of polar narcotics include nitrogen heterocycles (DiToro et al., 2000). Hence, in oil the heterocyclic aromatics may act as Type II narcotics.

Regardless of their Type I or Type II classification, all organic chemicals in a field mixture contribute to toxicity by narcosis (Deneer et al., 1988); therefore, mixtures of organic chemicals, such as found during an oil spill, would be expected to exhibit additive toxicity over a range of composition ratios (van Wezel et al., 1996). Toxic unit models have been applied to estimate the acute toxicity of PAH and other oil components (Swartz et al., 1995; DiToro et al., 2000; French-McCay, 2002). A toxic unit is the ratio of the measured concentration of a chemical and the corresponding effective concentration in the same medium. Assuming toxicity is additive, the toxic unit value for individual constituents can be summed to estimate acute toxicity of the mixture. DiToro et al. (2000) and French-McCay (2002) incorporated the critical body residue (i.e., lethal body burden) concept into the narcosis toxic unit model. The assumption for this toxicological model, known as the narcosis target lipid model (McGrath et al., 2004), is that mortality occurs when the concentration of narcotic chemicals in the target lipid reaches a threshold concentration. The acute toxicity threshold is assumed to be species specific.

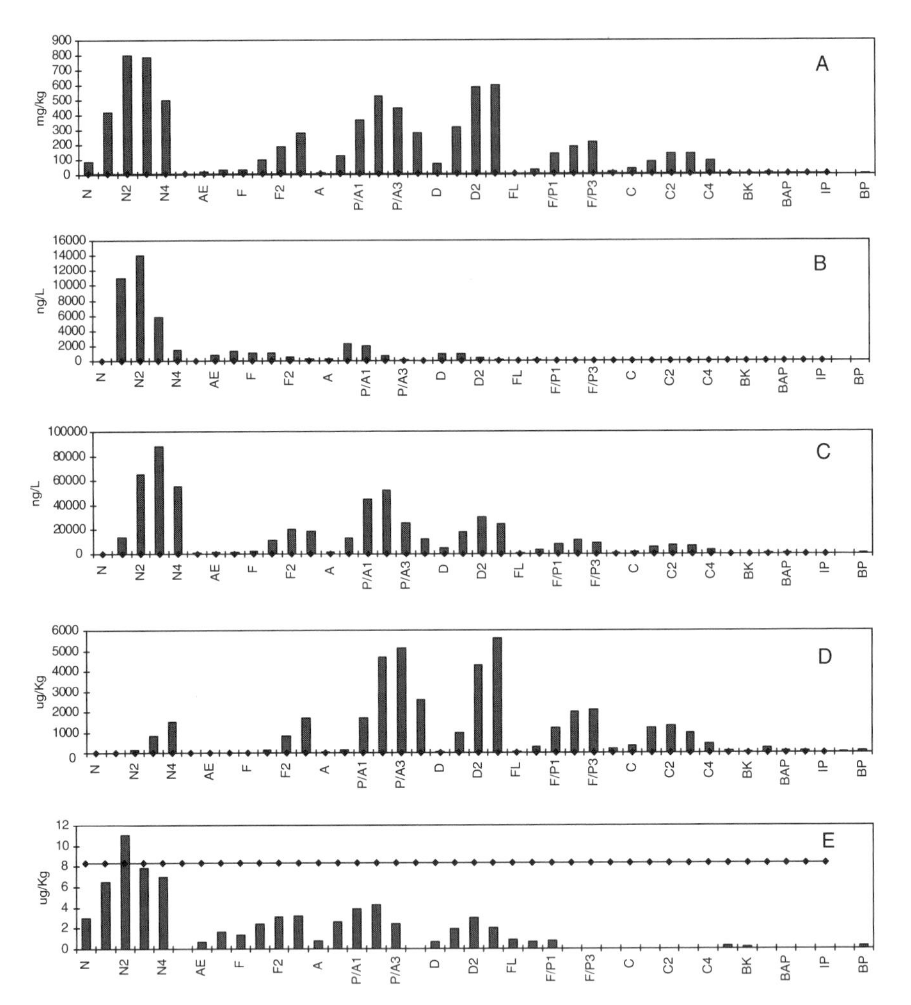

FIGURE 5-4 PAH histograms for: (A) mixed M/V *New Carissa* source oil "blend" (ET-2) collected from the beach adjacent to the vessel on 2/11/99; (B) dissolved- and (C) oil droplet-phase samples collected in the surf zone with the portable large volume water sampling system (PLVWSS) on 2/12/99; (D) mussels collected from the outside north jetty entrance to Coos Bay on 2/14/99; and (E) Dungeness crab collected inside Coos Bay midway up the main channel on 2/19/99. The diamonds connected by the horizontal line represent the sample-specific method detection limits. Note: Also provided is a complete list of analytes and abbreviations, in order, presented in Figure 5-4.

continued

Analytes	Abbreviation
Naphthalene	N
C1-Naphthalenes	N1
C2-Naphthalenes	N2
C3-Naphthalenes	N3
C4-Naphthalenes	N4
Biphenyl	BI
Acenaphthylene	AC
Acenaphthene	AE
Fluorene	F
C1-Fluorenes	F1
C2-Fluorenes	F2
C3-Fluorenes	F3
Anthracene	A
Phenanthrene	P
C1-Phenanthrene/Anthracenes	P/A1
C2-Phenanthrene/Anthracenes	P/A2
C3-Phenanthrene/Anthracenes	P/A3
C4-Phenanthrene/Anthracenes	P/A4
Dibenzothiophene	D
C1-Dibenzothiophenes	D1
C2-Dibenzothiophenes	D2
C3-Dibenzothiophenes	D3
Fluoranthene	FL
Pyrene	PYR
C1-Fluoranthene/Pyrenes	F/P1
C2-Fluoranthene/Pyrenes	F/P2
C3-Fluoranthene/Pyrenes	F/P3
Benzo(a)Anthracene	BA
Chrysene	C
C1-Chrysenes	C1
C2-Chrysenes	C2
C3-Chrysenes	C3
C4-Chrysenes	C4
Benzo(b)fluoranthene	BB
Benzo(k)fluoranthene	BK
Benzo(e)pyrene	BEP
Benzo(a)pyrene	BAP
Perylene	PER
Indeno(1,2,3-cd)pyrene	IP
Dibenzo(a,h)anthracene	DA
Benzo(g,h,i)perylene	BP

SOURCE: Data from Payne and Driskell, 2003; courtesy of the American Petroleum Institute.

The accuracy of toxic unit models is typically based on three assumptions: (1) all the constituents contributing to toxicity are known and measured; (2) effects concentrations of the constituents are known; and (3) chemical equilibrium exists between the organism and the exposure media (but see French-McCay, 2002). Clearly, under dispersed oil scenarios, whether in the laboratory or the field, these assumptions are not apt to be met. Nonetheless, the narcosis model may provide a better estimate of the potential acute effects of oil or dispersed oil than existing measures that rely on determining relationships between toxicity and mixtures of total volatiles, PAH, and/or TPH.

One advantage of the narcosis target lipid model is that it can and has been incorporated into oil fate models to allow estimation of toxicity to aquatic organisms (e.g., French-McCay, 2002, 2004; McGrath et al., 2003). For example, French-McCay (2002) developed an oil toxicity and exposure model, OilToxEx, as a submodel of the Spill Impact Model Application Program (SIMAP). In this model, oil toxicity is predicted by applying the narcosis target lipid model to the predicted concentrations of dissolved aromatic constituents of spilled oil. In a wide range of laboratory exposures with WAF, French-McCay (2002) found good agreement between the narcosis target lipid model predicted LC_{50}s and measured LC_{50}s. McGrath et al. (2003) used the narcosis target lipid model to estimate laboratory toxicity of different gasoline blends. Their model estimated the fate and effects of "hydrocarbon blocks," rather than tracking individual hydrocarbon components (e.g., individual aromatics). The hydrocarbon blocks represented pseudo-components with similar physical chemical properties (usually boiling point as reflected by distillate cut ranges; see Chapter 4). Their analysis indicated that reliable toxicity predictions could be achieved by modeling the fate and toxicity of the hydrocarbon blocks. The utility of this approach is being further explored to predict the fate and effects of spilled oil by incorporation into current models (e.g., GNU Network Object Model Environment) for use in pre-spill planning as well as real-time spill modeling. Nevertheless, more work needs to be done to link the additive compound-specific toxicity data with the component concentrations and mixtures within each hydrocarbon block or pseudo-component.

It should be noted that narcosis may not account for all the toxic effects due to exposure to oil or dispersed oil, particularly sublethal or long-term effects. Barron et al. (2004) evaluated the ability of four mechanism-based toxicity models, including narcosis, to predict chronic toxicity of oil to early life stage fish. They found that the narcosis model underpredicted the observed toxicity and appeared to be mechanistically inconsistent with many of the observed effects of early life stage toxicity in PAH-exposed embryos, including edema, deformities, and cardiovascular dysfunction.

Hence, in these chronic (16 to 35 days) exposures, narcosis appeared not to be the primary mode of action. In conclusion, narcosis models have utility in predicting acute mortality due to exposure to dispersed oil, but may underestimate toxicity in cases where petroleum compounds with non-narcotic modes of action are important components (e.g., alkyl phenanthrenes, heterocyclic aromatics) and where sublethal or delayed effects are manifested (Barron et al., 1999, 2004).

Photoenhanced Toxicity

A number of laboratory studies have indicated that toxicity due to PAH increases significantly (from 12 to 50,000 times) in exposures conducted under ultraviolet light, compared to exposures under the more typical conditions of fluorescent lights (e.g., Landrum et al., 1987; Ankley et al., 1994; Boese et al., 1997; Pelletier et al., 1997). This phenomenon, known as photoenhanced toxicity or phototoxicity, occurs through two mechanisms: photomodification and photosensitization (Neff, 2002; Figure 5-5). Both mechanisms result from the absorption of ultraviolet (UV) radiation by the conjugated double bonds of PAH, exciting them to the triplet state. With photomodification, the excited PAH molecule leads to the formation of highly reactive free radicals that oxidize to form products that are often more toxic than the parent PAH. As described earlier in Chapter 4, photomodification of PAH produces a wide variety of oxygenated products, including quinones, peroxides, and ketones, all of which are more water soluble than the parent PAH (Neff, 2002). Photosensitization occurs when the excited PAH transfers the energy to dissolved oxygen, forming reactive oxygen species. Because of the short-half life of these photoproducts in water, these reactions are only important when products bioaccumulate in the tissues of aquatic organisms (Newsted and Giesy, 1987) and attack cell membranes, bind DNA, or generate secondary radicals. Hence, photosensitization, the primary mechanism of photoenhanced toxicity, causes impacts that differ from the narcosis effects typically associated with PAH toxicity.

Photoenhanced toxicity has only recently received consideration in the assessment of risk associated with spilled oil (Pelletier et al., 1997; Ho et al., 1999; Barron and Ka'aihue, 2001; Duesterloh et al., 2002; Barron et al., 2004). This phenomenon has the potential to increase toxicity under spill scenarios where the opportunity for UV exposure is greatest, e.g., oil stranded on the shoreline, in a surface slick, or in shallow water. Because dispersants generally increase the water-column concentrations of dissolved and particulate petroleum hydrocarbons (including the photoactive compounds) relative to undispersed oil, photoenhanced toxicity of some PAH is an important consideration for evaluating toxicity associ-

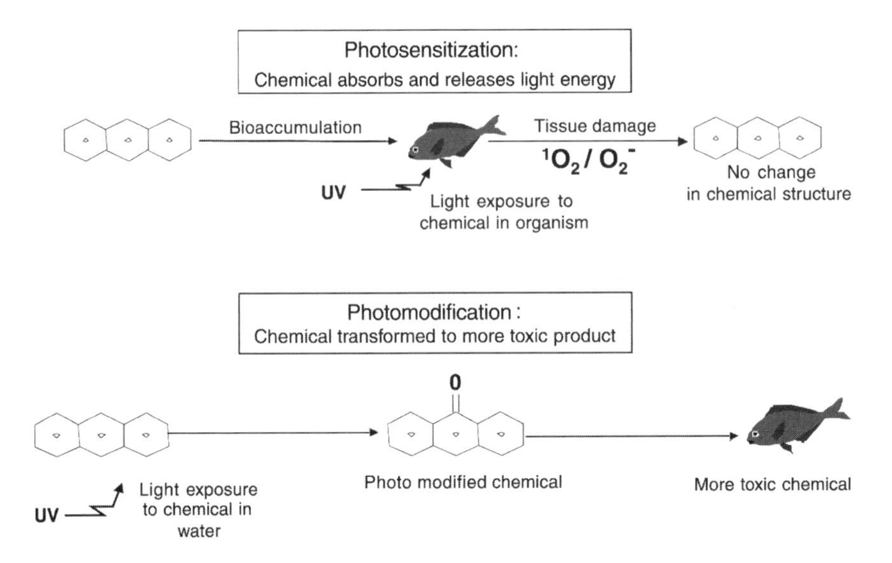

FIGURE 5-5 Mechanisms of photoenhanced toxicity.
SOURCE: Barron, 2000; courtesy of the Prince William Sound Regional Citizen's Advisory Council.

ated with water-column exposure to dispersed oil (Barron and Ka'aihue, 2001; Barron et al., 2004). Photoenhanced toxicity also has implications for the toxicological testing of spilled and dispersed oil. For example, Duesterloh et al. (2002) found that the toxicity of weathered Alaska North Slope crude oil for two calanoid copepod species was dramatically increased upon exposure of the copepods to natural sunlight. In this experiment, *Calanus marhallae* and *Metridia okhotensis* were exposed for 24 hr to low levels of oil in seawater and then exposed to different levels of natural sunlight for 3.8 to 8.2 hr. Toxicity to the copepods increased by up to 80 percent after exposure to UV in sunlight. Similarly, Pelletier et al. (1997) investigated phototoxicity in larvae and juveniles of the bivalve, *Mulinia lateralis,* and juvenile mysid shrimp, *Mysidopsis bahia,* exposed to WAF of several different petroleum products (No. 2 fuel oil, Arabian Light crude, Prudhoe Bay crude, No. 6 fuel oil). Large increases in toxicity (from 2 to 100-fold) in UV light exposures were seen in tests with Arabian Light crude, Prudhoe Bay crude, and No. 6 fuel oil, with the predominant increases found in heavier crudes corresponding to increases in the amount of higher-molecular-weight phototoxic PAH. In contrast, No. 2 fuel oil was highly toxic under both fluorescent and UV light. Finally, Barron et al. (2004) investigated the photoenhanced toxicity of weathered Alaska North Slope crude with and without dispersant (Corexit 9527) to eggs and larvae of the Pacific herring, *Clupea pallasi*. Brief exposure to sunlight (~ 2.5 hr per day for 2 days) increased toxicity from 1.5 to 48-fold over control lighting. In addition, the toxicity of chemically dispersed oil was similar to oil alone in the control treatment, but was significantly more toxic than oil alone in the treatments exposed to sunlight. Accumulation of even small amounts of PAH may make translucent organisms susceptible to toxicological effects if these animals are subsequently exposed to sunlight in the upper part of the water column. Organisms most susceptible to photoenhanced toxicity include translucent pelagic larvae and epibenthic or benthic organisms living in shallow water areas. This phenomenon may not be important for organisms that are opaque (e.g., adult fish, crabs) or avoid sunlight through vertical migration below the photic zone (Valiela, 1984).

Current dispersed oil testing protocols do not typically include exposure to natural sunlight as a factor in evaluating toxicity; thus they may underestimate toxicity for some species, and hence underestimate the "footprint" of toxicological effects on aquatic organisms in the field. Additional toxicological studies are needed to incorporate phototoxicity into effects models, including the identification of phototoxic compounds. Models can be used to overlay this information with expected species distribution in the water column to estimate potential impacts.

Toxicity of Chemically Versus Physically Dispersed Oil

A review of the recent literature (since the publication of the 1989 NRC report on oil dispersants) reveals no consensus in the evaluation of the relative toxicities of chemically dispersed and physically dispersed oil (Clark et al., 2001; Singer et al., 1998; Fingas, 2002a; Fucik et al,. 1994). Some of the inconsistency can be attributed to studies that have drawn conclusions about relative toxicity based on comparing nominal loading rates of oil and dispersant, not on measured concentrations of dissolved hydrocarbons (e.g., Epstein et al., 2000; Adams et al., 1999; Bhattacharyya et al., 2003; Gulec et al., 1997). Loading rate data are useful for comparing the toxicity of different oils when dispersed, different dispersants with the same oil, or sensitivity comparisons among species. However, this approach has limited utility in evaluating the relative toxicity of chemically dispersed versus untreated oil based on exposure to oil in the water column. The degree to which a dispersant facilitates dissolution of petroleum hydrocarbons into the water column will influence the resulting degree of toxicity observed. Many studies have found that the concentrations of PAH are higher in the chemically dispersed oil than in WAF for equal loading of oil. This is likely due to partitioning kinetics between the dispersed oil droplets and water. That is, the increased number of oil droplets and smaller droplet diameters increase the surface area to volume ratio such that more of the hydrocarbon components enter the dissolved phase. Consequently, it is essential to measure actual exposure concentrations to evaluate whether the bioavailability and toxicity of dispersed oil is greater than what would be expected based on the amount of oil in the water column.

Clark et al. (2001) tested three types of crude oil (Kuwait, weathered Kuwait, and Forties) and two dispersants (Corexit 9500 and 9527) in continuous and short-term spiked exposures using the early life stages of several marine species. They found that physically dispersed oil appears less toxic than chemically dispersed oil when LC_{50} is expressed as the nominal loading concentration (Figure 5-6), but when effects are based on the amount of oil measured in water (i.e., TPH), dose-response relationships are similar between chemically and physically dispersed oil.

Similarly, Ramachandran et al. (2004) measured induction of CYP1A (the liver enzyme ethoxyresorufin-*O*-deethylase or EROD) in rainbow trout to WAF and chemically dispersed oil (using Corexit 9500) made from three types of crude oil. They found that EROD activity was as much as 1,100 times higher in chemically dispersed oil treatments compared to WAF when results were expressed on percent (v/v) basis; however, when expressed as measured PAH concentrations, there was little difference between the EC_{50} values for EROD activity.

In contrast, Singer et al. (1998) concluded that the relative toxicity of CEWAF versus WAF was dependent on the test species, exposure time, and endpoint evaluated. In a series of tests, they evaluated the acute effects of untreated and dispersant-treated (Corexit 9527) Prudhoe Bay crude oil on early life stages of three Pacific marine species: red abalone, *Haliotis rufescens*, kelp forest mysid shrimp, *Holmesimysis costata*, and topsmelt, *Atherinops affinis*. Experiments were conducted using CROSERF spiked exposure protocols, including standard preparation of WAF and CEWAF. In addition to the standard toxicity test endpoints, Singer et al. (1998) evaluated initial narcosis in the exposures with *H. costata* and *A. affinis* by making behavioral observations during the first 6–7 hr of exposure and tallying the number of inactive and active animals. Narcosis was defined as those animals initially affected, but that recovered to an active state later in the exposure. Results are summarized in Table 5-4 (taken from Singer et al., 1998) and expressed as EC_{50} or LC_{50} values based on total hydrocarbon content ($THC_{(C7\text{-}C30)}$) measured at the beginning of the exposures. In tests with *H. rufescens* and *H. costata*, significant effects were seen in the CEWAF exposures at total hydrocarbon concentrations (THC) two to three times lower than in WAF tests (Table 5-4). In contrast, effects on mortality of the topsmelt, *A. affinis*, and initial narcosis were more severe in WAF exposures. Singer et al. (1998) suggest that a likely explanation for these results is compositional differences in dissolved petro-

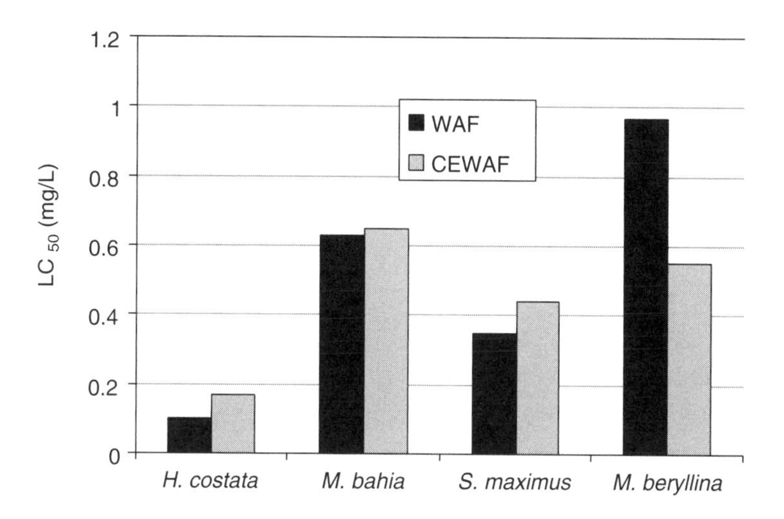

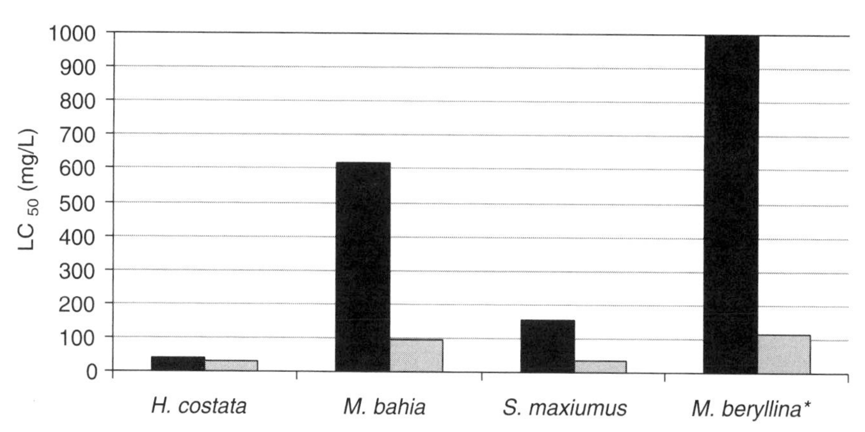

FIGURE 5-6 Comparison of expressing toxicity in terms of measured lethal concentrations (LC) of total petroleum hydrocarbons (TPH) or lethal loading (LL) concentrations based on nominal "oil added" values. Tests were constant 96-hour static-renewal tests with Kuwait oil and Corexit 9527 for the mysids (*Holmesimysis costata* and *Mysidopsis bahia*) and silversides (*Menidia beryllina*). Exposures of turbot (*Scophthalmus maximus*) were 48 hour exposures with Forties crude oil and Corexit 9500. Data expressed as LL imply that CEWAF is more toxic than WAF, but when expressed as measured TPH, toxicities are roughly equivalent.* The LL_{50} for *M. beryllina* exposed to WAF was 5,020 mg/L, but was not displayed for scaling purposes.
SOURCE: Data are from Clark et al., 2001.

TABLE 5-4 Results of Spiked Exposure Toxicity Tests Using Prudhoe Bay Crude Oil Alone and Combined with Corexit 9527 (O:D ratio = 10:1) from Singer et al., 1998 (Results are expressed as the EC or LC_{50} in mg/L of $THC_{(C7–C30)}$)

	WAF			CEWAF		
Species/Endpoint	Test 1	Test 2	Test 3	Test 1	Test 2	Test 3
Haliotis						
Larval abnormality	>34.03[a]	>46.99	>33.58	19.09	32.70	17.80
Holmesimysis						
96-h mortality	>34.68	>25.45	>28.55	10.54	10.75	10.83
Initial narcosis	11.31	11.58	15.90	11.07	>38.33	48.03
Atherinops						
96-h mortality	16.34	40.20	35.73	28.60	74.73	34.06
Initial narcosis	26.63	>48.22	31.76	>101.82	>140.97	>62.22

[a]EC/LC_{50} estimated to be above the highest test concentration.

leum hydrocarbons between CEWAF and WAF due to differences in mixing energy and loading rates used to prepare the exposure media. For example, WAF solutions were found to have a larger proportion of volatiles (96 percent) as compared to the CEWAF (67 percent). They conclude that different fractions of oil may drive toxicity in different types of solutions. Consequently, reporting toxicity based on only a few of the oil components may make comparisons across studies difficult (see Figure 5-6).

A similar conclusion was drawn by Fucik et al. (1994) in a series of tests comparing the toxicity of chemically dispersed oil, dispersant (Corexit 9527), and WAF to a variety of fish and invertebrate species and life stages from the Gulf of Mexico. Fucik et al. (1994) reported that the toxicity of dispersed oil was proportionately less than WAF when results were compared using a Toxicity Index (TI) applied to the measured TPH data. The TI expresses toxicity as a function of concentration and duration of exposure (e.g., ppm-h). Experiments included both static renewal and flow-through exposures in open containers that allowed significant volatilization of the petroleum constituents. To explain this result, Fucik et al. (1994) speculated that volatilization from dispersed oil was enhanced compared to WAF. Therefore, concentrations of benzene, toluene, ethylbenzene, and xylenes (BTEX) were higher in WAF, potentially enhancing toxicity in these exposures. Alternatively, they suggested that oil droplets or emulsions in the chemically dispersed oil may have lower bioavailability than the dissolved hydrocarbons. This explanation seems unlikely given recent studies suggesting that oil droplets may enhance uptake of petroleum hydrocarbons (Payne and Driskell, 2003; Ramachandran et al., 2004).

In conclusion, there is no compelling evidence that the toxicity of chemically dispersed oil is enhanced over physically dispersed oil if comparisons are based on measured concentrations of petroleum hydrocarbons in the water column. This conclusion is further discussed in the section on toxicological effects of dispersed oil on water column organisms. A similar conclusion was reached in the NRC (1989) review of oil dispersants. CROSERF testing protocols recommend analyzing total hydrocarbon content (composed of total petroleum hydrocarbons and volatile hydrocarbons) at a minimum, but also suggest in-depth investigations include the analysis of PAH (Singer et al., 2000; Table 5-5). The studies reviewed above clearly indicated that measuring fractional components of aqueous oil (e.g., TPH, total PAH, total volatiles) may not give the resolution necessary to adequately interpret toxicity test data. Consequently, it is recommended that chemical analyses in conjunction with toxicity tests should routinely include dissolved- and oil droplet-phase analyses of the full suite of parent and alkyl-substituted PAH and heterocyclics as well as the n-alkanes that typically comprise the THC. In addition, application of additive toxicity models for PAH and other petroleum constituents may facilitate the interpretation of toxicity test results.

Although acute toxicity studies do not indicate differences in the lethal or sublethal responses of organisms exposed to chemically dispersed or untreated oil, some studies have suggested that the bioaccumulation kinetics of PAH from dispersed oil may differ from those for undispersed oil. In a series of experiments, Wolfe et al. (1998a,b,c; 1999a; 2001) have investigated the bioavailability of naphthalene and phenanthrene in chemically dispersed oil versus WAF, including an assessment of uptake and depuration kinetics, to address the question of whether dispersants alter bioavailability of compounds. The premise of these experiments was that the bioavailability of dispersed oil may be enhanced due to interactions between dispersant, oil, and biological membranes, possibly as a result of dispersant-mediated changes in membrane permeability, osmoregulation, or other cellular mechanisms. Several experiments examined bioaccumulation of naphthalene as a model PAH by the microalga *Isochrysis galbana*. Naphthalene was selected because it has negligible dispersant facilitated solubility such that changes in bioavailability could be examined in the absence of differences in dissolved-phase concentrations between dispersed and untreated oil. In these experiments, algal cells were exposed to laboratory preparations of either WAF of Prudhoe Bay crude

TABLE 5-5 Recommended Target Analyte List for PAH from Singer et al. (2000)

Naphthalene	Fluoranthene
C-1 naphthalenes	Pyrene
C-2 naphthalenes	C-1 pyrenes
C-3 naphthalenes	C-2 pyrenes
C-4 naphthalenes	C-3 pyrenes
Biphenyl	C-4 pyrenes
Fluorene	Benzo(a,h)anthracene
C-1 fluorenes	Chrysene
C-2 flurorenes	C-1 chrysenes
C-3 fluorenes	C-2 chrysenes
Dibenzothiophene	C-3 chrysenes
C-1 dibenzothiophenes	C-4 chrysenes
C-2 dibenzothiophenes	Benzo(b)fluoranthene
C-3 dibenzothiophenes	Benzo(k)fluoranthene
C-4 dibenzothiophenes	Benzo(e)pyrene
Phenanthrene	Benzo(a)pyrene
C-1 phenanthrenes	Perylene
C-2 phenanthrenes	Indeno(g,h,i)pyrene
C-3 phenanthrenes	Dibenzo(a,h)anthracene
C-4 phenanthrenes	Benzo(1,2,3-cd)perylene

oil (PBCO) or dispersed oil mixture of PBCO and Corexit 9527 spiked with [U-^{14}C] naphthalene. Results suggest that dispersants enhanced the initial uptake of naphthalene by microalgae under a variety of temperature and salinity conditions. However, there were no differences in bioaccumulation as indicated by similarity in bioaccumulation factors between dispersed oil and WAF, suggesting that depuration rates were also enhanced. Wolfe et al. (1998a,b,c; 1999a,b; 2001) extended these experiments to a model food chain, including *I. galbana*, the rotifer *Brachionus plicatilis*, and larval topsmelt, *Atherinops affinis*. Direct aqueous exposures to phenanthrene and naphthalene were compared with aqueous plus dietary exposures. Depuration of phenanthrene by rotifers decreased significantly following dispersed oil exposures, while uptake and depuration of naphthalene by larval topsmelt significantly increased in both aqueous and dietary exposures to dispersed oil. These detailed and elegant experiments have enhanced our understanding of the bioaccumulation kinetics of dispersed oil PAH. These studies should be expanded to include other organisms and PAH. In addition, this model food chain could also be used to answer questions related to the importance of PAH uptake via the dissolved versus oil droplet phases.

EFFECTS ON BIOLOGICAL COMMUNITIES

In the sections that follow, the recent (post-1989) literature on the toxicological effects of chemically dispersed oil is reviewed by habitat type. A detailed review on dispersant toxicity studies pre-1989 was provided in NRC (1989). Besides avoiding duplication, for the most part these earlier studies are not included because many were based on comparisons using the older dispersant formulations and limited by the use of nominal exposures. Studies from freshwater systems are included where possible. It is noted, however, that the amount of literature concerned with dispersants and chemically dispersed oil effects on freshwater organisms is sparse, most likely a function of the fact that the most common U.S. dispersants, Corexit 9500 and 9527, have low efficacy in freshwater. Furthermore, the use of dispersants in freshwater is assumed to be unlikely because the increase in water-column burden of hydrocarbons would preclude their use in freshwater systems that provide a source of drinking water.

Water-Column Organisms

This section reviews the literature pertaining to dispersed oil effects on water column organisms, including larval stages of benthic organisms (Tables 5-6, 5-7, and 5-8). The review was limited by many studies that are still based on comparisons of nominal concentrations, despite the recommendation made in NRC (1989) that future studies include chemical analyses of the exposure media. One common technique is to measure TPH (and /or VOC and PAH) in the stock solutions and infer TPH levels upon serial dilutions of these solutions. While this is an improvement over the use of purely nominal values, it still limits the interpretation of the results unless some minimal and random sampling of test exposures provides confirmation that expected concentrations approximate measured concentrations. It is extremely important to provide an estimate of exposure based on measured concentrations when conducting toxicity tests.

In general, studies that concluded that chemically dispersed oil was more toxic were based on nominal loading of oil, not measured concentrations. For example, Clark et al. (2001) using three types of oil (variable loadings), two dispersants (Corexit 9500 and 9527), continuous and short-term spiked exposures, and early life stages of several marine organisms in 46 and 96 hr tests found that physically dispersed oil appears less toxic than chemically dispersed oil when LC_{50}s were expressed as nominal loading concentrations (see earlier in Chapter 5). When toxicity effects were based on measured TPH, no difference between chemically and physically dispersed oil was observed using continuous exposures. In an exposure study using freshwater fish, Pollino and Holdaway (2002b) con-

TABLE 5-6 Acute Effects of Chemically Dispersed Oil in Comparison to Physically Dispersed Oil in Water-Column Organisms (studies since 1989)

Species	Oil (D:O ratio)	Dispersant	Exposure (hr)	Type of Exposure (static/flow-through)
(1) Marine studies:				
MOLLUSCS				
Crassostrea gigas (Pacific oyster)	Kuwait (1:10)	Corexit 9527	48	constant
Crassostrea gigas	Kuwait (1:10)	Corexit 9527	48	spiked
Crassostrea gigas	Forties crude (1:10)	Corexit 9500	48	constant
Crassostrea gigas	Forties crude (1:10)	Corexit 9500	48	spiked
Crassostrea gigas	Medium fuel oil (1:10)	Corexit 9527	48	constant
Crassostrea gigas	Medium fuel oil (1:10)	Corexit 9527	48	spiked
Octopus pallidus (octopus)	BSC (1:50)	Corexit 9527	24	semi-static
Octopus pallidus	BSC (1:50)	Corexit 9527	48	semi-static
CRUSTACEANS				
Balanus amphitrite (barnacle)	Diesel oil (1:10)	Vecom B-1425	24	static
Balanus amphitrite	Diesel oil (1:10)	Vecom B-1425	48	static
Balanus amphitrite	Diesel oil (1:10)	Norchem OSD-570	24	static
Balanus amphitrite	Diesel oil (1:10)	Norchem OSD-570	48	static
Palaemon serenus (ghost shrimp)	BSC (1:10)	Corexit 9500	96	static (50% daily renewal)
Palaemon serenus	BSC (1:10)	Corexit 9527	96	static (50% daily renewal)
Palaemon elegans (prawn)	Middle East Crude Oil	Not disclosed	24	static
Allorchestes compressa (Amphipod)	BSC (1:10)	Corexit 9527	96	static (60% daily renewal)
Allorchestes compressa	BSC (1:10)	Corexit 9500	96	static (60% daily renewal)
Mysidopsis bahia (gulf mysid shrimp)	Kuwait (1:10)	Corexit 9527	96	constant
Mysidopsis bahia	Kuwait (1:10)	Corexit 9527	96	spiked
Mysidopsis bahia	Kuwait (W) (1:10)	Corexit 9527	96	constant

Endpoint	Oil Treatment Effect Conc. (LC_{50}) mg/L	Dispersed Oil Effect Conc. (LC_{50}) mg/L	Concentration Estimate[e]	Reference
larval mortality	NA	0.5	Initial TPH	Clark et al., 2001
larval mortality	NA	1.92	Initial TPH	Clark et al., 2001
larval mortality	NA	0.81	Initial TPH	Clark et al., 2001
larval mortality	NA	3.99	Initial TPH	Clark et al., 2001
larval mortality	>1.14	0.53	Initial TPH	Clark et al., 2001
larval mortality	>1.83	2.28	Initial TPH	Clark et al., 2001
hatchling mortality	0.51	3.11	Average TPH over 24 hr	Long and Holdaway, 2002
hatchling mortality	0.39	1.8	Average TPH over 24 hr	Long and Holdaway, 2002
larval mortality	NA	514	Initial nominal[a]	Wu et al., 1997
larval mortality	NA	48	Initial nominal[a]	Wu et al., 1997
larval mortality	NA	505	Initial nominal[a]	Wu et al., 1997
larval mortality	NA	71	Initial nominal[a]	Wu et al., 1997
mortality	258,000	3.6	Initial nominal	Gulec and Holdaway, 2000
mortality	258,000	8.1	Initial nominal	Gulec and Holdaway, 2000
mortality	83.5[b]	1.1[b]	Initial nominal	Unsal, 1991
mortality	311,000	16.2	Initial nominal	Gulec et al., 1997
mortality	311,000	14.8	Initial nominal	Gulec et al., 1997
mortality	0.63	0.65	Initial TPH	Clark et al., 2001
mortality	>2.93	17.2	Initial TPH	Clark et al., 2001
mortality	NA	0.11	Initial TPH	Clark et al., 2001

continues

TABLE 5-6 Continued

Species	Oil (D:O ratio)	Dispersant	Exposure (hr)	Type of Exposure (static/flow-through)
Mysidopsis bahia	Kuwait (W) (1:10)	Corexit 9527	96	spiked
Mysidopsis bahia	Forties crude (1:10)	Corexit 9500	96	constant
Mysidopsis bahia	Forties crude (1:10)	Corexit 9500	96	spiked
Mysidopsis bahia	AMC (W) (1:10)	Corexit 9500	96	spiked
Mysidopsis bahia	AMC (W) (1:10)	Corexit 9500	96	static (75% daily renewal), sealed
Mysidopsis bahia	ANS (1:10)	Corexit 9500	96	spiked
Mysidopsis bahia	ANS (1:10)	Corexit 9500	96	continuous
Mysidopsis bahia	VCO (1:10)	Corexit 9500	96	static (90% daily renewal), sealed
Mysidopsis bahia	VCO (1:10)	Corexit 9500	96	spiked
Mysidopsis bahia	PBCO (1:10)	Corexit 9500	96	spiked
Mysidopsis bahia	VCO (W) (1:10)	Corexit 9500	96	spiked
Mysidopsis bahia	KCO (1:10)	Corexit 9527	96	spiked
Mysidopsis bahia	KCO (1:10)	Corexit 9527	96	static daily renewal, sealed
Holmesimysis costata (kelp mysid shrimp)	Kuwait (1:10)	Corexit 9527	96	constant
Holmesimysis costata	Kuwait (1:10)	Corexit 9527	96	spiked
Holmesimysis costata	PBCO (1:10)	Corexit 9527	96	spiked
Americamysis (Holmesimysis) costata (kelp forest mysid)	PCBO (1:10)	Corexit 9500	96	spiked
Americamysis (Holmesimysis) costata	PCBO (W) (1:10)	Corexit 9500	96	spiked
CNIDARIANS				
Hydra viridissima (green hydra)	BSC (1:29)	Corexit 9527	96	static
Hydra viridissima	BSC (1:29)	Corexit 9500	96	static
FISH				
Clupea pallasi (Pacific herring)	Weathered ANS (1:25)	Corexit 9527	24	static

Endpoint	Oil Treatment Effect Conc. (LC_{50}) mg/L	Dispersed Oil Effect Conc. (LC_{50}) mg/L	Concentration Estimate[e]	Reference
mortality	>0.17	111	Initial TPH	Clark et al., 2001
mortality	NA	0.42	Initial TPH	Clark et al., 2001
mortality	NA	15.3	Initial TPH	Clark et al., 2001
larval mortality	26.1–83.1	56.5–60.8	Initial TPH	Fuller and Bonner, 2001
larval mortality	0.56–0.67	0.64–0.65	Initial TPH	Fuller and Bonner, 2001
larval mortality	8.21	5.08	Initial THC	Rhoton et al., 2001
larval mortality	2.61	1.4	Initial THC	Rhoton et al., 2001
larval mortality	0.15–0.4	0.50–0.53	Average TPH	Wetzel and van Fleet, 2001
larval mortality	0.59–0.89	10.2–18.1	Average TPH	Wetzel and van Fleet, 2001
larval mortality	>6.86	15.9	Average TPH	Wetzel and van Fleet, 2001
larval mortality	>0.63–>0.83	72.6–120.8	Average TPH	Wetzel and van Fleet, 2001
mortality	>2.9	17.7	Initial TPH	Pace et al., 1995
mortality	0.78	0.98	Initial TPH	Pace et al., 1995
mortality	0.1	0.17	Initial TPH	Clark et al., 2001
mortality	>2.76	1.8	Initial TPH	Clark et al., 2001
juvenile mortality	>25.45–>34.68	10.54–10.83	Initial THC[c]	Singer et al., 1998
early-life stage mortality	14.23–>17.5	9.46–14.40	Initial THC[c]	Singer et al., 2001
early-life stage mortality	0.951–>1.03	5.72–33.27	Initial THC[c]	Singer et al., 2001
mortality	0.7	9	Initial stock TPH	Mitchell and Holdaway, 2000
mortality	0.7	7.2	Initial stock TPH	Mitchell and Holdaway, 2000
larval mortality	~0.045	0.199	Initial tPAH	Barron et al., 2004

continues

TABLE 5-6 Continued

Species	Oil (D:O ratio)	Dispersant	Exposure (hr)	Type of Exposure (static/flow-through)
Cyprinodon variegatus (sheepshead minnow)	No. 2 fuel oil (1:1 to 1:10)	Omniclean	96	static
Cyprinodon variegatus	AMC (W) (1:10)	Corexit 9500	96	spiked
Cyprinodon variegatus	AMC (W) (1:10)	Corexit 9500	96	static (75% daily renewal), sealed
Atherinops affinis (topsmelt)	PBCO (1:10)	Corexit 9527	96	spiked
Atherinops affinis	PBCO (1:10)	Corexit 9500	96	spiked
Atherinops affinis	PBCO (W) (1:10)	Corexit 9500	96	spiked
Scophthalamus maxiumus (turbot)	Kuwait (1:10)	Corexit 9527	48	constant
Scophthalamus maxiumus	Kuwait (1:10)	Corexit 9527	48	spiked
Scophthalamus maxiumus	Forties (1:10)	Corexit 9500	48	constant
Scophthalamus maxiumus	Forties (1:10)	Corexit 9500	48	spiked
Menidia beryllina (Inland silveride)	Kuwait (1:10)	Corexit 9527	96	constant
Menidia beryllina	Kuwait (1:10)	Corexit 9527	96	spiked
Menidia beryllina	Kuwait (W) (1:10)	Corexit 9527	96	constant
Menidia beryllina	Kuwait (W) (1:10)	Corexit 9527	96	spiked
Menidia beryllina	Forties (1:10)	Corexit 9500	96	constant
Menidia beryllina	Forties (1:10)	Corexit 9500	96	spiked
Menidia beryllina	PBCO (1:10)	Corexit 9500	96	spiked
Menidia beryllina	ALC (W) (1:10)	Corexit 9500	96	spiked
Menidia beryllina	ALC (W) (1:10)	Corexit 9500	96	static (75% daily renewal), sealed
Menidia beryllina	PBCO (W) (1:10)	Corexit 9500	96	spiked
Menidia beryllina	ANS (1:10)	Corexit 9500	96	spiked
Menidia beryllina	ANS (1:10)	Corexit 9500	96	continuous
Menidia beryllina	PBCO (1:10)	Corexit 9500	96	spiked
Menidia beryllina	PBCO (1:10)	Corexit 9500	96	continuous
Menidia beryllina	VCO (1:10)	Corexit 9500	96	static (90% daily renewal), sealed
Menidia beryllina	VCO (1:10)	Corexit 9500	96	spiked
Menidia beryllina	PBCO (1:10)	Corexit 9500	96	spiked

Endpoint	Oil Treatment Effect Conc. (LC_{50}) mg/L	Dispersed Oil Effect Conc. (LC_{50}) mg/L	Concentration Estimate[e]	Reference
larval mortality	94	~ 80–165[d]	Nominal initial mg/L	Adams et al., 1999
larval mortality	>5.7–6.1	31.9–39.5	Initial TPH	Fuller and Bonner, 2001
larval mortality	3.9–4.2	>9.7–10.8	Initial TPH	Fuller and Bonner, 2001
larval mortality	16.34–40.20	28.6–74.73	Initial THC	Singer et al., 1998
early life stage mortality	9.35–12.13	7.27–17.70	Initial THC	Singer et al., 2001
early life stage mortality	>1.45–>1.60	16.86–18.06	Initial THC	Singer et al., 2001
mortality	NA	2	Initial TPH	Clark et al., 2001
mortality	NA	16.5	Initial TPH	Clark et al., 2001
mortality	0.35	0.44	Initial TPH	Clark et al., 2001
mortality	>1.33	48.6	Initial TPH	Clark et al., 2001
mortality	0.97	0.55	Initial TPH	Clark et al., 2001
mortality	>1.32	6.45	Initial TPH	Clark et al., 2001
mortality	0.14	1.09	Initial TPH	Clark et al., 2001
mortality	>0.66	10.9	Initial TPH	Clark et al., 2001
mortality	NA	0.49	Initial TPH	Clark et al., 2001
mortality	NA	9.05	Initial TPH	Clark et al., 2001
early life stage mortality	11.83	32.47	Initial THC	Singer et al., 2001
larval mortality	>14.5–32.3	24.9–36.9	Initial TPH	Fuller and Bonner, 2001
larval mortality	4.9–5.5	1.5–2.5	Initial TPH	Fuller and Bonner, 2001
early life stage mortality	NA	20.28	Initial THC	Singer et al., 2001
larval mortality	26.36	12.22	Initial THC	Rhoton et al., 2001
larval mortality	15.59	12.42	Initial THC	Rhoton et al., 2001
larval mortality	>19.86	12.29	Initial THC	Rhoton et al., 2001
larval mortality	14.81	4.57	Initial THC	Rhoton et al., 2001
larval mortality	<0.11	0.68	Average TPH	Wetzel and van Fleet, 2001
larval mortality	0.63	2.84	Average TPH	Wetzel and van Fleet, 2001
larval mortality	>6.86	18.1	Average TPH	Wetzel and van Fleet, 2001

continues

TABLE 5-6 Continued

Species	Oil (D:O ratio)	Dispersant	Exposure (hr)	Type of Exposure (static/flow-through)
Menidia beryllina	VCO (W) (1:10)	Corexit 9500	96	spiked
Menidia beryllina	ANS (W) (1:10)	Corexit 9500	96	continuous
Menidia beryllina	ANS (W) (1:10)	Corexit 9500	96	spiked
Sciaenops ocellatus (Red drum)	VCO (1:10)	Corexit 9500	96	spiked
Macquaria novemaculeata (Australian bass)	BSC (1:10)	Corexit 9500	96	static (50% daily renewal)
Macquaria novemaculeata	BSC (1:10)	Corexit 9527	96	static (50% daily renewal)
Macquaria novemaculeata	BSC (1:50)	Corexit 9527	96	static daily renewal
(2) Freshwater studies:				
CNIDARIANS				
Hydra viridissima (green hydra)	BSC (1:29)	Corexit 9527	96	static
Hydra viridissima	BSC (1:29)	Corexit 9500	96	static
FISH				
Melanotaenia fluviatilis (crimson-spotted rainbowfish)	BSC (1:50)	Corexit 9500	24	static, daily renewal
Melanotaenia fluviatilis	BSC (1:50)	Corexit 9500	48	static, daily renewal
Melanotaenia fluviatilis	BSC (1:50)	Corexit 9500	72	static, daily renewal
Melanotaenia fluviatilis	BSC (1:50)	Corexit 9500	96	static, daily renewal
Melanotaenia fluviatilis	BSC (1:50)	Corexit 9527	48	static, daily renewal
Melanotaenia fluviatilis	BSC (1:50)	Corexit 9527	72	static, daily renewal
Melanotaenia fluviatilis	BSC (1:50)	Corexit 9527	96	static, daily renewal

[a]Nominal; concentrations refer to the quantity of dispersant:diesal mixture.
[b]Percent of stock solution.
[c]THC, total hydrocarbon content of C_7 to C_{30} compounds.
[d]Depending on dispersant concentration from 1:1 to 1:10 dispersant to oil ratio.
[e]Effects concentrations based on initial chemical quantiations (measured or nominal).

Endpoint	Oil Treatment Effect Conc. (LC_{50}) mg/L	Dispersed Oil Effect Conc. (LC_{50}) mg/L	Concentration Estimate[e]	Reference
larval mortality	>1.06	30.8	Average TPH	Wetzel and van Fleet, 2001
larval mortality	0.79	0.65	Initial THC	Rhoton et al., 2001
larval mortality	>1.13	18.89	Initial THC	Rhoton et al., 2001
larval mortality	0.85	4.23	Average TPH	Wetzel and van Fleet, 2001
larval mortality	465,000	14.1	Initial nominal	Gulec and Holdaway, 2000
larval mortality	465,000	28.5	Initial nominal	Gulec and Holdaway, 2000
mortalilty			Initial TPH on stocks	Cohen and Nugegoda, 2000
mortality	0.7	9	Initial stock TPH	Mitchell and Holdaway, 2000
mortality	0.7	7.2	Initial stock TPH	Mitchell and Holdaway, 2000
embryo mortality	4.48	2.62	Initial stock TPH	Pollino and Holdaway, 2002b
embryo mortality	3.38	1.94	Initial stock TPH	Pollino and Holdaway, 2002b
embryo mortality	2.1	1.67	Initial stock TPH	Pollino and Holdaway, 2002b
embryo mortality	1.28	1.37	Initial stock TPH	Pollino and Holdaway, 2002b
embryo mortality	3.38	2.92	Initial stock TPH	Pollino and Holdaway, 2002b
embryo mortality	2.1	1.25	Initial stock TPH	Pollino and Holdaway, 2002b
embryo mortality	1.28	0.74	Initial stock TPH	Pollino and Holdaway, 2002b

NOTE: THC, summation of total hydrocarbon content C_6 to C_{36}; (W), weathered; ANS, Alaska North Slope crude oil; PBCO, Prudhoe Bay crude oil; BSC, Bass Strait crude oil; ALC, Arabian light crude; VCO, Venezuelan medium crude oil.

TABLE 5-7 Sublethal Effects of Chemically Dispersed Oil in Comparison to Physically Dispersed Oil in Water-Column Organisms (studies since 1989)

Species	Life Stage	Oil	Dispersant (D:O ratio)	Exposure (hr)	Type of Exposure (Static/Flow-through)	Endpoint
(1) Marine studies:						
CRUSTACEANS						
Holmesimysis costata (kelp mysid shrimp)	Adult	PBCO	Corexit 9527 (1:10)	96	spiked-flow through	initial narcosis
Balanus amphitrite (barnacle)	Larvae	Diesel oil	Vecom B-1425 (1:10)	24	static	phototaxis inhibition
Balanus amphitrite (barnacle)	Larvae	Diesel oil	Vecom B-1425 (1:10)	48	static	phototaxis inhibition
Balanus amphitrite (barnacle)	Larvae	Diesel oil	Norchem OSD-570 (1:10)	24	static	phototaxis inhibition
Balanus amphitrite (barnacle)	Larvae	Diesel oil	Norchem OSD-570 (1:10)	48	static	phototaxis inhibition
MOLLUSCS						
Haliotis rufescens (red abalone)	Adult	PBCO	Corexit 9527 (1:10)	48	spiked-flow through	larval abnormality
FISH						
Atherinops affinis (topsmelt)	Adult	PBCO (variable)	Corexit 9527 (1:10)	96	spiked-flow through	initial narcosis
Clupea pallasi (Pacific herring)	embryos/ larvae	ANS (W)	Corexit 9257 (1:25)	24 (larval),[a] 96 (eggs)[a]	static. Daily renewal (for egg studies)	hatching time

Oil Treatment Effect Conc. (EC_{50}) mg/L	Dispersed Oil Effect Conc.	Concentration Estimate[c]	Comments	Reference
11.31–15.90	111.07–48.03	Initial THC		Singer et al., 1998
NA	LOEC; 400[b]	Initial nominal	No oil alone comparison.	Wu et al., 1997
NA	LOEC; 60L[b]	Initial nominal		Wu et al., 1997
NA	LOEC; 400[b]	Initial nominal		Wu et al., 1997
NA	LOEC; 80[b]	Initial nominal		Wu et al., 1997
> 33.58–>46.99	17.81–32.70	Initial THC		Singer et al., 1998
16.34–40.20	>62.22–>140.97	Initial THC		Singer et al., 1998
NA	NA	Initial tPAH	1 μm filtering of WAF/DO. Similar toxicity WAF & DO in control and UVA treatments but DO more toxic in sunlight.	Barron et al., 2003

continues

TABLE 5-7 Continued

Species	Life Stage	Oil	Dispersant (D:O ratio)	Exposure (hr)	Type of Exposure (Static/Flow-through)	Endpoint
Clupea pallasi	Embryos/ larvae	ANS (W)	Corexit 9257 (1:25)	24 (larval),[a] 96 (eggs)[a]	static. Daily renewal (for egg studies)	Hatching success
Clupea pallasi	Embryos/ larvae	ANS (W)	Corexit 9257 (1:25)	24 (larval),[a] 96 (eggs)[a]	static. Daily renewal (for egg studies)	Larval abnormalities
Macquaria novemaculeata (Australian bass)	Juvenile	BSC	Corexit 9527 (1:30)	96	constant flow-through (2% of stock prepared daily)	Cytochrome C oxidase (CCO)
Macquaria novemaculeata	Juvenile	BSC	Corexit 9527 (1:30)	96	constant flow-through (2% of stock prepared daily)	Lactate dehydrogenase (LDH)
Macquaria novemaculeata	Juvenile	BSC	Corexit 9527 (1:30)	96	constant flow-through (2% of stock prepared daily)	Oxygen consumption rate
Menidia beryllina (Inland silversides)	Embryonic/ larval	No. 2 Fuel Oil	Corexit 7664 (1:40) and 9527 (1:50)	240	static	Teratogenic endponts
Salmo salar (Atlantic salmon)	Immature	BSC	Corexit 9527 (1:50)	144 (plus 29 days recovery)	constant flow-through (1% of stock WAF)	Serum sorbitol dehydrogenase (SDH; indicator of liver damage)

Oil Treatment Effect Conc. (EC_{50}) mg/L	Dispersed Oil Effect Conc.	Concentration Estimate[c]	Comments	Reference
NA	NA	Initial tPAH	1 μm filtering of WAF/DO. Similar toxicity WAF & DO in control and UVA treatments but DO more toxic in sunlight.	Barron et al., 2003
NA	NA	Initial tPAH	1 μm filtering of WAF/DO. Similar toxicity WAF & DO in control and UVA treatments but DO more toxic in sunlight.	Barron et al., 2003
NA	NA	Initial TPH on stocks	Stimulated activity if DO cf WAF in gills; in livers stimulated in both WAF and DO WAF. DO WAF concentrations >5x higher cf. WAF	Cohen et al., 2001a
NA	NA	Initial TPH on stocks	LDH activity higher in DO WAF cf WAF. DO WAF concentrations >5x higher cf. WAF	Cohen et al., 2001a
NA	NA	Initial TPH on stocks	Oxygen consumption higher in DO WAF cf WAF. DO WAF concentrations >5x higher cf. WAF	Cohen et al., 2001a
NA	NA	Initial THC on stocks	WAF effect only at 100% stock solution; WAF 7664 effects at 1% stock and WAF 9527 at 10%.	Middaugh and Whiting, 1995
NA	NA	Initial TPH	No change with any treatment.	Gagnon and Holdaway, 1999

continues

TABLE 5-7 Continued

Species	Life Stage	Oil	Dispersant (D:O ratio)	Exposure (hr)	Type of Exposure (Static/Flow-through)	Endpoint
Salmo salar	Immature	BSC	Corexit 9527 (1:50)	144 (plus 29 days recovery)	constant flow-through (1% of stock WAF)	Hepatic EROD activity
Cyprinodon variegatus (sheepshead minnow)	0–24 h old fry	No. 2 Fuel oil	Omniclean (1:1 to 1:10)	168 (ELS)	static	Biomass
ALGAE						
Scenedesmus armatus (chlorococcal alga)	NA	No. 2 Fuel oil	DP 105 (1:20)	24	static	Variety of growth and reproductive endpoints
Isochrysis galbana	NA	PBCO	Corexit 9527 (1:100)	24	static	HSP60
ECHINODERM						
Coscinasterias muricata (eleven-armed asteroid)	Adult	BSC	Corexit 9500 (1:10)	96	Daily static renewal	Alkaline phosphatase activity (AP), cytochrome P450, behavioral assays
ROTIFERA						
Brachionus plicatilis (rotifer)	Adult	PBCO	Corexit (1:50)	8 to 24	static	Heat-shock 60

Oil Treatment Effect Conc. (EC_{50}) mg/L	Dispersed Oil Effect Conc.	Concentration Estimate[c]	Comments	Reference
NA	NA	Initial TPH	Induction of EROD by 2 days in WAF and DO WAF—induction levels higher and more persistent in DO WAF.	Gagnon and Holdaway, 2000
NA	25	Initial nominal	EC_{50}s reported as nominal mixed (oil and/or dispersant) mg/L values. Oil/dispersant mixtures equal or more toxic than oil alone.	Adams et al., 1999
NA	NA	Initial nominal	No clear difference between O and DO mixes. Nominal exposures.	Zachleder and Tukaj, 1993
NA	NA	Initial nominal	No differnce between WAF or DO	Wolfe et al., 1999
NA	NA	Initial PAH	tPAH in stocks WAF 1.8mg/L and dispersed oil 3.5 mg/L. AP no differences. P450 decreased in dispersed oil cf control or WAF. WAF and dispersed oil impacted behavior.	Georgiades et al., 2003
NA	NA	Initial nominal	8 h significant elevations in HSP60 in WAF, only elevated in DO exposures in unfed exposures.	Wheelock et al., 2002

continues

TABLE 5-7 Continued

Species	Life Stage	Oil	Dispersant (D:O ratio)	Exposure (hr)	Type of Exposure (Static/Flow-through)	Endpoint
(2) Freshwater studies:						
CNIDARIANS						
Hydra viridissima (green hydra)	Adult	BSC	Corexit 9527 (1:29)	168	static renewal	population growth rate
Hydra viridissima (green hydra)	Adult	BSC	Corexit 9500 (1:29)	168	static renewal	population growth rate
FISH						
Salmar salmar (rainbow trout)	Juvenile	Mesa sour crude (W)	Corexit 9500 (1:20)	48	static daily renewal	EROD activity (CYP1A induction)
Salmar salmar	Juvenile	Tera Nova	Corexit 9500 (1:20)	48	static daily renewal	EROD activity (CYP1A induction)
Salmar salmar	Juvenile	Scotian light	Corexit 9500 (1:20)	48	static daily renewal	EROD activity (CYP1A induction)
Melanotaenia fluviatilis (Australian crimson-spotted rainbowfish)	Adult	BSC	Corexit 9500 (1:50)	72	50% daily static renewal	EROD activity
Melanotaenia fluviatilis	Adult	BSC	Corexit 9500 (1:50)	72	50% daily static renewal	Citrate synthase activity
Melanotaenia fluviatilis	Adult	BSC	Corexit 9500 (1:50)	72	50% daily static renewal	LDH activity

Oil Treatment Effect Conc. (EC_{50}) mg/L	Dispersed Oil Effect Conc.	Concentration Estimate[c]	Comments	Reference
>0.6	0.6	Initial stock TPH		Mitchell and Holdaway, 2000
>0.6	4	Initial stock TPH		Mitchell and Holdaway, 2000
0.00072	0.0006	Initial TPH and PAH	CYP1A induction x106 in CEWAF (if expressed as % v/v ratio)	Ramachandran et al., 2004
0.0018	0.0015	Initial TPH and PAH	CYP1A induction x1116 in CEWAF (if expressed as % v/v ratio)	Ramachandran et al., 2004
0.00156	0.002	Initial TPH and PAH	CYP1A induction x6 in CEWAF (if expressed as % v/v ratio)	Ramachandran et al., 2004
NA	NA	Initial (daily averages) TPH	Higher activity cf controls in males at 0.8, 2.6, & 7.8 mg/L TPH WAF and in males and females at 14.5 mg/L TPH DCWAF.	Pollino and Holdaway, 2003
NA	NA	Initial (daily averages) TPH	Higher activity cf controls at 2.6 & 7.8 mg/L TPH WAF and 1.4 & 14.5 mg/L TPH DCWAF.	Pollino and Holdaway, 2003
NA	NA	Initial (daily averages) TPH	Higher activity cf controls at 7.8 mg/L TPH WAF and 14.5 mg/L TPH DCWAF.	Pollino and Holdaway, 2003

continues

TABLE 5-7 Continued

Species	Life Stage	Oil	Dispersant (D:O ratio)	Exposure (hr)	Type of Exposure (Static/Flow-through)	Endpoint
Melanotaenia fluviatilis	Adult	BSC	Corexit 9500 (1:50)	72	50% daily static renewal	Plasma estradiol/ testosterone; GSI and histopathology
Melanotaenia fluviatilis	Adult	BSC	Corexit 9500 (1:50)	72	50% daily static renewal	Egg production, % hatch and larval lengths

[a]Followed by UV exposures and assessment of combined effects of PAH accumulation and UV exposure.
[b]Represents mg/l value of oil and/or dispersant mixture.
[c]Effects concentrations based on initial chemical quantiations (measured or nominal).

TABLE 5-8 Dispersed Oil Effects on Water Column Organisms—Field Studies

Species	Treatment	Nominal/ Measured Concentrations	Results	Reference
Plankton, bioassays (*Daphnia*, rainbow trout, and microtox)	*O:* NWC *D:* Corexit 9550 (1:10 D/O ratio) *Details:* Fen lake plots, monitored 29 days before exposure and 30 days post-exposure *Response:* plankton counts, metabolic rate, aqueous microbial counts, bioassays (*Daphnia*, rainbow trout, and microtox)	Measured (fluorescence in field); TPH in lab	Bioassays no toxicity for O or DO plots No change in phyto- or zoo-plankton density, planktonic biomass, metabolic rates, or microbial populations with O or DO plots	Brown et al., 1990

NOTE: O, oil; D, dispersant; DO, chemically dispersed oil; NWC, Norman Wells Crude Oil.

Oil Treatment Effect Conc. (EC_{50}) mg/L	Dispersed Oil Effect Conc.	Concentration Estimate[c]	Comments	Reference
NA	NA	Initial (daily averages) TPH	No significant differences between WAF or DC WAF.	Pollino and Holdaway, 2002a
NA	NA	Initial (daily averages) TPH	No significant differences betweenWAF or DC WAF (high variability), although DC WAF exposure caused cessation in egg production at 14.5 mg/L.	Pollino and Holdaway, 2002a

NOTE: ANS, Alaska North Slope Crude Oil; BSC, Bass Strait Crude Oil; PBCO, Prudhoe Bay Crude Oil; (W), weathered.

cluded that 96-hr LC_{50}s for WAF and chemically dispersed oil were similar for both first- and second-generation fish based on measured TPH concentrations. It should be noted that a complex preparation of the chemically dispersed oil using Corexit 9527 and 9500 was used. The chemically dispersed oil was prepared by mixing oil and water for 24 hr, removing crude oil from the top, and then applying the dispersant to this oil. The chemically dispersed oil was then prepared by adding 1 mL of this mixture to 1L of WAF.

Singer et al. (1998) evaluated the acute effects of untreated and dispersant-treated (Corexit 9527) Prudhoe Bay crude oil on early life stages of three Pacific marine species: the red abalone, *Haliotis rufescens*, a kelp forest mysid shrimp, *Holmesimysis costata*, and the topsmelt, *Atherinops affinis* and concluded that CEWAF versus WAF toxicity was dependent upon test species and exposure time (also see earlier in Chapter 5). Results were expressed as measured THC concentrations and it was observed that WAF was more toxic at early time points (<1 hr), but in tests with *H. rufescens* and *H. costata* significant effects were seen in the CEWAF exposures at THC concentrations two to three times lower than in WAF tests (Table 5-4). Cohen and Nugegoda (2000) exposed fish to Bass Straight crude oil and Corexit 9527 and found that the chemically dispersed oil was more toxic than WAF, based on a comparison of measured TPH values. As noted previously (see earlier section in this chapter on toxicity of chemically versus physically dispersed oil), these results are likely due to compositional differences in dissolved petroleum hydrocarbons in chemically dispersed oil compared to WAF and argue for more detailed chemical evaluations of exposure. Other studies that indicate an enhanced acute toxicity from dispersed oil on a variety of marine and freshwater organisms are listed in Table 5-6, but are not discussed because they employed nominal exposures.

Since the NRC (1989) recommendation for increased investigations of chronic and sublethal effects of dispersed oil, many studies have been undertaken (sublethal studies summarized in Table 5-7). Many endpoints including molecular targets through behavioral responses have been assessed in a variety of species from phytoplankton to various early life stages of common nearshore benthic and water-column species. Again, several of these studies report nominal exposures (e.g., all of the phytoplankton reports, which demonstrate no effect of chemically dispersed oil versus WAF), although the majority of studies do evaluate at least TPH. Ramachandran et al. (2003) measured induction of hepatic CYP1A in juvenile rainbow trout in WAF and chemically dispersed oil (using Corexit 9500) using three types of crude oil. They found that CYP1A expression (measured as EROD activity) was as much as 1,100 times higher in the CEWAF exposures compared with WAF when results were expressed on a percent (v/v) basis; however, when expressed as measured PAH concentrations there was little difference between the EC_{50} values for EROD activity. Similarly, Cohen et al. (2001a,b) using juvenile fish exposed to Bass Straight crude oil and Corexit 9527 found that chemically dispersed oil increased the response in many of the biochemical indicators examined (e.g., cytochrome C oxidase). Barron et al. (2004) demonstrated that CEWAF and WAF toxicity were similar in exposed fish eggs and larvae. Other studies have demonstrated mixed responses (depending on metrics chosen) or decreased effects of chemically dispersed oil compared to WAF in both marine and freshwater species (e.g., Pollino and Holdaway, 2003; Gagnon and Holdaway, 2000; Wheelock et al., 2002; Georgiades et al., 2003).

Intertidal and Subtidal Habitats

These habitats include benthic invertebrates and plants inhabiting subtidal and intertidal areas in both hard and soft-bottom environments, as well as intertidal wetlands. Under most deepwater spill scenarios (>10 m), use of dispersants is thought to present minimal risk to benthic subtidal communities because water-column concentrations of petroleum hydrocarbon will be sufficiently dilute (McAuliffe et al., 1981; Mackay and Wells, 1983). In shallow-water systems, these organisms are more likely to be exposed to and affected by dispersed rather than floating oil. Consequently, increased impacts on subtidal benthic resources may be one of the environmental trade-offs of using dispersants. Intertidal areas, such as salt marshes and mangroves, are often considered sensitive areas because they serve as habitat for many adult, juvenile, and larval organisms. Hence, if valuable resources exist in the intertidal area, dispersing oil before it reaches this habitat may be preferable. In terms of short-term effects, an extensive evaluation of the relative acute sensitivities of benthic and water-column species to a variety of chemicals, including PAH, suggests that the toxicity of dispersed oil to benthic organisms would be similar to that on water-column organisms (DiToro et al., 1991). However, this evaluation does not consider the potential for long-term exposure to oil that may occur as a result of the persistence of oil in sediments, particularly in low-energy areas with minimal flushing. Thus, in order to adequately evaluate the potential effects on subtidal and intertidal temperate communities in shallow water systems, the persistence and behavior of dispersed oil versus untreated oil in benthic sediments and on the shoreline should be assessed. Field studies conducted in the 1980s still constitute much of what is known about these fate and effects processes and are summarized below.

In 1981, a field study in Long Cove, Searsport, Maine compared the fate and effects of dispersed and undispersed crude oil on nearshore temperate habitats (Gilfillan et al., 1986). The cove was divided into three areas: a control, dispersed oil (using Corexit 9527), and untreated oil. The spill of 250 gallons of untreated oil was released during high tide in water approximately 1.5 to 2.0 m deep. The oil was allowed to coat the beach and after two tidal cycles, oil was cleansed from the beach using conventional methods. The dispersed oil (10:1, O:D) was mixed and released into approximately 2.5 to 3.0 m. The deepest samples were taken near the center of the cove, in approximately 18 m depth. The treated oil quickly dispersed into the water column, reaching concentrations of 15–20 ppm near the bottom. However, this short-term exposure appeared to have little effect on the benthic community in this treatment. On the other hand, significant amounts of oil remained in the intertidal sediments exposed to untreated oil, but not in sediments exposed to the dispersed oil. In addition, hydrocarbons were found in clams and mussels near the untreated oil site, but were not detected in similar species collected at the dispersed oil site. Finally, effects on infaunal benthic communities were found in the untreated oil site but not in the area exposed to dispersed oil. Researchers attributed these differences to the greater persistence of undispersed oil in the intertidal sediments.

Similar results were seen in the Baffin Island Oil Spill Project (BIOS) initiated in 1980 (Blackall and Sergy, 1981). This large-scale field project consisted of four bays, two of which received either 94 barrels of untreated, partly weathered crude oil released on the surface or an underwater release of oil and dispersant (10:1). The untreated oil caused no immediate effects on benthic organisms, but some intertidal amphipods and larval fish were affected by physical coating. Oil concentrations in the top 1 m of water ranged from 0.01 to 2.8 ppm. In the dispersed oil treatment, concentrations of oil on the bottom (approximately 10 m) ranged from approximately 50 ppm to a high of 167 ppm. Benthic organisms appeared stressed in this treatment, most likely due to narcotic effects. However, systematic monitoring of benthic populations demonstrated that exposure to dispersed oil did not cause large-scale mortality. After one year, there was no statistical difference in benthic community composition between the dispersed oil treatment and the control bays. As in the Searsport study, the persistence of dispersed oil in subtidal sediments was much less (approaching background after 1 year) than at the untreated oil site. However, in this study there was no attempt to recover oil from the untreated oil site; hence, amounts of residual oil were likely higher than would have occurred had some recovery been attempted.

Michel and Henry (1997) evaluated PAH uptake and depuration by oysters after use of dispersants on a shallow water oil spill in El Salvador (see Box 5-3). Because the PAH levels dropped to nearly background within three weeks after application of dispersant, the authors concluded that the subtidal sediments in the spill site did not contain residual oil and therefore did not constitute a continuing source of oil to coastal resources. Studies in which the sediments were a major reservoir for spilled oil have reported elevated levels of PAH in oysters for months to years after the spill (Neff and Haensly, 1982; Blumer et al., 1970). Because most of the oil in the El Salvador spill was dispersed there was no opportunity to compare uptake and depuration of dispersed oil versus untreated oil. Thus, it was not possible to determine if the use of dispersants increased the amount of oil that reached benthic habitats. However, a qualitative comparison of PAH measurements in oysters collected during other oil spills where dispersants were not applied, does not suggest any dramatic difference in uptake (Michel and Henry, 1997). The SERF in Corpus Christi, Texas, was used in a series of mesocosm experiments to evaluate the ecological effects of shorelines impacted by oil and chemically dispersed oil (Coelho et al., 1999; Fuller et al., 1999; Bragin et al., 1999). Simulated beaches were constructed in experimental wave tanks (described in detail in Chapter 3) with fine sand. Treatments included artificially weathered Arabian medium crude oil, oil premixed with Corexit 9500, and controls. Six liters of oil or oil-dispersant mixture were poured onto the surface of the tanks. After an initial mixing period of one hour, fresh sea water was circulated continuously through the wave tanks to simulate tides with a 12-hour period. A variety of organisms (fiddler crabs, polychaete worms, amphipods, fish, and oysters) were exposed *in situ* in the wave-tank mesocosms or *ex-situ* in laboratory toxicity tests. In the oil-only treatment, the TPH concentrations in water peaked at 15,360 μg/L at 6 hr and then declined to a concentration of 2,948 μg/L at 24 hr (Coelho et al., 1999). The resulting total PAH concentrations in fish (*Cyprinodon variegatus*) and oysters (*Crassostrea virginica*) in the wave tanks at 24 hr were 8,420 and 8,590 μg/g, respectively. In the dispersed oil treatment, the TPH concentrations in water peaked at one hour at 48,580 μg/L and declined to 5,258 μg/L after 24 hr. The total PAH concentrations in fish and oysters were 18,440 and 3,550 μg/g, respectively after 24 hr. The similarity in PAH concentrations in oysters under the two treatments may be related to the oil-only exposure being limited to certain phases of the tidal cycle. As has been documented in field studies, sediment concentrations of TPH in the dispersed oil treatments were very low compared to the oil-only treatment, a consequence of the untreated oil becoming trapped in the mesocosm wave tank (Coelho et al., 1999). Interpretation of toxicological evaluations was confounded, in some instances, by unacceptable control mortality. However, in general, results suggested comparable toxicity of chemically and physically dispersed oil in these mesocosm experiments (Fuller et al., 1999; Bragin et al., 1999).

In general, the available information from field and mesocosm studies seems to indicate that dispersants will reduce the persistence of oil in subtidal and intertidal sediments compared to untreated oil. Consequently, there may be a trade-off between short-term acute effects due to increased concentrations of petroleum hydrocarbons in the water column countered by the reduction in long-term chronic exposure to petroleum hydrocarbons from stranded oil. However, this conclusion is based on limited information, and the interactions between dispersed oil and sediments are still poorly understood. For example, Ho et al. (1999) found that toxicity of sediments in the vicinity of the *North Cape* spill (a spill that had incredibly high physical dispersion of home heating oil) lasted for more than 6 months in some areas. Sediments in this study were fine grained, unlike those in the SERF mesocosms that were sandy. Consequently, a focused series of experiments should be conducted to quantify the final fate of chemically dispersed oil droplets compared to undispersed oil, including an evaluation of the interaction with a broader range of sediment types.

BOX 5-3
Case Study: Acajutla, El Salvador

Spilled Oil Type/Volume/Conditions. An estimated 400 ± 100 barrels of a blended crude oil called Venezuela Recon was released about 1 km offshore at the mooring buoy off the Refineria de Acajutla, El Salvador on 23 June 1994. Venezuela Recon is a 50:50 blend of a heavy Venezuelan crude and light, intermediate products such as naphtha and gas oil. It appeared much like a black diesel. Properties were: API gravity of 34.9; viscosity of 4.38 cSt; and pour point of –15°C. It would be readily dispersible.

Physical and Biological Setting. The spill affected open, exposed coastline consisting of rocky shores and sand beaches. Water depths were 4–6 m over mixed sand and rock bottom. Winds were high during the spill, but calm during dispersant applications over the next few days. There are inshore fisheries both for finfish (by boat) and for benthic oysters attached to rock outcrops (by free diving).

Dispersant Application. Thirty barrels of Corexit 9527 were applied over a 3-day period, for an application rate of 1:13. Applications followed guidelines in the facility's oil spill contingency plan. Dispersant was first applied on 24 June within 12–15 hr after the spill by fixed wing aircraft and workboats. Some Corexit 7664 was applied from shore to oil in the surf zone. Small nearshore slicks were treated with Corexit 9527 sprayed by workboats for two more days. On the morning of 27 June, no visible slicks were reported.

Wildlife

One of the widely held assumptions concerning the use of dispersants is that chemically dispersion of oil will dramatically reduce the impacts to seabirds and aquatic mammals, primarily by reducing their exposure to petroleum hydrocarbons (e.g., French-McCay, 2004). Evaluating the validity of this assumption is critical because it is often a key factor in the decision on whether or not to use dispersants on a particular spill. Unfortunately, little is known about the effects of dispersed oil on wildlife, especially aquatic mammals. Oil can affect wildlife through a combination of effects: toxicity due to ingestion of oil or contaminated prey; inhalation of petroleum vapors; and loss of thermoregulatory capacity due to physical oiling of feathers and fur. In addition, adults that become oiled may transfer oil from their plumage to their more sensitive eggs or

Monitoring Results. *Effectiveness:* Monitors conducting visual observations during overflights reported that the application was highly effective. The small amount of oil that stranded onshore was removed manually. *Effects:* Because of concern over potential impacts of the spill and dispersant use on fisheries, a monitoring plan was developed. Fishermen were queried to determine if they had encountered any oil on their nets or catch or any dead organisms. No encounters were reported. Commercial fishermen were hired to free-dive for oysters at four locations (included two background locations). Whole oysters (including the gut) were analyzed for PAH to fingerprint the oil and monitor for the presence and bioavailability of oil to benthic resources at 7, 28, 185, and 280 days post-spill (though there was another small spill reported just prior to the 185 day sampling event).

Two samples of oysters from the area where the oil was dispersed in 4–6 m of water contained total PAH of 147 and 164 ppm, dry weight, compared to background levels less than 1.0 ppm. The PAH patterns indicated that the oil in the oysters was slightly weathered whole oil. Since the oysters had been exposed to clean water for at least five days, it is likely that they were already depurating the oil and the oil measured represents a body burden rather than oil in the digestive glands. Four weeks post-spill, PAH levels in oysters from these areas decreased by 94–98 percent. Half-lives for 2- and 3-ringed PAH were calculated to range from 2.8 to 4.7 days, and 4- to 6-ringed PAH ranged from 3.7 to 30 days. These values were similar to results of laboratory studies. These studies showed that dispersed oil did reach benthic communities when dispersed in 4–6 m of water in open-water conditions. Uptake by oysters was rapid, and depuration was complete within 28 days.

SOURCE: Summary based on Michel and Henry (1997).

hatchlings—refined oil is highly toxic to avian embryos. The limited available information suggests comparable toxicity of dispersed and untreated oil to seabirds and mammals. A literature review by Peakall et al. (1987) concluded that, from the toxicological perspective, the effects of oil and chemically dispersed oil on seabirds were similar, based on sublethal responses at the biochemical and physiological level. Similarly, studies on the effects of oil on the hatching success of bird eggs (summarized in NRC, 1989) also indicated that toxicities of oil and dispersed oil were similar.

Hence, the main concern for the impacts of dispersed oil and dispersants is in the physical loss of insulative properties of the feathers and fur of wildlife when coated with oil, which in turn can lead to hypothermia, stress, starvation, and ultimately death of the animal. The effect of external oiling on the thermal insulation of plumage has been shown to be dependent on the amount of water that is absorbed into the plumage as a function of the amount of oil exposure. Peakall et al. (1987) derived a mathematical model to estimate the amount of dispersed oil to which seabirds would be exposed. The risk of exposure to oil is dependent on the behavioral characteristics of birds. Because the purpose of dispersants is to drive oil into the water column, only those activities that cause seabirds to submerge, such as feeding, would lead to an increased exposure to oil. Based on their modeling analysis, Peakall et al. (1987) concluded that there is no significant exposure of birds to oil in the water column, rather, the highest exposure occurs when the bird dives or returns to the water-oil surface. They concluded that the assumption that dispersing oil benefits seabirds depends on the efficiency of the dispersion. However, several later evaluations have challenged this assumption, asserting that exposure to even small amounts of organic petroleum compounds and surfactants may result in adverse effects to birds and potentially bird populations (Jenssen, 1994; Briggs et al., 1996; Stephenson, 1997).

The waterproof properties of feathers and their value as thermal insulators are due to their composition and their structure. The keratin of feathers is inherently water repellant. In addition, the lattice structure and contour of feathers promote the shedding of water droplets from the surface of the feather (Stephenson, 1997). Thus, it is reasonable to predict that any factors that compromise the integrity of the plumage, such as exposure to oil or dispersants, will affect thermoregulation and result in a physiological cost to the animal. Similar effects would be expected in aquatic mammals, such as otters, that rely on water-repellant fur to maintain normal thermal regulation (Jenssen, 1994).

As noted previously, very few studies have evaluated the effects of dispersed oil on thermoregulation. Lambert et al. (1982) compared metabolic rates of mallards exposed to Prudhoe Bay crude oil and Corexit 9527. They found higher metabolic rates in birds exposed to dispersant, presumably due to increased energy expended to maintain a normal body temperature. Jenssen and Ekker (1991) reported that a much smaller volume of chemically treated oil compared to crude oil was required to cause significant effects on plumage insulation and thermoregulation in eiders. Because dispersants are surface active agents that reduce water surface tension, they may also increase the wettability of bird feathers and hence disrupt their insulation properties (Stephenson, 1997). Stephenson and Andrews (1997) concluded that adult bird feathers could be wetted when the surface tension of water is reduced below a certain threshold. In addition, Stephenson (1997) indicates that a multitude of surface-active organic contaminants, including petroleum compounds and detergents, may have detrimental effects on aquatic birds due to alterations in water surface tension. Application of chemical dispersants during an oil spill may lower the amount of oil to which a bird or aquatic mammal is exposed while at the same time increasing the potential loss of the insulative properties of feathers or fur through reduction of surface tension at the feather/fur-water interface. Clearly, more studies are needed to address the uncertainties associated with the impacts of dispersants and dispersed oil on wildlife. A similar conclusion was also reached by NRC (1989), and very few studies have been conducted since that initial recommendation.

Microbial Communities

During the decision-making process an important factor to be considered is whether degradation of the spilled oil will be enhanced or inhibited using dispersants, thereby affecting the ultimate fate of the oil. As discussed in Chapter 4, there is no conclusive evidence demonstrating either the enhancement or the inhibition of microbial biodegradation when dispersants are used. Studies specifically addressing the toxic effects of dispersants or dispersed oil on microorganisms are limited and effects are often inferred from inhibited rates of oil biodegradation (see Chapter 4 and Table 5-9). To determine toxic effects to bacterial populations as a result of dispersant use, consideration should be given as to the transport mechanism involved for oil uptake by the particular species under study. Transport mechanisms include uptake from the dissolved phase or via a direct contact mechanism. Addition of dispersants can alter the concentration of dissolved phase hydrocarbons and interfere with normal bacteria-oil droplet attachment mechanisms (Zhang and Miller, 1994) as discussed in Chapter 4. These changes could result in enhanced or decreased exposure of the bacteria to particular hydrocarbons, which may be either advantageous or detrimental (toxic) to the microbe. There are few studies that directly examine routes of exposure and toxicity to microorganisms.

Inhibition of biodegradation rates may be caused by a variety of factors, including toxicity, though it could also result from the fact that the dispersant may substitute for the oil as the carbon source. However, it is also possible that an increased concentration of dispersed oil (or dispersant) could cause temporary toxic effects to natural microbial populations. Studies of biodegradation rates that report changes in bacterial growth (numbers) or uptake of glucose as indicators of toxic effects should be interpreted with caution. Many other factors could be limiting, such as nutrients and other growth factors. Extrapolating data from laboratory tests is difficult because hydrocarbon degradation rates are often several orders of magnitude higher compared with *in-situ* rates. Conversely, any toxic or inhibitory effects are also likely to be magnified in the laboratory setting (NRC, 1989).

Studies addressing specific toxicity issues in microbial communities are very limited, with the majority being an indirect observation from biodegradation studies using enhanced or inhibited growth of microbial populations. For example, Linden et al. (1987), in a microcosm system aimed at modeling the littoral ecosystem of the Baltic Sea, demonstrated elevated numbers of water-borne heterotrophic bacteria after 30 hr in dispersed oil treatments relative to oil alone. After 7 days post-exposure, the differences between treatments were not significant. This study indicates no toxic effect to the microbial population as a whole with the use of dispersants; however, growth as measured by bacterial counts may mask selective toxicity to some bacterial strains concordant with elevations in numbers of tolerant or specific hydrocarbon degrading strains. It should be noted that a 100-fold increase of C_{16}-specific organisms was observed after 30 hr in the dispersant-oil treatment compared with oil alone (Linden et al., 1987). A similar elevation in bacterial numbers in response to chemical (Corexit 9500) versus physical dispersion was observed by MacNaughton et al. (2003), again measured by total bacterial counts. Some dispersant studies have demonstrated that when microbial processes are inhibited, rates of oil decomposition decline (see Chapter 4; NRC, 1989; Mulyono et al., 1994; Varadaraj et al., 1995).

Although there are a few studies specifically on microbial toxicity, none examined natural marine microbial populations. George et al. (2001) indirectly addressed the toxicity of oil and oil plus dispersant treatments to microbes by determining effects on the intestinal flora of rats and the mutagenic potential of these mixtures using an assay on bacteria (see below). The reasoning behind this study was to determine the adverse health effects of cleanup options on marine mammals. It was hypothesized that even low levels of oil (with or without dispersant) may cause toxic effects following ingestion due to the alteration in gastrointestinal tract metabolic processes. The rat was used as a model organism to determine if co-administration of Corexit 9527 enhanced oil toxicity or mutagenicity. The study demonstrated that oil exposure reduced several cecal microflora populations (see Table 5-9 and 5-10), and Corexit alone reduced the lactose-fermenting enterobacteria Conversely, the oil plus dispersant treatment increased the lactose fermenting group with no changes in other bacterial populations. It should be noted that these data were derived from only three rats. In test treatments, the authors found that both dispersants (Corexit 9500 and 9527) were mutagenic in various strains of *Salmonella typhimurium* (employed for the Ames histidine reversion bioassay), using dilutions up to 1:1,000, but weathered Nigerian crude oil was not mutagenic. No data were available for the dispersed oil mixture. A similar study also found Corexit 9527 alone to be toxic in the Microtox assay with an EC_{20} of 1 ppm (Poremba and Gunkel, 1990). Although both studies demonstrated the toxic effects of dispersant, dispersed oil was not investigated. Because these studies examined a single laboratory species exposed to relatively high levels of dispersant, the potential effects on natural mixed, marine bacterial populations cannot be assessed.

There are a multitude of implications regarding the effects of dispersant and dispersed oil on microbial communities. A lack of toxicity is often inferred in studies that show increases in numbers of bacteria. However, this may not accurately reflect the entire microbial community because elevations in some bacterial (tolerant) species may mask the inhibition (toxicity) of other types. A lack of inhibition observed at the community levels does not necessarily indicate the absence of toxicity. Elevated numbers of bacteria may also reflect an indirect enhancement if dispersant or dispersed oil is toxic to bacteriovores (Lee et al., 1985). The removal of the bacterial grazers would also cause elevated bacterial counts, although these would probably be temporary. Alterations in bacterial species composition may have severe consequences for the ecosystem as a whole. In addition, elevated numbers of bacteria may result in toxic effects to other forms of life. For example, elevated bacterial numbers may deplete oxygen levels in benthic substrates, resulting in indirect toxic effects to organisms inhabiting this environment. Additionally, some microbial pathways may lead to transformation of the oil into more toxic byproducts. The impact of dispersants and/or dispersed oil on gut microflora, particularly in relation to ingestion by marine mammals, has been discussed above. Because of their importance in aquatic systems, targeted toxicity studies should be conducted to address the effects of dispersant and dispersed oil on the composition and metabolic activities of mixed microbial populations representing marine (or estuarine/freshwater) communities.

Coral Reefs

Compared with other test species, data on the effects of dispersants and/or chemically dispersed oil and comparisons with physically dispersed oil on coral species are even more limited. The majority of research was conducted in the 1970s and 1980s, and these studies (field and laboratory based) have been adequately discussed and summarized in NRC (1989). Many of the early studies were conducted by researchers at the Bermuda Biological Station (e.g., Cook and Knap, 1983; Dodge et al., 1984, 1995; Knap, 1987; Knap et al., 1983, 1985; Wyers, 1985; Wyers et al., 1986) who conducted an extensive series of laboratory and field based studies on the effects of dispersants (e.g., Corexit 9527 and BP1100WD) and dispersed oil (Arabian light crude) on the brain coral *Diploria strigosa*. These studies were based on 6 to 24 hr exposures followed by recovery in clean seawater. They found no significant differences between the oil and the

TABLE 5-9 Detail of Studies Addressing Effects of Dispersant/ Dispersed Oil on Microbial Populations

Microbial sps./Community	Dispersant/Oil (D:O ratio)	Metrics Used
Indigenous mixed microbial population	D; Corexit 8666, Gamlen Sea Clean, GH Woods degreaser, Formula 11470, Sugee 2 O; Arabian Crude (1:1)	Bacterial no. (growth; drop-plate method). Species diversity.
	D; Corexit 8666, Shell oil herder #3, Smith oil herder O; Crude oil	CO_2 evolution
Arthrobacter simplex *Candida tropicalis*	D; ONGC-1, ONGC-2, ONGC-3, ONGC-4 O; Saudi Arabian Crude, Bombay high crude (1:5)	Growth (turbidity)
Indigenous mixed bacterial population	D; IB 2/80, IB 1/80, IB 11/80, IB 12/80, IB 13/80, BP 1100WD, BP 1100 O; Saudi Arabian Crude (1:1)	Bacterial no. (spread plate method)
Mixed population	D; Corexit 9500 O; Forties crude (W), ANS (W) (1:10)	Bacterial no.
Mixed culture of oil degrading bacteria	D; 15 FW dispersants O; Newman-wells (D:O various)	Bacterial no.'s CO_2 evolution
Photobacterium phosphoreum	D; E09, DK50, DK 160 O; Ekofisk crude (± W) (1:100–10,000)	Microtox assay (loss of bacterial bioluminescence indicates toxicity)
Rat intestinal bacterial mixed population *Salmonella typhimimurium* (mutagenicity study)	D; Corexit 9527, Corexit 9500 O; Weathered Bonnie light Nigerian crude oil	Bacterial no.'s Species diversity Bacterial enzymes quantitation
Natural flora (from pond)	D; Corexit 9550 O; Forties North Sea (1:10)	No. heterotrophic bacteria, plus 4 specific-species counts
Acinetobacter calcoaceticus, *Photobacterium phosphoreum* and *Serratia marioruba* *P.phosphoreum* (microtox test)	D; Finasol OSR-5, Corexit 9527 (plus biosurfactants and other synthetic surfactants) O; none	Bacterial no.'s Bacterial bioluminescense (microtox test)
Natural flora (enclosed ecosystem—SEAFLUXES)	D; Corexit 9527 O; Prudhoe Bay Crude Oil (1:10) (No oil alone test)	Heterotrophic bacterial production (thymidine incorporation) Direct counts (epifluor. microscope) Bacterial biomass (electron micros.)

Finding	Reference
Increased no.'s with D alone Elevated no.'s in DO c.f. O alone Changes in species diversity with DO (genus level).	Mulkins-Phillips & Stewart, 1974
Increased CO_2 evolution in DO c.f. O	
D non-toxic (growth). Increased growth DO c.f. O alone	Bhosle & Mavinkurve, 1984
Only D toxic was IB 2/80. No difference in growth with D c.f. DO. O alone toxic.	Bhosle & Row, 1983
Bacterial no.'s increase with DO c.f. O (forties). ANS study, DO bacterial no.'s initial elevation (quick colonization), no difference c.f. O alone at later time-points	[a]MacNaughton et al., 2003
Changes in no.'s and species diversity is D dependent, some toxic, others no-effect or increase growth	[b]Foght et al., 1987
Decreased toxicity of DO c.f. O. High levels of D toxic.	[c]Poremba, 1993
Treatment changes in bacterial enzyme activities. Oil reduction of microflora in 3 populations; D alone 1 reduction and DO slight elevation (1 population) Species composition changes. D toxic to *S. typhimum* (O alone not).	George et al., 2001
30 hr-increase bacterial no.'s in DO c.f. O; no differences at 7 days C_{16}-organisms 100x in DO c.f. O, other species same	Linden et al., 1987
No inhibition of growth, some elevated. EC_{20} Corexit and Finasol at 1mg/L	[d]Poremba, 1993
Elevated bacterial production by D and highest in DO test. Toxicity to bacteriovors	Lee et al., 1985

continues

TABLE 5-9 Continued

Microbial sps./Community	Dispersant/Oil (D:O ratio)	Metrics Used
Soil bacteria; mixed microbial population	D; Corexit 9550 O; Arabian crude, Louisiana crude (1:5)	Gross metabolic capacity (CO_2, CH_4)
Pond natural bacterial population *Salmonella typhimimurium* (mutagenicity study) *Spirillum volutans* (toxicity test)	D; Corexit 9527 O; Fresh Norman Wells Crude	General biomass (microscope enumeration and ATP levels), heterotrophic plate count, MPN

NOTE: ANS, Alaskan North Slope crude oil; ATP, adenosine triphosphate; D, dispersant; DO, chemically dispersed oil; FW, freshwater; MPN, most probable numbers; O, oil; W, weathered.

[a]Biodegradation study with indirect toxicity observations.
[b]Freshwater study.

TABLE 5-10 *Cecal microflora* Effects Following 5 Weeks of Nigerian Crude Oil and Corexit 9527 Treatment of F344 Rats

Microflora Population	Selective Medium	Control[a]	Oil	Corexit	Oil + Corexit
Enterocci	KF	4.72	0.00[b]	4.90	4.74
Lactose-fermenting enterobacteria	MacConkey +	3.25	0.00[b]	2.59[b]	4.10[b]
Lactose-nonfermenting enterobacteria	MacConkey –	4.92	0.00[b]	4.71	4.90
Total anaerobic count	Blood agar	8.46	8.32	8.39	8.42
Obligately anaerobic Gram-negative rods	VK	8.19	8.12	8.13	8.24
Lactobactilli	Rogosa	7.73	7.81	7.78	7.64

[a]Male Fischer 344 rats were gavaged for 5 weeks with Nigerian crude oil (1:20) with and without Corexit 9527 (1:50). The cecum was removed from each animal, homogenized under CO_2, and diluted and plated anaerobically on selective media for enumeration. Results are an average from three rats.
[b]Significant at $p < 0.05$, one-way ANOVA.

SOURCE: modified from George et al., 2001.

Finding	Reference
No inhibition, some elevations (temporary)	[e]Nyman, 1999
No toxicity/mutagenicity of O or DO, slight short-term effects, i.e., O decreased no.'s but DO elevated no.'s (7 days)	Dutka & Kwan, 1984

[c]Dispersants alone.
[d]Using Microtox toxicity test bacteria.
[e]Soil study.

dispersed oil treatments using an array of biometrics including tentacle extension, mucus production, pigmentation loss, tissue swelling, and skeletal growth. Any stress effects were transient and recovery occurred within one week post-exposure. However, they did note reduced photosynthesis of the zooxanthellae (symbiotic algae) within the coral resulting from 8 hr exposure to 19 ppm dispersed oil, whereas this was not apparent in treatments with either oil or dispersant alone. Carbon fixation and lipid synthesis recovered to normal levels within 24 hr.

One of the more robust and extensive studies on early life stages of corals was undertaken by Negri and Heyward (2000). They exposed *Acropora millepora* eggs and sperm to WAF (heavy crude oil) and chemically dispersed oil (using Corexit 9527; dispersant to oil ratio at1:100 and 1:10) or dispersant alone for 4 hr and assessed fertilization rates. They found no inhibition of fertilization at >0.165 ppm THC in WAF exposures (>10 percent dilution of stock WAF) but significant inhibition for exposure to dispersed oil (1:10 DOR) at 0.0325 ppm (equal to a 1 percent dilution). Exposure concentrations were estimates based on measured concentrations of THC in the stock solutions used to make the dilutions. Dispersants alone resulted in significant inhibition (final dilution of 0.1 percent), although at a lower magnitude than dispersed oil at the same dispersant concentrations. Although fertilization in this species appeared to be relatively insensitive to naturally dispersed oil droplets, crude oil and dispersant alone inhibited larval metamorphosis, with the greatest inhibition observed when larvae were exposed to chemically dispersed oil. Metamorphosis was inhibited at 0.0824 ppm THC and 0.0325 ppm THC for crude oil and chemically dispersed oil (1:10 DOR), respectively. The authors concluded that there may be additive toxicity of dispersants and oil and recommended that the timing of spawning events be considered in management decisions on dispersant use in coralline environments. However, as noted previously, without evaluation of specific chemical constituents in the various exposures regimes, conclusions regarding relative toxicity of chemically dispersed versus physically dispersed oil are tenuous.

A study by Epstein et al. (2000) investigated the toxicity of five third-generation dispersants to early life stages of coral. Planula larvae of stony coral (*Stylophora pistillata*) and soft coral (*Heteroxenia fuscesense*) were exposed to varying concentrations of WAF, chemically dispersed oil (1:10, DOR), and dispersants alone (0.5–500 ppm) using short-term (2–96 hr) bioassays. WAF treatments resulted in a concentration-dependent reduction in planulae settlement, but no mortality. All the tested dispersants also decreased settlement rates, even at the lowest tested concentrations (0.5 ppm). In addition, larval survival at 50 and 500 ppm after 96 hr was completely or significantly reduced in most of the dispersants tested. Chemically dispersed oil exposures resulted in a dramatic increase in acute toxicity to both coral species larvae. In addition, the authors reported that dispersants and dispersed oil treatments caused larval morphological deformations, loss of normal swimming behavior, and rapid tissue degeneration. Interpretations of physically versus chemically dispersed oil toxicities in this study are hampered by the use of nominal exposures.

A recent study investigating the effects of dispersant and dispersed oil by Shafir et al. (2003) using coral nubbins of the hard coral *Stylophora pistillata* exposed to water-soluble fractions (WSF), dispersant, and chemically dispersed oil for 24 hr (static exposures) followed by recovery for long-term assessments in clean seawater. No mortality was observed at any of the WSF concentrations, but extensive mortality was observed with dispersant alone (at 24 hr all doses including 1 percent stock dilution) with a delayed enhanced mortality occurring at the 0.1 percent concentration after 6 days. Survivorship of chemically dispersed oil exposed corals was similar to that described for dispersant alone.

The Tropical Oil Pollution Investigations in Coastal Systems (TROPICS) field experiments are particularly useful in evaluating the impacts and trade-offs of dispersants and dispersed oil on corals, seagrasses, and mangroves (Ballou et al., 1987, 1989; Dodge et al., 1995). In these field experiments in Panama, corals were exposed to oil and chemically dispersed oil for relatively short periods (1–5 days) followed by extensive monitoring for 1–10 years post-exposure (see Box 5-4). Sites were monitored repeatedly in the first two years, and at two later dates (ten years final). At the untreated oil site no significant impacts to corals were observed at any of the time points (Dodge et al., 1995). At the dispersed oil site, corals were exposed to higher concentrations of oil (i.e., 24 hr averages of 5.1 ppm vs. 0.14 ppm at the untreated oil site). Significant impacts to the coral reef were observed and at two-years post-exposure these included reduced coral coverage and reduced growth in two hard coral species (*Agaricia tennuifolia* and *Porites porites*) with no reduction in two other species (*Montastrea annularis* and *Acropora cervicornis*). However, at the 10 year monitoring time point, recovery was complete and comparable to pre-spill conditions and conditions at the control site (Dodge et al., 1995).

Another field experiment using two oil exposure regimes was conducted in the Arabian Gulf by LeGore et al. (1983, 1989). Exposures consisted of oil alone (Arabian light crude), dispersant alone (Corexit 9527), and oil/dispersant mixtures with analysis of water chemistry. The two series of experiments consisted of a 24 hr or 5 day (120 hr) exposure period. The authors concluded that coral growth appeared to be unaffected by exposure to the toxicants, although some *Acropora* sp. exposed to the dispersed oil for 5 days did exhibit delayed, but minor effects, that became apparent only during the relatively cold and stressful winter season.

Corals are particularly susceptible to PAH dissolved in seawater or adsorbed to particles because the layer of tissue covering the coral skeleton is thin (approximately 100 μm; Peters et al., 1997). Also, coral tissue is rich in lipids (high lipid/protein ratios), facilitating the direct uptake and bioaccumulation of lipophilic chemicals, including PAH found in oil (Peters et al., 1981). Indeed, it has been observed that oil is quickly and readily bioaccumulated in coral tissues and is slow to depurate, possibly reflecting inefficient contaminant metabolism or lack of detoxification pathways (see Shigenaka, 2001). Long residence times of PAH were indicated by high PAH concentrations found in oiled corals (up to 50 mg hydrocarbon g lipid^{-1}) from Panama as long as 5 months after the original spill (Burns and Knap, 1989). A laboratory study by Kennedy et al. (1992) demonstrated a linear uptake rate of benzo(a)pyrene in corals and their zooxanthellae. Accumulated levels were slowly eliminated with 38–65 percent of the accumulated benzo(a)pyrene remaining after 144 hr depuration (recovery) in clean seawater (Kennedy et al., 1992). This rapid uptake and slow depuration may be of particular relevance to oil toxicity mechanisms in corals. Many studies have shown that a brief exposure to oil may not result in immediate death to coral species (acute oil toxicity), but induces mortality over an extended period of time (delayed effects) (see Shigenaka, 2001 for a summary). On a similar theme, Fucik et al. (1984) suggested that acute toxicity is probably not a good indicator of oil impact, stating that it is much more likely that adverse effects to coral species would be manifested at sublethal levels.

One relatively unstudied hypothesis that could explain delayed effects is that most of the toxicity is derived from exposure to the UV radiation in sunlight (see earlier section on Phototoxicity in this chapter). This phenomenon may be of particular relevance in explaining the high toxicity of accumulated oil in corals, species that are slow to depurate PAH.

BOX 5-4
Case Study: TROPICS, Panama

Spilled Oil Type/Volume/Conditions. In 1984, a field oil experiment called the Tropical Oil Pollution Investigations in Coastal Systems (TROPICS) was conducted in Panama. The objective of the TROPICS experiment was to evaluate the relative impacts of oil and dispersed oil on mangroves, seagrasses, and corals. Exposure concentrations were targeted to be as high as 50 ppm, in a worst-case scenario, with dispersants applied to oil directly over corals.

Physical and Biological Setting. Sheltered shallow area near Bocas del Toro, Panama (Figures 5-7 and 5-8). Mature mangroves with extensive seagrass beds (water depth average about 40 cm), and coral reefs (water depth average 60 cm).

Oil and Dispersed Oil Application. The oil, or dispersed oil, was applied inside boomed areas 30 m wide and 30 m deep, extending across all three habitats. The pre-mixed dispersed oil (4.5 barrels) was released over a 24-hour period so that the dispersed oil concentrations would stay elevated over the exposure period. The untreated whole oil (6 barrels) was released in two periods over the 24 hr, at an application rate of 1 liter/m^2. After one more day, the remaining floating oil was removed with sorbents.

Monitoring Results. *Water Column Monitoring:* Oil concentrations at each treatment site (oil or dispersed oil) were monitored continuously for 24 hr using a field fluorometer that was calibrated to convert fluorescence into the concentration of physically and chemically dispersed oil. Discrete and unfiltered water samples were collected for chemical analysis by gas chromatography (GC). In comparing the oil concentrations in the water as measured by both approaches, the field fluorometer readings were 3 times higher that the GC concentration for samples from the dispersed oil site, and they were 17 times higher than the samples from the undispersed oil site. Therefore, the oil concentrations as measured in the discrete water samples by GC were used to calculate the oil exposures because these results are more quantitative.

Not only are corals in high-light environments, they are translucent and seek high intensity light environments (by regulating pigments or altering their position with respect to the sun) to foster the symbiotic relationship with photosynthetic algae.

An additional stress for corals may be attributed to the physical toxicity of oil droplets. It has been observed that oil droplets adhere to the surface of the coral, which results in a complete breakdown of the under-

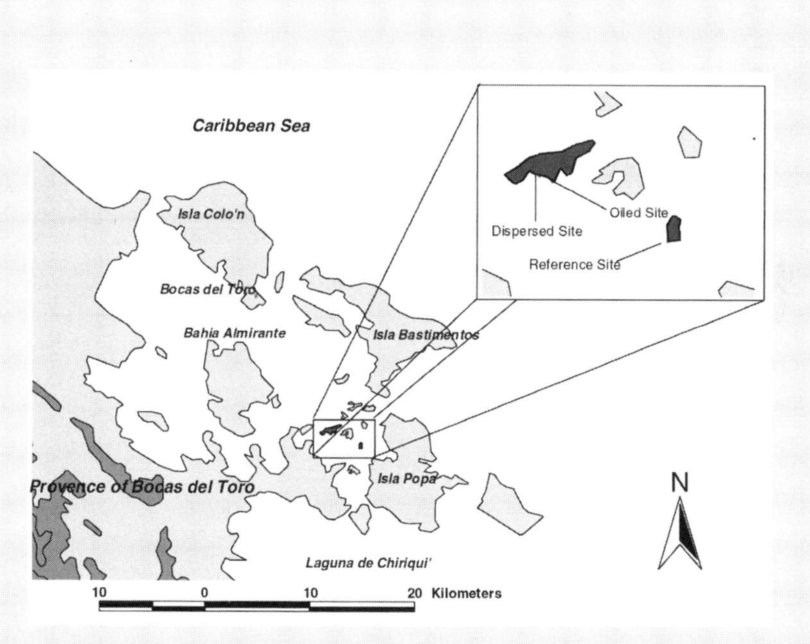

FIGURE 5-7 Case study: (TROPICS, Panama) Map of TROPICS study sites near Bocas del Toro, Panama.
SOURCE: Ward et al., 2003; courtesy of the American Petroleum Institute.

Effects: The sites were monitored five times in the first two years and once in 1994, ten years later. At the oil-only site, the corals were exposed to a 24-hour average of 0.14 ppm and a 48-hour average of 0.14 ppm. No significant impacts to corals were observed during any monitoring period.

At the dispersed oil site, the corals were exposed to a 24-hour average of 5.1 ppm (with a 1 hr maximum of 14.8 ppm) and 1.6 ppm at 48 hr. The average exposure over the 48-hour period was 3.4 ppm. At these expo-

FIGURE 5-8 Case study: (TROPICS, Panama) Aerial view of whole oil and dispersed oil sites.
SOURCE: Coastal Science Associates, Southern Affiliate, Incorporated.

sures, there were significant impacts to the shallow coral reef communities. Impacts observed at two years post-exposure included: reduced coverage by the major categories of all organisms (30 percent), hard corals (10 percent), all animals (30 percent), and plants (10 percent); reduced growth of the two most important hard coral species (*Agaricia tennuifolia* and *Porites porites*) but not two others (*Montastrea annularis* and *Acropora cervicornis*); and mortality of binding sponges. Studies conducted ten years post-exposure showed full recovery of coral coverage to levels equal those present pre-spill at the dispersed site and equal to conditions at the non-oiled control site.

Dispersed oil concentrations over the shallower seagrass (*Thalassia testudinum*) habitat were five times higher than over the coral habitat, av-

lying tissues (Johannes, 1975). Again this phenomenon may be of direct relevance in interpreting physically versus chemically dispersed oil toxicities. NRC (1989) stated that the smaller droplets in chemically dispersed oil did not adhere to the corals, in contrast to the larger, physically dispersed oil droplets, some of which were found on coral a few weeks after

eraging 22 ppm over 24 hr with a maximum of 70 ppm as measured in discrete water samples analyzed by GC. Even at these high exposures (the maximum likely oil concentrations), no negative effects were observed for plant survival, growth rates, or leaf blade area at the dispersed oil treatment site compared to the non-oiled reference site.

Untreated, whole oil caused significant impacts to mangrove habitats with high levels of defoliation and 17 percent mortality of adult mangroves after 2 years. After 10 years, mangrove mortality increased to 46 percent and some subsidence of the sediment surface was observed at the oiled site. After 18 years, the oiled site started to show some recovery as new trees replaced the dead trees (Figure 5-9; Ward et al., 2003). This field experiment clearly demonstrates the trade-offs associated with dispersant use in shallow tropical settings.

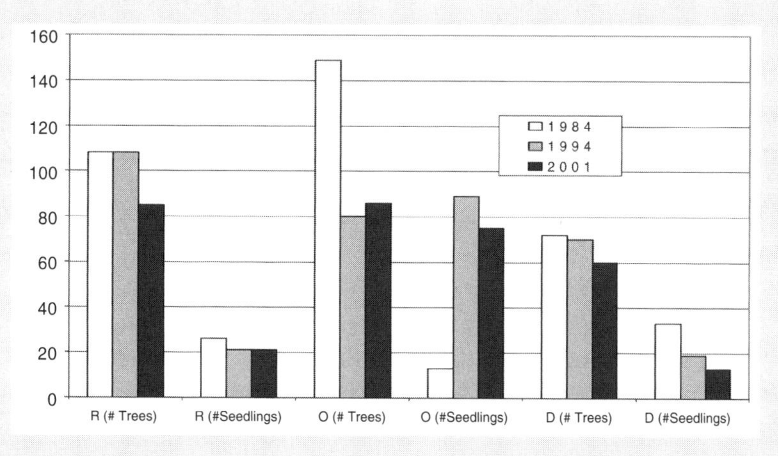

FIGURE 5-9 Results of 18 years of monitoring impacts to mangroves in Panama as part of TROPICS. Histograms reflect mangrove tree or seedling population counts (1984–2001) from whole oil (Site O) and dispersed oil (Site D) compared to a reference site (Site R). SOURCE: Ward et al., 2003; courtesy of the American Petroleum Institute.

SOURCE: Summary compiled from Ballou et al. (1987), Dodge et al. (1995), and Ward et al. (2003).

exposure to oil. In addition, a common stress response to oil pollution that has repeatedly been observed in coral species is the excessive production of mucus (see Shigenaka, 2001). This protective response can reduce the bioaccumulation of chemical contaminants by binding them in this lipid-rich mucus matrix that is ultimately "sloughed off" (or eaten by grazing fish) the surface of the coral, so protecting the underlying tissues. It is unclear whether chemically dispersed droplets or physically dispersed droplets or accumulation of dissolved components could alter this response. The excessive production of mucus takes energy away from normal cellular processes potentially reducing the overall health and fitness of the coral. In the case of chronic oil pollution events, such as continued leaching from mangrove sediments, excess mucus production could ultimately lead to coral death.

In conclusion, recent studies of coral larvae clearly demonstrate impacts of dispersants and dispersed oil on corals and, because of their life history and habitat characteristics, these species may be especially susceptible (Table 5-11). Consequently, decisions concerning dispersant use should take coral toxicity studies into consideration. In addition, laboratory studies are needed to estimate the relative contribution of dissolved- and particulate-phase oil to toxicity in representative coral species. Because corals typically experience high levels of natural sunlight, these toxicity tests should include an evaluation of delayed effects and photo-enhanced toxicity.

Mangroves

Few reports have been published that address the use of dispersants in treating oil spills close to mangroves. Early work by Getter and Ballou (1985) used an experimental spill at a site in Panama and concluded that dispersant use reduced the overall impact of oil on mangroves. This was a long-term project (10 years), but lacked replication of study sites (Dodge et al., 1995). In order to investigate the types of oil spill responses that might reduce the impact of oil spills and to address the need for more relevant information on the effects of oil spills on mangroves, Duke, Burns and co-workers carried out a number of field trials to assess the benefits of two remediation strategies for mangrove forests (see Burns et al., 1999; Duke and Burns, 1999; Duke et al., 1998a,b,c, 1999, 2000). These experiments were aimed at bridging the gap between surveys of real spill incidents (e.g., Volkman et al., 1994; Duke et al., 1997, 1998c) and those obtained from seedling laboratory experiments (Lai and Lim, 1984; Wardrup, 1987; Duke et al., 1998a). Field experiments, named the Gladstone trials, investigated the effects of different oils and remediation strategies on mangroves over both short and long-term time scales (1995–1998) utilizing a variety of replicated trials. One study compared the effects of dispersant (Corexit 9527) or bioremediation (aeration plus nutrients) strategies on a controlled spill using pre-weathered (24 hr) Gippsland light crude oil. It should be noted that the dispersant Corexit 9527 was premixed and weathered with the oil mixture before application. There were no differences observed between oil alone and dispersed oil treatments on resident fauna. Death of mangrove trees, however, was significantly lower in the plots treated with dispersant, similar to data previously obtained from laboratory and field studies (Duke et al., 1998a,c; Duke and Burns, 1999). With oil alone, long-term impacts on the fauna and little sign of recovery of trees led the authors to conclude that dispersion of spilled oil before it reaches mangroves should be considered for reducing the long-term impact of oil on mangrove habitat. It was interesting to note that the use of Corexit 9527 resulted in no difference in the amount of oil absorbed by the sediments, the penetration of oil to depth, or the weathering patterns of the oil over time.

IMPROVING THE USE OF INFORMATION ABOUT EFFECTS IN DECISIONMAKING

As discussed in Chapter 2, the ultimate decision regarding the use of dispersants in spill response generally rests upon answering the question as to whether use of dispersants will reduce the overall impact (Figure 2-4 in Chapter 2) by reducing the effects on some specific and sensitive species or habitat, without causing unacceptable harm to another specific and sensitive species or habitat. This decision represents a trade-off that will be dictated by a range of ecological, social, and economic values associated with the potentially affected resources. When spills occur offshore, where the potential magnitude and duration of impacts on organisms in the water column or seafloor can be assumed to be minimal, a decision to use dispersant can be made with information that is generally available. As the capability to deploy dispersants offshore increases, however, the capability to use dispersants in nearshore and shallower water settings will also increase. At the present time, the current understanding of the risk of dispersant use to shallow water or benthic species during a given spill is typically not adequate to allow rapid and confident decisionmaking. Several factors contribute to this uncertainty.

The rate of processes controlling the ultimate fate of dispersed oil is poorly understood. Of particular concern is the fate of dispersed oil in areas with high suspended solids and areas of low flushing rates. There is insufficient information to determine how chemically dispersed oil interacts with suspended sediments, as well as biotic components of aquatic systems, both short- and long-term, compared to naturally dispersed oil. **Relevant state and federal agencies, industry, and appropriate international partners should develop and implement a focused series of experiments to quantify the weathering rates and final fate of chemically dispersed oil droplets compared to undispersed oil.** Results from these experiments could be integrated with results from biological exposures

TABLE 5-11 Toxicity Studies of Chemically Dispersed Oil (or Dispersant Alone) to Coral Species in Laboratory and Field Studies (since 1988)

Species	Oil (D:O ratio)	Dispersant	Exposure
Coral reef (primarily *Porites porites* and *Agaricia tennuifolia*)	PBCO (1:20)	Commercial nonionic glycol ether-based	24 hr continuous release
Acropora spp. (growth), variety of corals visually assessed	Arabian light crude (1:20)	Corexit 9527	24 hr and 120 hr exposures plus 1 year recovery. Growth assessed.
Acropora palmata, Montastrea annularis, Porites porites	Oil (W) not detailed (1:10)	12 D including Corexit 9527, Corexit 9550, Finasol OSR7	DO and O, 6–10 hr, 2 week recovery and delayed assessments in clean SW.
Larvae of *Stylophora pistillata* and *Heteroxenia fuscescense*	Egyptian crude (1:10)	Inipol IP-90, Petrotech PTI-25, Biosolve, Bioreico R-93, Emulgal C-100	WSF (of O), DO WAF and D (5–500 ppm). 2–96 hr, static
Acropora millepora (eggs and larvae)	Heavy crude oil (1:10/100)	Corexit 9527	WAF, DO and D alone. Exposures; 4 hr fertilization assays (FA), 24 hr larval metamorphosis assay (LM); static
Stylophora pistillata (adult)	Egyptian crude (1:10)	Emulgal C-100	WSF (of O), D and DO WAF. 24 hr, static with recovery in clean SW.

NOTE: D, dispersant; DO, chemically dispersed oil; D:O, dispersant:oil ratio; HC, Hydrocarbon concentration (ppb); O, oil; PBCO, Prudhoe Bay Crude Oil; SW, seawater; TPH, total petroleum hydrocarbons; WAF, water-accommodated fraction; WSF, water soluble fraction.

comparing uptake of dissolved, colloidal, and particulate oil to provide a comprehensive model of the fate of dispersed oil in aquatic systems.

There is insufficient understanding of the actual concentrations and temporal/spatial distributions and behavior of chemically dispersed oil from field settings (from either controlled experiments or actual spills). Data from field studies (both with and without dispersants) are needed to validate models, provide real-world data to improve knowledge of oil fate and effects, and fulfill other information needs. **Relevant state and federal agencies, industry, and appropriate international partners should develop and implement steps to ensure that future wave-tank or spill-of-opportunity studies (or during the Natural Resource Damage**

Response	Comments	Reference
DO decrease in coral cover—complete elimination of *A. tennuifolia*.	Continuous field measurement of TPH and C_1-C_{10} hydrocarbons	[a]Ballou et al., 1989
Delayed sublethal impacts in all plots (bleaching); DO 120 hr exposure plots recovery less. No difference in growth rates.	HC concentrations measured over time (to 120 hr)	[a]Legore et al., 1989
Mortality was D dependent.	Nominal exposures	Thorhaug et al., 1989
Varied with exposure—from unsuccessful larval settlement to death. D toxic, DO WAF more toxic cf. WSF (and D alone).	Nominal exposures (dilutions of stocks)	Epstein et al., 2000
FA; WAF no effect. DO slight more toxic c.f. D alone. LM; DO more toxic cf. WAF, D toxic but at higher levels cf. [D] in DO.	Measured THC mg/L in stocks. Nominal concentrations calculated for dilutions.	Negri and Heyward, 2000
No death in WSF. D alone (1% or >) very toxic within 24 hr, delayed death (day 6) at 0.1%. DO WAF similar to D alone.	Nominal exposures (dilutions of stocks)	Shafir et al., 2003

[a]Field study.

Assessment investigations of oil spills that are not treated with dispersants) implement a field program to measure both dissolved-phase PAH and particulate/oil-droplet phase PAH concentrations for comparison to PAH thresholds measured in toxicity tests and predicted by computer models for oil spill fate and behavior. Accomplishing this will require the development and implementation of detailed plans (including preposition of sufficient equipment and human resources) for rapid deployment of a well-designed monitoring plan for actual dispersant applications in the United States. The RRT Region 6 Spill of Opportunity Monitoring Plan for dispersant application in the Gulf of Mexico should be finalized and implemented at the appropriate time. In addition, consideration should be given to long-term monitoring of sensitive habitats and species (e.g., mangroves, corals, sea grasses) after dispersant application to assess chronic effects and long-term recovery. These data will be valuable in validating the assumptions associated with environmental trade-offs of using dispersants.

One of the widely held assumptions concerning the use of dispersants is that chemical dispersion of oil will dramatically reduce the impacts of oil to seabirds and aquatic mammals, primarily by reducing their exposure to petroleum hydrocarbons. Evaluating the validity of this assumption is critical, because it is often a key factor in the decision on whether or not to use dispersants on a particular spill (e.g., in the ecological risk assessment workshop analyses). In addition, populations of waterfowl and some aquatic mammals may be higher in nearshore and estuarine areas; therefore, validating this assumption becomes even more important. Unfortunately, there is very little available information on the effects of dispersed oil on wildlife, especially aquatic mammals. Of additional concern is the effect of dispersed oil and dispersants on the waterproof properties of feathers and their role as thermal insulators. One of the recommendations of the NRC (1989) report was that **studies be undertaken "to assess the ability of fur and feathers to maintain the water-repellency critical for thermal insulation under dispersed oil exposure conditions comparable to those expected in the field."** This recommendation is reaffirmed because of the importance of this assumption in evaluating the environmental trade-offs associated with the use of oil dispersants in nearshore and estuarine systems and because it has not been adequately addressed.

The primary assumption for models predicting acute toxicity of physically and chemically dispersed oil is additive effects of dissolved-phase aromatic hydrocarbons. However, the possibility of photoenhanced toxicity and particulate/oil droplet phase exposure is generally not considered. A number of laboratory studies have indicated toxicity due to PAH increases significantly (from 12 to 50,000 times) for sensitive species in exposures conducted under ultraviolet light (representative of natural sunlight), compared to those conducted under the more traditional laboratory conditions of fluorescent lights. In addition, the toxicity tests typically do not consider delayed acute or sublethal effects. Consequently, current testing protocols may significantly underestimate toxicity for some species. For example, corals appear to be particularly sensitive to dispersants and dispersed oil due to the potential for photoenhanced toxicity and delayed effects. Similarly, toxicological effects due to increased exposure to oil from smothering, ingestion, or enhanced uptake are not explicitly considered in exposure models. Better understanding of these variables will decrease the uncertainty associated with predicting ecological effects of dispersed oil. **Relevant state and federal agencies, industry,**

and appropriate international partners should develop and implement a series of focused toxicity studies to: (1) provide data that can be used to parameterize models to predict photoenhanced toxicity; (2) estimate the relative contribution of dissolved and particulate oil phases to toxicity with representative species, including sensitive species and life stages; and (3) expand toxicity tests to include an evaluation of delayed effects. Detailed chemical analyses should accompany these tests, including characterization of dissolved and particulate oil composition and concentrations, as well as bioaccumulation. By refining our understanding of these variables, and incorporating them into decision-making tools, such as fate and effects models and risk rankings, the ability of decisionmakers to estimate the impacts of dispersants on aquatic organisms will be enhanced.

6

Research Priorities to Support Dispersant Use Decisionmaking

The primary response methods for oil spills in the United States consist of the deployment of mechanical on-water containment and recovery systems, such as booms and skimmers. Under the Oil Pollution Act of 1990 (OPA 90), the U.S. Coast Guard (USCG) passed rules for vessel and facility response plans that specified the minimum equipment capabilities for oil containment and recovery for likely and maximum spill volumes. Mechanical recovery is not always effective, thus OPA 90 also called for national and regional response teams to develop guidelines for other on-water response strategies, specifically the use of chemical dispersants and *in-situ* burning. Regional Response Teams have designated areas and conditions where dispersants and *in-situ* burning may be considered as appropriate response strategies when mechanical recovery is determined to be insufficient to protect sensitive resources. There are three types of approvals for dispersant use: case-by-case approval, expedited approval, and pre-approval.

In early 2002, the U.S. Coast Guard proposed changes to the rules for oil spill response capabilities to include minimum capabilities for dispersant application in all zones where dispersant use has been pre-approved. Thus, availability of dispersant application assets will no longer be a limiting factor in the decision-making process. Instead, other factors, such as effectiveness and effects, will be the major drivers of the decision whether or not dispersants should be used. Furthermore, the ready availability of dispersants and their application systems is expected to result in an increase in situations where dispersant use may be considered to combat oil spills, even in nearshore settings. Timeliness of the decision-making process is even more critical for nearshore spills where oil can quickly threaten highly sensitive resources.

Oil spills occur under a very wide range of conditions; thus decisions about appropriate response strategies have to be made for each event. Even for spills that meet pre-approval guidelines for dispersant use, the Federal On-Scene Coordinator (FOSC) has to make a series of decisions to determine whether or not dispersants would be effective and appropriate. Where dispersant use is not pre-approved, many more decisions must be made and the decision-making process becomes that much more difficult. One of the primary objectives of this study was to identify gaps in the information needed to support decisionmaking regarding appropriate use of dispersants. The decision-making framework used most frequently in the United States, and shown in Figure 2-4 in Chapter 2, is used to organize and prioritize the recommendations made in Chapters 3, 4, and 5. Many of the recommendations are relevant to more than one question in the decision-making process, but each recommendation is matched to the question it most strongly supports.

In response to the statement of task, the committee reviewed and evaluated existing information and ongoing research regarding the efficacy and effects of dispersants as an oil spill response technique. The statement of task specifically directed the committee to address "how laboratory and mesoscale experiments could inform potential controlled field trials and what experimental methods are most appropriate for such tests." All experiments, whether conducted at the bench-top or field scale, represent an attempt to measure or otherwise quantify the contribution of one variable among many that interact to dictate a specific outcome. Depending on the scale of the experiment, many variables may need to be held constant so that the change in outcome can be mapped against variation in a single variable of interest. Unfortunately, this simplification, if not taken under consideration, can reduce the realism of tests to the point that the results are of limited application to real world situations. For example, when bench-top tests are conducted, temperature, salinity, and other natural variables are set and held constant to compare the effectiveness of different dispersants and oil combinations. At the other end of the spectrum, when field tests are carried out, researchers have very limited control over environmental conditions, thus each test result is specific to a narrow set of environmental conditions. If the test is conducted under conditions typical of most spills, some extrapolation is possible; however, spills occur over such a wide range of settings that a relatively large number of tests at a variety of geographic locations would be necessary to identify the range of settings in which the dispersants would be effective. Even if this were possible, many of the most important variables (e.g., concentration of dissolved phase constituents or dispersed oil droplets) are extremely difficult to measure under field conditions. For example, one of the most challenging aspects of tank tests involves determination of a mass balance. Trying to obtain a mass balance in the field would be, at best, very difficult. Laboratory and wave-tank experiments and monitoring of spills of opportunity have the potential to answer key questions, if properly designed and conducted. Laboratory and mesocosm studies are most appropriate to test specific hypotheses on specific mechanisms or processes. Field tests are expensive, are difficult to control, offer many challenges to achieving good scientific measurements, and represent only a single set of conditions which are not readily controlled.

Serious consideration should be given to determining the true value and potential contribution of field testing. The body of work completed to date has provided important, but still limited, understanding of many aspects of the efficacy of dispersants in the field and behavior and toxicity of dispersed oil. Developing a robust understanding of these key processes and mechanisms to support decisionmaking in nearshore environments will require taking dispersant research to the next level. Many factors will need to be systematically varied in settings where accurate measurements can be taken. It is difficult to envision the appropriate role of field testing in a research area that has yet to reach consensus on standard protocols for mesocosm testing. The greater complexities (and costs) of carrying out meaningful field experiments suggest that greater effort be placed, at least initially, on designing and implementing a thorough and well-coordinated bench-scale and mesocosm research program. Such work should lead to more robust information about many aspects of dispersed oil behavior and effects. When coupled with information obtained through more vigorous monitoring of actual spills (regardless of whether dispersants are used effectively in response), this experimental work should provide far greater understanding than is currently available. Upon completion of the work discussed below, the value of further field-scale experiments may become obvious. In any case, such field-scale work would certainly be better and more effectively designed and executed than is currently possible. **Future field-scale work, if deemed necessary, should be based on the systematic, coordinated bench-scale and wave-tank testing discussed throughout this report and summarized in the remainder of this chapter.**

The committee also realizes that funding for oil spill research is very limited and this situation is not expected to change significantly in the future. Therefore, it is important to prioritize the research recommendations according to how they will improve the ability to make informed and science-based decisions on whether dispersants are the appropriate response tool for a specific spill. As stated previously, the committee's efforts have focused on questions arising from the potential use of dispersants in shallow, nearshore settings. Each of the questions in Figure 2-4 is listed below, followed by the research priorities needed to support the decisions.

D.1 WILL MECHANICAL RESPONSE BE SUFFICIENT?

Given that the use of dispersants does not remove oil from the environment, dispersant use is only considered when the answer to this question is "no," which then leads decisionmakers to evaluate the appropriateness of dispersant use. Although a subset of factors that may limit the effectiveness of mechanical response are readily evaluated (e.g., spill occurs too far from shore to allow safe and effective mechanical spill response), other factors require an ability to forecast environmental conditions at the spill site or along the projected path of the surface slick. Thus efforts to support real-time, tactical decisions regarding mechanical response will indirectly and directly support real-time decisionmaking regarding dispersant use.

D.2 IS THE SPILLED OIL OR REFINED PRODUCT KNOWN TO BE DISPERSIBLE?

As discussed in Chapter 3, there are two parts to this question of chemical effectiveness: (1) Is the freshly spilled oil dispersible? and (2) If the oil is initially dispersible, how long before it becomes non-dispersible? Currently, responders use "rules of thumb," experimental results for specific oils, and past experience to determine if an oil will be dispersible. Generally, it is possible for experienced and knowledgeable responders to predict whether a specific oil is initially dispersible. However, the more difficult determination is the length of time remaining until the oil weathers to the point that it is no longer dispersible. The state of the practice is to use visual observations to estimate the degree of emulsification or to conduct a first application of dispersant and see if the surface slick disperses. Such observations provide a qualitative approach to whether an oil has dispersed; however, they will not begin to answer the quantitative question as to the effectiveness of dispersant application. There is significant confusion among decisionmakers on how to interpret existing data on dispersant effectiveness. Results from experiments designed to answer one type of question are inappropriately used to answer a different question or to predict behavior under very different conditions (e.g., laboratory results are inappropriately used to estimate field effectiveness). The key areas of research include studies that will allow better predictions, using simple models, of the weathering processes that limit dispersant effectiveness for different oil types and environmental conditions.

NOAA, the Environmental Protection Agency (EPA), the Department of the Interior (including MMS and USGS), USCG, relevant state agencies, industry, and appropriate international partners should work together to develop and fund a research program to identify the mechanisms and rates of weathering processes that control the chemical effectiveness of dispersants. The research program should include both bench-scale tests and wave-tank experiments. Because of the limited funds and high costs of wave-tank experiments, it is essential that wave-tank studies be well coordinated. Agencies and industry should work together to establish an integrated research plan that focuses on collecting information about key aspects of dispersant use in a scientifically robust, but environmentally meaningful context. This new work will require systematic analysis using rigorous experimental design and execution, making use of standard chemical and other measurement techniques carried out by trained, certified personnel. Specific recommendations for these experiments are listed below.

- Research should be conducted in laboratory and wave-tank systems to investigate those parameters that control oil dispersability, including oil rheology and chemistry, dispersant rheology and dispersant chemistry, and dispersant-oil ratio. Past research on these topics often did not include measurement of important system characteristics (e.g., energy input to experimental systems) and response variables (e.g., oil droplet-size distributions), and the results are often contradictory. Future studies should conform to accepted standards of experimental design (discussed in Chapters 3, 4, and 5) that support statistical analysis of the data.
- Experimental bench-scale tests should be used to characterize the energy dissipation rates that prevail over a range of operating conditions to determine the functional relationship between variables for a range of oil viscosities and weathering states. Furthermore, evaluation of chemical effectiveness should always include measurement of the droplet-size distribution of the dispersed oil.
- The relationships between energy dissipation rates and chemical effectiveness should be determined for a variety of oil-dispersant combinations, including a range of oil viscosities and weathering states. Oil dispersant chemical efficiency tests should be designed to collect data that can be used in fate and transport modeling.
- Wave-tank studies should be designed to specifically address the chemical treatment of weathered emulsions of water-in-oil. Oil mass balances should be reported. In addition, the droplet-size distribution of the dispersed oil should be measured and reported.

D.3 ARE SUFFICIENT CHEMICAL RESPONSE ASSETS (I.E., DISPERSANT, EQUIPMENT, AND TRAINED PERSONNEL) AVAILABLE TO TREAT THE SPILL?

Under the proposed U.S. Coast Guard rulemaking, assets to treat spills in U.S. waters with chemical dispersant will be available within 12 hours after the spill in areas that have pre-approval plans. This increased availability of chemical response assets will likely result in more frequent consideration of dispersants as a response option for all spills, including those closer to shore and in shallow waters. If dispersant application becomes a required capability, it will be necessary to implement methods and procedures to ensure the readiness of response equipment and supplies for dispersant use, similar to the requirements for mechanical response equipment.

D.4 ARE THE ENVIRONMENTAL CONDITIONS CONDUCIVE TO THE SUCCESSFUL APPLICATION OF DISPERSANT AND ITS EFFECTIVENESS?

This question addresses environmental and operational effectiveness. Currently, it is not possible to predict the overall field effectiveness of dispersants for a spill event, a critical aspect of the trade-off analysis. Resource trustees need to be able to evaluate the benefits of reduced loadings of oil on shoreline habitats and smaller slicks that threaten water-surface resources compared to increased risks from dispersed oil plumes on water-column and benthic resources. Both potential risks and potential benefits depend on the effectiveness of dispersant application, particularly in nearshore settings. Better information is needed to determine the window of opportunity and percent effectiveness of dispersant application for different oil types and environmental conditions. Currently, dispersant effectiveness is a user input to fate and transport models, but potential effectiveness should be estimated by a physical-chemical efficiency model that integrates all of the complex processes controlling oil weathering and oil entrainment into the water column. Furthermore, there is no standard definition of field effectiveness and how it should be reported.

Relevant state and federal agencies, industry, and appropriate international partners should develop and fund a research program that provides the data necessary to predict, through modeling of the chemical, environmental, and operational conditions, the overall effectiveness of a dispersant application, specifically including conditions representative of nearshore physical settings. Two general types of modeling efforts and products should be recognized: (1) output intended to support decisionmaking during preplanning efforts; and (2) output intended to support emergency response to provide "rough-cut" outputs in hours. Detailed and specific recommendations are discussed at length in Chapters 3 and 4. The research program should consider the following issues:

- Energy-dissipation rates should be determined for wave tanks over the range of operating conditions that will be used in dispersant effectiveness tests. The wave conditions used in dispersant effectiveness tests should represent a specific environment of interest. It may be necessary to conduct experiments over a range of energy dissipation rates to adequately represent the spill environment.
- The design of wave-tank dispersant-effectiveness studies should test specific hypotheses regarding factors that may influence operational effectiveness. These factors include oil properties expected to prevail under spill-response conditions such as water-in-oil emulsification and the potential for heterogeneity in the rheological properties of the floating oil (e.g., formation of a "skin" that resists dispersant penetration).
- Tanks test studies should be conducted to determine the ability of mechanical recovery methods to retrieve oil that has been ineffectively treated with dispersant and re-floated oil. A more complete understanding of how dispersant use may subsequently limit mechanical recovery, if the dispersant is ineffective, could greatly reduce concern about the reliance on operational testing of dispersant effectiveness during early phases of spill response.
- Experiments should be designed to provide data on the rate and consequences of surfactant wash-out for both dispersed oil droplets that re-coalesce and surface slicks that were treated under calm conditions. Coalescence and resurfacing of dispersed oil droplets should be studied in flumes or wave tanks with high water-to-oil ratios (to promote leaching of surfactant into the water column) as a function of mixing time and energy dissipation rates.
- Evaluation of dispersant effectiveness in wave-tank tests should include measurement of oil concentrations on the water surface, in the water column, lost to the atmosphere, and on wave-tank surfaces. Oil mass balances should be reported. In addition, the droplet-size distribution of the dispersed oil should be measured and reported.

Currently, protocols for monitoring effectiveness of dispersants (e.g., Specialized Monitoring of Advanced Response Technologies [SMART]) are for guidance only and reflect the concern that use of dispersants not be unnecessarily postponed until monitoring assets can be put in place. Often, monitoring resources cannot be mobilized within the timeframe of emergency dispersant applications. SMART protocols have not been updated since the first design. However, it is very important to document actual oil concentrations under dispersed oil slicks, to validate dispersed oil model predictions and document actual exposures to sensitive resources. Monitoring would also support evaluation of the effectiveness of dispersant applications. **NOAA and USCG should develop updated SMART protocols and consider adding a detailed standard operating procedure (including instrument calibrations and data quality objectives) for each sampling and analytical module.**

D.5 WILL THE EFFECTIVE USE OF DISPERSANTS REDUCE THE IMPACTS OF THE SPILL TO SHORELINE AND WATER SURFACE RESOURCES WITHOUT SIGNIFICANTLY INCREASING IMPACTS TO WATER-COLUMN AND BENTHIC RESOURCES?

This trade-off analysis is the most difficult step of the process because of the lack of quantitative tools to predict the fate and effects of the dispersed oil plume and the benefits associated with less surface oil. It is also one of the most critical questions needing answer for adequate and appropriate decisionmaking. As discussed in Chapter 2, current ecological risk assessment (ERA) workshops on dispersant decisionmaking use a very qualitative approach that is difficult to apply to nearshore conditions where the potential impacts are not easily characterized. For instance, in offshore settings, one might reasonably assume that there is only a small likelihood that organisms on the seafloor may be exposed to significant concentrations of dispersed oil. Such assumptions may not be reasonable in some nearshore settings. Resource trustees need better information on the likely exposure regime, the mechanisms of toxicity of dispersed oil, and appropriate endpoints.

Oil trajectory models for dispersed oil plumes could be valuable tools to predict exposure, but they are incomplete in terms of their representation of the natural physical process involved, verification of the codes, and validation of the output from these models in an experimental setting or during an actual spill. As discussed in Chapter 4, the ability of models to predict the concentrations of dispersed oil and dissolved aromatic hydrocarbons in the water column with sufficient accuracy to aid in spill decisionmaking has yet to be fully determined.

As discussed in Chapter 4, one of the most significant weaknesses in correlating laboratory-scale and mesoscale experiments with conditions in the field results from a lack of understanding of the turbulence regime in all three systems. Likewise, one of the biggest uncertainties in computer modeling of oil spill behavior (with and without dispersant addition) comes from specifying horizontal and vertical diffusivities. It is very difficult to integrate all interacting transport and fate processes and oil properties to predict how much oil will be found in specific areas during an actual oil spill without the use of models.

Oil trajectory and fate models used to predict the behavior of dispersed oil should be improved, verified, and then validated in an appropriately designed experimental setting or during an actual spill. These models should meet the needs of both planning and real-time decisionmaking in complex nearshore settings. Key elements of the model improvements include:

- Studies should be conducted to quantify horizontal and vertical diffusivities and the rate of energy dissipation in the field (shear in vertical dimension, variations in the vertical diffusivity as a function of depth, sea-surface turbulence, etc.) under a variety of sea states that can be used as inputs into models (to improve the physical components of dispersed oil behavior), as well as to better design laboratory and mesocosm systems that may be suitable for estimating dispersant performance at sea.
- Models should include advective transport of entrained oil droplets, and the model codes and results should be validated in flume/tank studies and open-ocean (spill-of-opportunity) tests.
- The results of studies to better understand the processes that lead to formation of water-in-oil emulsions should be reflected in the improved models.
- Model output should include the concentrations of dissolved and dispersed oil, expressed as specific components or pseudo-components that can be used to support toxicity analysis.
- Once the improved models are available, sensitivity analyses should be conducted based on three-dimensional, oil-component, transport, and fate models, and the necessary databases developed (evaporation, dissolution, degradation, etc.) so that oil concentrations and fate can be used in decisionmaking.

One of the major concerns with use of dispersants in nearshore settings is the dispersed oil interaction with suspended particulate matter and the ultimate fate of the droplets.

Relevant state and federal agencies, industry, and appropriate international partners should develop and fund a focused series of experiments to quantify the weathering rates and final fate of chemically dispersed oil droplets compared to undispersed oil. Results from these experiments could be integrated with results from biological exposures comparing uptake of dissolved and particulate oil to provide a comprehensive model of the fate of dispersed oil in aquatic systems.

As discussed in Chapters 2 and 5, information regarding potential adverse effects on sensitive organisms or habitats involves a variety of factors, including seasonal variation in the presence, absence, and number of organisms at the spill site or along a projected path of a surface slick or dispersed oil plume; the sensitivity of the species to various toxic components in crude oil or refined products; and some estimation of the time needed for a given population to recover from acute exposure.

Relevant state and federal agencies, industry, and appropriate international partners should conduct a series of transport and fate modeling and associated biological assessments with and without dispersants, and develop operational envelopes of the dispersant use (e.g., for what oil types and volumes; when, where, and what type of water bodies) for planning prior to actual oil spills. Dissemination of these modeling efforts also provides scientific knowledge and intuition to make rational decisions for dispersant use.

Models are envisioned as a key tool to support decisionmaking on the appropriateness of dispersant applications. These models can be improved with the results of laboratory and wave-tank studies, but they need to be validated by comparing model results with actual field data. **Relevant state and federal agencies and industry should develop and implement detailed plans (including preposition of sufficient equipment and human resources) for rapid deployment of a well-designed monitoring effort for actual dispersant applications in the United States.** The plans should include measurement of total petroleum hydrocarbon (TPH) and polynuclear aromatic hydrocarbon (PAH) concentrations in both dissolved phase and particulate/oil-droplet phase for comparison to TPH and PAH concentrations predicted by computer models for oil spill fate and behavior. The spill-of-opportunity monitoring planning for dispersant application in the Gulf of Mexico (Aurand et al., 2004) should be finalized and implemented at the appropriate time.

Relevant state and federal agencies, industry, and appropriate international partners should develop and implement a series of focused toxicity studies to: (1) provide data that can be used to parameterize models to predict photo-enhanced toxicity; (2) estimate the relative contribution of dissolved and particulate oil phases to toxicity with representative species; and (3) include an evaluation of delayed effects. Detailed chemical analyses should accompany these tests, including characterization of dissolved and particulate oil composition and concentrations, as well as bioaccumulation. Increased understanding of these variables, and effective incorporation of them into decision-making tools, such as fate and effects models and risk rankings, will enhance the ability of decisionmakers to estimate the impacts of dispersants on aquatic and benthic organisms. **To this end, every effort should be taken to ensure that the spill response research community continues to monitor developments in the broad field of ecotoxicology, as various applications of**

increased understanding of toxicological effects, on various time scales, at the population and community-level may be of significant value to dispersant decisionmaking (see Chapter 5 for more details).

In addition, consideration should be given to long-term monitoring of sensitive habitats and species (e.g., mangroves, corals, sea grasses) after dispersant application to assess chronic effects and long-term recovery. These data will be valuable in validating the assumptions associated with environmental trade-offs of using dispersants.

The 1989 report recommended that studies be undertaken "to assess the ability of fur and feathers to maintain the water-repellency critical for thermal insulation under dispersed oil exposure conditions comparable to those expected in the field." This committee re-affirms this recommendation because of the importance of this assumption in evaluating the environmental trade-offs associated with the use of oil dispersants in nearshore and estuarine systems and because it has not been adequately addressed.

References

Aamo, O.M., M. Reed and P. Daling. 1993. A laboratory based weathering model: PC version for coupling to transport models. Pp. 617–626 in *Proceedings of the Sixteenth Arctic and Marine Oil Spill Program (AMOP) Technical Seminar*, Calgary, Canada. Environment Canada, Ottawa, Ontario, Canada.

Aamo, O.M., M. Reed and K. Downing. 1997. Oil Spill Contingency and Response (OSCAR) Model System: Sensitivity Studies. Pp. 429–438 in *Proceedings of the 1997 International Oil Spill Conference*, Fort Lauderdale, Florida. American Petroleum Institute, Washington, D.C.

Adams, G.G., P.L. Klerks, S.E. Belanger and D. Dantin. 1999. The Effect of the Oil Dispersant Omni-Clean on the Toxicity of Fuel Oil No. 2 in Two Bioassays With the Sheepshead Minnow *Cypriodon variegates*. *Chemosphere* 39:2141–2157.

Addassi, Y.M. Sowby, H. Parker-Hall and W. Robberson. In press. Establishment of Dispersant Use Zones in the State of California: A consensus approach for marine waters 200 nautical miles from shore. *Proceedings of the 2005 International Oil Spill Conference*, Miami Beach, Florida. American Petroleum Institute, Washington, D.C.

Agrawal, Y.C., E.A. Terray, M.A. Donelan, P.A. Hwang, A.J. Williams, W.M. Drennan, K.K. Kahma and S.A. Kitaigorodskii. 1992. Enhanced dissipation of kinetic energy beneath surface waves. *Nature* 359:219–220.

Alaska Department of Environmental Conservation. 1989. Letter dated March 28, 1989 from the State of Alaska to the U.S. Coast Guard. State of Alaska, Department of Environmental Conservation, Office of the Commissioner, Juneau.

Alaska Department of Environmental Conservation. 1993. *The Exxon Valdez Oil Spill: Final Report State of Alaska Response.* State of Alaska, Department of Environmental Conservation, Office of the Commissioner, Juneau.

Alaska Oil Spill Commission. 1990. *The Wreck of the Exxon Valdez.* Final report of the State of Alaska Oil Spill Commission. State of Alaska, Office of the Governor, Juneau.

Allen, A.A. 1988. Comparison of response options for offshore oil spills. Pp. 289–306 in *Proceedings of the Eleventh Arctic and Marine Oilspill Program (AMOP) Technical Seminar*, Edmonton, Alberta, Canada. Environment Canada, Ottawa, Ontario, Canada.

Allen, A.A. and J.R. Payne. 2001. *Liverpool Bay crude oil discharge tests at OHMSETT Facilities—December 2–6, 2000.* A report prepared for BHP Petroleum Limited, Spiltec, and Payne Environmental Consultants, Incorporated. Payne Environmental Consultants, Incorporated, Encinitas, California.

American Society of Civil Engineers (ASCE) Task Committee on Modeling of Oil Spills of the Water Resources Engineering Division. 1996. State-of-the Art Review of Modeling Transport and Fate of Oil Spills. *Journal of Hydraulic Engineering* 122(11):594–609.

American Society for Testing and Materials (ASTM). 1992. *Standard Guide for Oil Spill Dispersant Application Equipment: Boom and Nozzle Systems.* Designation: F1413-92. American Society for Testing and Materials, Philadelphia, Pennsylvania.

American Society for Testing and Materials (ASTM). 1993. *Standard Practice for Calibrating Oil Spill Dispersant Application Equipment: Boom and Nozzle Systems.* Designation: F1460-93. American Society for Testing and Materials, Philadelphia, Pennsylvania.

American Society for Testing and Materials (ASTM). 1996. *Standard Guide for Use of Oil Spill Dispersant Application Equipment During Spill Response: Boom and Nozzle Systems.* Designation: F1737-96. American Society for Testing and Materials, Philadelphia, Pennsylvania.

American Society for Testing and Materials (ASTM). 2000. *Standard Test Method for Laboratory Oil Spill Dispersant Effectiveness Using the Swirling Flask.* Designation: F2059-00. American Society for Testing and Materials, Philadelphia, Pennsylvania.

Ankley, G.T., S.A. Collyard, P.D. Monson and P.A. Kosian. 1994. Influence of ultraviolet light on the toxicity of sediments contaminated with polycyclic aromatic hydrocarbons. *Journal of Environmental Toxicology and Chemistry* 13:1791–1796.

Aquatic Testing Laboratories. 1994. *Abalone larval development short term toxicity test for oil spill cleanup agents.* Laboratory report. Aquatic Testing Laboratories, Ventura, California.

Aunaas, T., A. Olsen and K.E. Zachariassen. 1991. The effects of oil and oil dispersants on the amphipod *Gammarus-oceanicus* from arctic waters. *Polar Research* 10(2)619–630.

Aurand, D., G. Coelho, J. Clark and G. Bragin. 1999. Goals, objectives and design of a mesocosm experiment on the environmental consequences of nearshore dispersant use. Pp. 629–643 in *Proceedings of the Twenty-Second Arctic and Marine Oilspill Program (AMOP) Technical Seminar,* Calgary, Canada. Environment Canada, Ottawa, Ontario, Canada.

Aurand, D., L. Walko and R. Pond. 2000. *Developing Consensus Ecological Risk Assessments: Environmental Protection in Oil Spill Response Planning A Guidebook.* United States Coast Guard. Washington, D.C.

Aurand, D.V., G.M. Coelho and A. Steen. 2001. Ten years of research by the U.S. oil industry to evaluate the ecological issues of dispersant use: An overview of the past decade. Pp. 429–434 in *Proceedings of the 2001 International Oil Spill Conference,* Tampa, Florida. American Petroleum Institute, Washington, D.C.

Aurand, D., M. Hitchings, L. Walko, J. Clark, J. Bonner, C. Page, R. Jamail and R. Martin. 2001. *Justification for the Proposed Texas General Land Office "Spill of Opportunity" Testing Program.* EM&A Preliminary Report 01-08A. Ecosystem Management & Associates, Incorporated, Lusby, Maryland.

Aurand, D., M. Hitchings, L. Walko, J. Clark, J. Bonner, C. Page, R. Jamail and R. Martin. 2004. *Texas General Land Office "Spill of Opportunity" Dispersant Demonstration Project Description.* EM&A Final Report 01-08A. Ecosystem Management & Associates, Incorporated, Lusby, Maryland.

Baca, B.J. and C.D. Getter. 1984. The toxicity of oil and chemically dispersed oil to the seagrass *Thalassia testudinum.* Pp. 314–323 in *Oil Spill Chemical Dispersants: Research, Experience, and Recommendations,* T.E. Allen, Ed. American Society for Testing and Materials, Philadelphia, Pennsylvania.

Baker, J.M., J.H. Cruthers, D.I. Little, J.H. Oldham and C.M. Wilson. 1984. Comparison of the fate and ecological effects of dispersed and non-dispersed oil in a variety of marine habitats. Pp. 239–279 in *Oil Spill Chemical Dispersants: Research, Experience, and Recommendations*, T.E. Allen, ed. American Society for Testing and Materials, Philadelphia, Pennsylvania.

Ballou, T.G., R.E. Dodge, S.C. Hess, A.H. Knap and T.D. Sleeter. 1987. *Effects of a Dispersed and Undispersed Crude Oil on Mangroves, Seagrasses, and Corals.* American Petroleum Institute Publication Number 4460. American Petroleum Institute, Washington, D.C.

Ballou, T.G., S.C. Hess, R.E. Dodge, A.H. Knap and T.D. Sleeter. 1989. Effects of untreated and chemically dispersed oil on tropical marine communities: A long-term field experiment. Pp. 447–454 in *Proceedings of the 1989 International Oil Spill Conference*, San Antonio, Texas. American Petroleum Institute, Washington, D.C.

Barnea, N. and R. Laferriere. 1999. SMART: Scientific monitoring of advanced response technologies. Pp.1265–1267 in *Proceedings of the 1999 International Oil Spill Conference*, Seattle, Washington. American Petroleum Institute, Washington, D.C.

Barron, M.G. 2000. *Potential for photenhanced toxicity of spilled oil in Prince William Sound and Gulf of Alaska waters.* Contract No. 602.00.1. Report prepared for the Prince William Sound Regional Citizen's Advisory Council. Prince William Sound Regional Citizens' Advisory Council, Anchorage, Alaska.

Barron, M.G. and L. Ka'aihue. 2001. Potential for photoenhanced toxicity of spilled oil in Prince William Sound and Gulf of Alaska waters. *Marine Pollution Bulletin* 43:86–92.

Barron, M.G. and L. Ka'aihue. 2003. Critical evaluation of CROSERF test methods for oil dispersant toxicity testing under subarctic conditions. *Marine Pollution Bulletin* 46:1191–1199.

Barron, M.G., T. Podrabasky, S. Ogle and R.W. Ricker. 1999. Are aromatic hydrocarbons the primary determinant of petroleum toxicity to aquatic organisms? *Aquatic Toxicology* 46:253–268.

Barron M.G., M.G. Carls, R. Heintz and S.D. Rice. 2004. Evaluation of fish early life stage toxicity models of chronic embryonic exposures to complex polycyclic aromatic hydrocarbon mixtures. *Toxicological Sciences* 78:60–67.

Bassin, N.J. and T. Ichiye. 1977. Flocculation behavior of suspended sediments and oil emulsions. *Journal of Sedimentary Petrology* 47:671–677.

Beaupoil, C. and D. Nedelec. 1994. *Etude de la toxicite du produit de lavage Corexit® 9500 vis-a-vis de la crevette blanche Palaemonetes varians.* Laboratoire de Biologie Marine, Concarneau, France.

Becker, K.W., M.A. Walsh, R.J. Fiocco and M.T. Curran. 1993. A new laboratory method for evaluating oil spill dispersants. Pp. 507–510 in *Proceedings of the 1993 International Oil Spill Conference*, Tampa, Florida. American Petroleum Institute, Washington, D.C.

Belk, J.L., D.J. Elliott and L.M. Flaherty. 1989. The comparative effectiveness of dispersants in fresh and low salinity waters. Pp. 333–336 in *Proceedings of the 1989 International Oil Spill Conference*, San Antonio, Texas. American Petroleum Institute, Washington, D.C.

Belluck, D. 1993. Defining Scientific Procedural Standards for Ecological Risk Assessment. Pp. 440–450 in *Environmental Toxicology and Risk Assessment: 2nd Volume*, ASTM STP 1216, J.W. Gorsuch, F.J. Dwyer, C.G. Ingersoll and T.W. LaPoint, eds., American Society for Testing and Materials, Philadelphia, Pennsylvania.

Belore, R. 2003. Large wave tank dispersant effectiveness testing in cold water. Pp. 381–385 in *Proceedings of the 2003 International Oil Spill Conference*, Vancouver, Canada. American Petroleum Institute, Washington, D.C. [Online] Available at: http://www.slross.com/publications/IOSC.htm [February 11, 2005].

Berger, D. and D. Mackay. 1994. The evaporation of viscous or waxy oils—When is a liquid-phase resistance significant? Pp. 77–92 in *Proceedings of the Seventeenth Arctic and Marine*

Oil Spill Program (AMOP) Technical Seminar, Vancouver, British Columbia, Canada. Environment Canada, Ottawa, Ontario, Canada.

Bertuccioli, L., G.I. Roth, J. Katz, and T.R. Osborn. 1999. Turbulence measurements in the bottom boundary layer using particle image velocimetry. *Journal of Atmospheric and Oceanic Technology* 16(11):1635–1646.

Bhattacharyya, S., P.L. Klerks and J.A. Nyman. 2003. Toxicity to freshwater organisms from oils and oil spill chemical treatments in laboratory microcosms. *Environmental Pollution* 122:205–215.

Bhosle, N.B. and S. Mavinkurve. 1984. Effects of dispersants on microbial growth and biodegradation of crude oil. *Mahasagar-Bulletin of the National Institute of Oceanography* 17(4):233–238.

Bhosle, N.B. and A. Row. 1983. Effect of dispersants on the growth of indigenous bacterial population and biodegradation of crude oil. *Indian Journal of Marine Sciences* 12(3):194–196.

Blackall, P.J. and G.A. Sergy, 1981. The BIOS project-frontier oil spill countermeasures research. Pp. 167–172 in *Proceedings of the 1981 International Oil Spill Conference*, Atlanta, Georgia. American Petroleum Institute, Washington, D.C.

Blondina, G.J., M.L. Sowby, M.T. Ouano, M.M. Singer and R.S. Tjeerdema. 1997. A modified swirling flask efficacy test for oil spill dispersants. *Spill Science and Technology Bulletin* 4(3):177–185.

Blondina, G.J., M.M. Singer, I. Lee, M.T. Ouano, M. Hodgins, R.S. Tjeerdema and M.L. Sowby. 1999. Influence of salinity on petroleum accommodation by dispersants. *Spill Science and Technology Bulletin* 5(2):127–134.

Blum, D.J. and R.E. Speece. 1990. Determining chemical toxicity to aquatic species, *Environmental Science and Technology* 24:284–293.

Blumer, M., G. Souza and J. Sass. 1970. Hydrocarbon pollution of edible shellfish by an oil spill. *Marine Biology* 5:195–202.

Bobra, M. 1990. A study of the formation of water-in-oil emulsions. Pp. 87–117 in *Proceedings of the Thirteenth Arctic and Marine Oil Spill Program (AMOP) Technical Seminar*, Edmonton, Alberta, Canada. Environment Canada, Ottawa, Ontario, Canada.

Bobra, M. 1991. Water-in-Oil Emulsification: A Physicochemical Study. Pp. 483–492 in *Proceedings of the 1991 International Oil Spill Conference*, San Diego, California. American Petroleum Institute, Washington, D.C.

Bobra, M. 1992. *A Study of the Evaporation of Petroleum Oils.* Report No. EE-135. Report to Environmental Emergencies Science Division of Environment Canada. Environment Canada, Ottawa, Ontario, Canada.

Bocard, C., G. Castaing, J. Ducreux, C. Gatellier, J. Croquette and F. Merlin. 1987. PROTECMAR: The French experience form a seven-year dispersant offshore trials programme. Pp. 225–229 in *Proceedings of the 1987 International Oil Spill Conference*, Baltimore, Maryland. American Petroleum Institute, Washington, D.C.

Boehm, P.D. 1987. Transport and Transformation Processes Regarding Hydrocarbon and Metal Pollutants in Offshore Sedimentary Environments. Pp. 233–287 in *Long-Term Environmental Effects of Offshore Oil and Gas Development*, D.F. Boesch and N.N. Rabalais, eds. Elsevier Applied Science, New York, New York.

Boehm, P.D. and D.L. Fiest. 1980. Surface water column transport and weathering of petroleum hydrocarbons during the IXTOC-I blowout in the Bay of Campeche and their relation to surface oil and microlayer compositions. Pp. 267–338 in *Proceedings of the Symposium on Preliminary Results from the September 1979 Researcher/Pierce IXTOC-I Cruise*, Key Biscayne, Florida. National Oceanic and Atmospheric Administration Publications Office, Boulder, Colorado.

Boehm, P.D. and D.L. Fiest. 1982. Subsurface distribution of petroleum from an offshore well blowout. The IXTOC-I blowout, Bay of Campeche. *Environmental Science and Technology* 16(2):67–74.

Boese, B.L., J.O. Lamberson, R.C. Swartz and R.J. Ozretich. 1997. Photoinduced toxicity of fluoranthene to seven marine benthic crustaceans. *Archives of Environmental Contamination and Toxicology* 32:389–393.

Bonner, J.S., C.A. Page and C.B. Fuller. 2003. Meso-scale testing and development of test procedures to maintain mass balance. *Marine Pollution Bulletin* 47:406–414.

Box, G.E.P., W.G. Hunter and J.S. Hunter. 1978. *Statistics for Experimenters: An Introduction to Design, Data Analysis, and Model Building.* John Wiley & Sons, Incorporated, New York, New York.

Bradbury, S., R.Carlson and T. Henry. 1989. Polar narcosis in aquatic organisms. *Aquatic Toxicology and Hazard Assessment* 12:59–73.

Braddock, J.F. and Z.D. Richter. 1998. *Microbial degradation of aromatic hydrocarbons in marine sediments.* Final Report OCS Study MMS 97-0041. Institute of Arctic Biology, University of Alaska, Fairbanks.

Bragg, J.R. and E.H. Owens. 1994. Clay-oil flocculation as a natural cleansing process following oil spills: Part 1—Studies of shoreline sediments and residues from past spills. Pp. 1–23 in *Proceedings of the Seventeenth Arctic and Marine Oilspill Program (AMOP) Technical Seminar*, Vancouver, British Columbia, Canada. Environment Canada, Ottawa, Ontario, Canada.

Bragg, J.R. and E.H. Owens. 1995. Shoreline cleansing by interactions between oil and fine mineral particles. Pp. 216–227 in *Proceedings of the 1995 International Oil Spill Conference*, Long Beach, California. American Petroleum Institute, Washington, D.C.

Bragg, J.R. and S.H. Yang. 1995. Clay–oil flocculation and its effects on the rate of natural cleansing in Prince William Sound following the Exxon Valdez oil spill. Pp. 178–214 in *Exxon Valdez Oil Spill—Fate and Effects in Alaskan Waters*, P.G. Wells, J.N. Butler, J.S. Hughes eds., American Society for Testing and Materials, Philadelphia, Pennsylvania.

Bragin, G., G. Coelho, E. Febbo, J. Clark and D. Aurand. 1999. Coastal oilspill simulation system comparison of oil and chemically dispersed oil released in near-shore environments: Biological effects. Pp. 671–683 in *Proceedings of the Twenty-Second Arctic and Marine Oilspill Program (AMOP) Technical Seminar*, Calgary, Alberta, Canada. Environment Canada, Ottawa, Ontario, Canada.

Brandvik, J.J. and P.S. Daling. 1990. Statistical experimental design optimization of dispersant's performance. Pp. 243–254 in *Proceedings of the Thirteenth Arctic and Marine Oilspill Program (AMOP) Technical Seminar*, Edmonton, Alberta, Canada. Environment Canada, Ottawa, Ontario, Canada.

Brandvik, J.J., P.S. Daling and K. Aareskjold. 1991. *Chemical dispersibility testing of fresh and weathered oils—an extended study with eight oil types.* DIWO Report No. 12. SINTEF Group, Trondheim, Norway.

Brandvik, J.J., M.D. Moldestad and P.S. Daling. 1992. Laboratory testing of dispersants under arctic conditions. Pp. 123–134 in *Proceedings of the Fifteenth Arctic and Marine Oilspill Program (AMOP) Technical Seminar*, Edmonton, Alberta, Canada. Environment Canada, Ottawa, Ontario, Canada.

Brandvik, J.J., P.S. Daling, A. Lewis and T. Lunel. 1995. Measurements of dispersed oil concentrations by *in-situ* UV fluorescence during the Norwegian experimental oil spill with Sture blend. Pp. 519–535 in *Proceedings of the Eighteenth Arctic and Marine Oilspill Program (AMOP) Technical Seminar*, Edmonton, Alberta, Canada. Environment Canada, Ottawa, Ontario, Canada.

Brandvik, P.J., T. Strom-Kristiansen, A. Lewis, P.S. Daling, M. Reed, H. Rye and H. Jensen. 1996. The Norwegian Sea trial 1995 offshore testing of two dispersant systems in simulation of an underwater pipeline leakage, a summary paper. Pp. 1395–1416 in *Proceedings of the Nineteen Arctic and Marine Oil Spill Program (AMOP) Technical Seminar*, Edmonton, Alberta, Canada. Environment Canada, Ottawa, Ontario, Canada.

Briceno, J., W.J. McKee, J.R. Clark and D.D. Whiting. 1992. *Relative sensitivity of Gulf of Mexico species and national test species in acute toxicity tests with dispersants. Poster presentation.* Paper presented at the Thirteenth Annual Meeting of the Society of Environmental Toxicology and Chemisty (SETAC). SETAC, North America, Pensacola, Florida.

Bridie, A.L, Th.H. Wanders, W. Zegveld and H.B. Van der Heijde. 1980a. *The Formation, Prevention and Breaking of Sea-Water-in-Crude-Oil Emulsions: "Chocolate Mousse."* Paper presented at International Research Symposium, Chemical Dispersion of Oil Spills, Toronto, Canada. University of Toronto, Institute for Environmental Studies, Toronto, Ontario, Canada.

Bridie, A.L., Th.H. Wanders, W. Zegveld and H.B. Van der Heijde. 1980b. Formation, Prevention and Breaking of Sea Water in Crude Oil Emulsions: "Chocolate Mousses." *Marine Pollution Bulletin* 2:343–348.

Briggs, K.T., S.H. Yoshida and M.E. Gershwin. 1996. The influence of petrochemicals and stress on the immune system of seabirds. *Regulatory Toxicology and Pharmacology* 23:145–155.

Brochu, C., E. Pelletier, G. Caron and J.E. Desnoyers. 1986. Dispersion of crude oil in seawater: The role of synthetic surfactants. *Oil & Chemical Pollution* 3:257–279.

Broecker, W.S. and T.-H. Peng. 1982. *Tracers in the Sea.* Lamont-Doherty Geological Observatory, Palisades, New York.

Brooks, J.M., D.A. Weisenburg, R.A. Burke, M.C. Kenicutt and B.B. Bernard. 1980. Gaseous and volatile hydrocarbons in the Gulf of Mexico following the IXTOC-I blowout. Pp. 53–88 in *Proceedings of the Symposium on Preliminary Results from the September 1979 Researcher/Pierce IXTOC-I Cruise*, Key Biscayne, Florida. National Oceanic and Atmospheric Administration Publications Office, Boulder, Colorado.

Brovchenko, I., A. Kuschan, V. Maderich, M. Shlakhtun, V. Koshebutsky and M. Zheleznyak. 2003. Model of oil spill simulation in the Black Sea. Pp. 101–112 in *Proceedings of the Third Conference on Oil Spills, Oil Pollution and Remediation*, Istanbul, Turkey. Bo-aziçi University, Institute of Environmental Sciences, Istanbul, Turkey.

Brown, C.E., M.F. Fingas, R.H. Goodman, J.V. Mullin, M. Choquet and J.-P. Monchalin. 2000. Progress in achieving airborne oil slick thickness measurement. Pp. 493–498 *Proceedings of the Twenty-Third Arctic and Marine Oil Spill Program (AMOP)* Technical Seminar, Vancouver, British Columbia, Canada. Environment Canada, Ottawa, Ontario, Canada.

Brown, H.M. and R.H. Goodman. 1987. *The Dispersion of Alaska North Slope Oil in Wave Basin Tests.* Report to Alaska Clean Seas and Esso Imperial Oil. Alaska Clean Seas, Prudhoe Bay, Alaska, and Esso Imperial Oil, Calgary, Alberta, Canada.

Brown, H.M. and R.H. Goodman. 1988. Dispersant tests in a wave basin—four years of experience. Pp. 501–514 in *Proceedings of the Eleventh Arctic and Marine Oilspill Program (AMOP) Technical Seminar*, Edmonton, Alberta, Canada. Environment Canada, Ottawa, Ontario, Canada.

Brown, H.M., R.H. Goodman and G.P. Canevari. 1987. Where has all the oil gone? Dispersed oil detection in a wave basin and at sea. Pp. 307–312 in *Proceedings of the 1987 International Oil Spill Conference*, Baltimore, Maryland. American Petroleum Institute, Washington, D.C.

Brown, H.M., J.S. Goudey, J.M. Foght, S.K. Cheng, M. Dale, J. Hoddinott, L.R. Quaife and D.W.S. Westlake. 1990. Dispersion of spilled oil in freshwater systems: field trial of a chemical dispersant. *Oil and Chemical Pollution* 6:37–54.

Bruheim, P. and K. Eimhjellen. 2000. Effects of non-ionic surfactants on the uptake and hydrolysis of fluorescein diacetate by alkane-oxidizing bacteria. *Canadian Journal of Microbiology* 46(4):387–390.

Bruheim, P., H. Bredholt and K. Eimhjellen. 1997. Bacterial degradation of emulsified crude oil and the effect of various surfactants. *Canadian Journal of Microbiology* 43:17–22.

Bruheim, P., H. Bredholt and K. Eimhjellen. 1999. Effects of surfactant mixtures, including Corexit 9527, on bacterial oxidation of acetate and alkanes in crude oil. *Applied and Environmental Microbiology* 65(4):1658–1661.

Burbank, D.C. 1977. *Environmental studies of Kachemak Bay and lower Cook Inlet—Volume III. Circulation studies in Kachemak Bay and lower Cook Inlet.* Alaska Department of Fish and Game, Anchorage, Alaska.

Burns, K.A. and A.H Knap. 1989. The Bahia Las Minas oil spill: Hydrocarbon uptake by reef building corals. *Marine Pollution Bulletin* 20(8):391–398.

Burns, K.A., S. Codi, C. Pratt and N.C. Duke. 1999. Weathering of hydrocarbons in mangrove sediments: testing the effects of using dispersants to treat oil spills. *Organic Geochemistry* 30:1273–1286.

Burridge, T.R. and M. Shir. 1995. The comparative effects of oil, dispersants, and oil/dispersant conjugates on germination of marine macroalga *Phyllospora comosa* (Fucales: Phaeophyta). *Marine Pollution Bulletin* 31:446–452.

Bury, S.J. and C.A. Miller. 1993. Effect of micellar solubilization on biodegradation rates of hydrocarbons. *Environmental Science and Technology* 27:104–110.

Butler, J.N. 1975. Evaporative weathering of petroleum residues: the age of pelagic tar. *Marine Chemistry* 3:9–21.

Butler, J.N., B.F. Morris and T.D. Sleeter. 1976. The fate of petroleum in the open ocean. Pp. 287–297 in *Sources, effects, and sinks of hydrocarbons in the aquatic environment*. The American Institute of Biological Sciences, Washington, D.C.

Byford, D.C., P.J. Green and A. Lewis. 1983. Factors influencing the performance and selection of low-temperature dispersants. Pp. 140–150 in *Proceedings of the Sixth Annual Arctic Marine Oilspill Program (AMOP) Technical Seminar*, Edmonton, Alberta, Canada. Environment Canada, Ottawa, Ontario, Canada.

Byford, D.C., P.R. Laskey and A. Lewis. 1984. Effect of low temperature and varying energy input on the droplet size distribution of oils treated with dispersants. Pp. 208–228 in *Proceedings of the Seventh Annual Arctic Marine Oilspill Program (AMOP) Technical Seminar*. Environment Canada. Ottawa, Ontario, Canada

Canevari, G.P. 1969. The role of chemical dispersants in oil cleanup. Pp. 29–51 in *Oil on the Sea*, D.P. Hoult, Ed. Plenum Press, New York, New York.

Canevari, G.P. 1984. A review of the relationship between the characteristics of spilled oil and dispersant effectiveness. Pp. 87–93 in *Oil Spill Chemical Dispersants: Research, Experience, and Recommendations*, T.E. Allen, Ed. American Society for Testing and Materials. Philadelphia, Pennsylvania.

Canevari, G.P., P. Calcavecchio, R.R. Lessard, K.W. Becker and R.J. Fiocco. 2001. Key parameters affecting the dispersion of viscous oil. Pp. 479–483 in *Proceedings of the 2001 International Oil Spill Conference*, Tampa, Florida. American Petroleum Institute, Washington, D.C.

Champ, M.A., ed. 2000. Special issue: Langmuir circulation and oil spill modeling. *Spill Science and Technology Bulletin* 6(3/4):207–275.

Chandrasekar, S., G. Sorial and J.W. Weaver. 2003. Determining dispersant effectiveness data for a suite of environmental conditions. *Proceedings of the 2003 International Oil Spill Conference*, Vancouver, Canada. American Petroleum Institute, Washington, D.C.

Choquet, M., R. Heon, G. Vaudreuil, J.-P. Monchalin, C. Padioleau and R.H. Goodman. 1993. Remote thickness measurement of oil slicks on water by laser-ultrasonics. Pp. 531–536 in *Proceedings of the 1993 International Oil Spill Conference*, Tampa, Florida. American Petroleum Institute, Washington, D.C.

Chow, V.T. 1988. *Open Channel Hydraulics.* McGraw-Hill, Incorporated, New York, New York.

Churchill, P.F. and S.A. Churchill. 1997. Surfactant-enhanced biodegradation of solid alkanes. *Journal of Environmental Science and Health, A* 32(1):293–306.

Clark, J.R., G.E. Bragin, R.J. Febbo and D.J. Letinski. 2001. Toxicity of physically and chemically dispersed oils under continuous and environmentally realistic exposure conditions: Applicability to dispersant use decisions in spill response planning. Pp. 1249–1255 in *Proceedings of the 2001 International Oil Spill Conference*, Tampa, Florida. American Petroleum Institute, Washington, D.C.

Clayton, J.R., Jr., J.R. Payne and J.S. Farlow. 1993. *Oil Spill Dispersants: Mechanisms of Action and Laboratory Tests.* CRC Press, Incorporated, Boca Raton, Florida.

Cloutier, D., C.L. Amos, P.R. Hill and K. Lee. 2002. Oil erosion in an annular flume by seawater of varying turbidities: a critical bed shear stress approach. *Spill Science and Technology Bulletin* 8(1):83–93.

Coelho, G.M. and D.V. Aurand, Eds. 1996. *Proceedings of the Fifth Meeting of the Chemical Response to Oil Spills: Ecological Effects Research Forum.* Ecosystem Management and Associates, Purcellville, Virginia.

Coelho, G.M. and D.V. Aurand, Eds. 1997. *Proceedings of the Sixth Meeting of the Chemical Response to Oil Spills: Ecological Effects Research Forum.* Ecosystem Management and Associates, Purcellville, Virginia.

Coelho, G., D. Aurand and D.A. Wright. 1999. Biological uptake analysis of organisms exposed to oil and chemically dispersed oil. Pp. 685–694 in *Proceedings of the Twenty-Second Arctic and Marine Oilspill Program (AMOP) Technical Seminar*, Calgary, Alberta, Canada. Environment Canada, Ottawa, Ontario, Canada.

Cohen, A.M. and D. Nugegoda. 2000. Toxicity of three oil spill remediation techniques to the Australian bass *Macquaria novemaculeata. Ecotoxicology and Environmental Safety* 47(2): 178–185.

Cohen, A.M., D. Nugegoda and M.M. Gagnon. 2001a. Metabolic Responses of Fish Following Exposure to Two Different Oil Spill Remediation Techniques. *Ecotoxicology and Environmental Safety* 48:306–310.

Cohen, A.M., D. Nugegoda and M.M. Gagnon. 2001b. The Effect of Different Oil Spill Remediation Techniques on Petroleum Hydrocarbon Elimination in Australian Bass (*Macquaria novemaculeata*). *Archives of Environmental Contamination and Toxicology* 40: 264–270.

Cohen, A., M.M. Gagnon and D. Nugegoda. 2003. Biliary PAH metabolite elimination in Australian bass, *Macquaria novemaculeata,* following exposure to bass straight crude oil and chemically dispersed crude oil. *Bulletin of Environmental Contamination and Toxicology* 70(2):394–400.

Conover, R.J. 1971. Some relations between zooplankton and Bunker C oil in Chedabucto Bay following the wreck of the tanker *Arrow. Journal of the Fisheries Research Board Canada* 28:1327–1330.

Cook, C.B. and A.H. Knap, 1983. The effects of crude oil and chemical dispersant on photosynthesis in the brain coral, *Diploria strigosa. Marine Biology* 78:21–27.

Cormack, D., B.W.J. Lynch and B.D. Dowsett. 1987. Evaluation of dispersant effectiveness. *Oil and Chemical Pollution* 3:87–103.

Coutou, E.I. Castritis-Catharios and M. Moraitou-Apostolopoulo. 2001. Surfactant-based oil dispersant toxicity to developing nauplii of Artemia: effects on ATPase enzymatic system. *Chemosphere* 42:959–964.

CRC. 1967. *CRC Handbook of Chemistry and Physics: A Ready-Reference Book of Chemical and Physical Data.* Forty-eighth edition. R.C. Weast and S.M Selby, eds. The Chemical Rubber Company, Cleveland, Ohio.

Crowell, M.J. and P.A. Lane. 1988. *The Effects of Crude Oil and the Dispersant Corexit 9527 on the Vegetation of a Nova Scotian Saltmarsh: Impacts After Two Growing Seasons.* Report no. EE-103, Environment Canada, Ottawa, Ontario, Canada.

Csanady, G.T. 1973. *Turbulent Diffusion in the Environment.* Reidel Publishing Company, Boston, Massachusetts.

Daling, P.S. 1988. A study of the chemical dispersibility of fresh and weathered crudes. Pp. 481–499 in *Proceedings of the Eleventh Arctic and Marine Oilspill Program (AMOP) Technical Seminar*, Edmonton, Alberta, Canada. Environment Canada, Ottawa, Ontario, Canada.

Daling, P.S. and P.J. Brandvik. 1989. *The effects of photolysis on oil slicks.* SINTEF Group, Trondheim, Norway.

Daling, P.S. and R. Lichtenthaler. 1987. Chemical dispersion of oil. Comparison of the effectiveness results obtained in laboratory and small-scale field tests. *Oil & Chemical Pollution* 3:87–103.

Daling, P.S., D. Mackay, N. Mackay and P.J. Brandvik. 1990a. Droplet size distributions in chemical dispersion of oil spills: towards a mathematical model. *Oil and Chemical Pollution* 7:173–198.

Daling, P.S., P.J. Brandvik, D. Mackay and O. Johansen. 1990b. Characterization of crude oils for environmental purposes. Pp. 119–138 in *Proceedings of the Thirteenth Arctic and Marine Oil Spill Program (AMOP) Technical Seminar*, Edmonton, Alberta, Canada. Environment Canada, Ottawa, Ontario, Canada.

Daling, P.S., O.M. Aamo, A. Lewis and T. Strom-Kritiansen. 1997. SINTEF/IKU Oil-Weathering Model: Predicting Oil's Properties at Sea. Pp. 297–307 in *Proceedings of the 1997 International Oil Spill Conference*, Fort Lauderdale, Florida. American Petroleum Institute, Washington, D.C.

Daling, P.S., P.J. Brandvik and M. Reed. 1998. *Dispersant experience in Norway: Dispersant effectiveness, monitoring, and fate of dispersed oil. Dispersant Application in Alaska: A Technical Update.* Prince William Sound Oil Spill Recovery Institute, Cordova, Alaska.

Dames and Moore. 1978. *Drilling fluid dispersion and biological effects study for the Lower Cook Inlet C.O.S.T. well.* Report prepared for the Atlantic Richfield Company. Dames and Moore Group, Los Angeles, California.

Davies, L., F. Daniel, R.P.J. Swannell and J.F. Braddock. 2001. *Biodegradability of chemically-dispersed oil.* A report prepared for the Minerals Management Service, Alaska Department of Environmental Conservation, and the United States Coast Guard. Minerals Management Service, Herndon, Virginia.

Delaune, R.A., C.J. Smith, W.H. Patrick, J.W. Fleeger and M.D. Tolley. 1984. Effect of oil on salt marsh biota: Methods for restoration. *Environmental Pollution* 36(3)207–227.

Delvigne, G.A.L. 2002. Physical appearance of oil in oil-contaminated sediment. *Spill Science and Technology Bulletin* 8(1)55–63.

Delvigne, G.A.L. and C.E. Sweeney. 1988. Natural dispersion of oil. *Oil and Chemical Pollution* 4:281–310.

Delvigne, G.A.L., J.A. Roelvink and C.E. Sweeney. 1986. Research on vertical turbulent dispersion of oil droplets and oiled particles—Literature review. OCS Study MMS 86-0029. Report to Minerals Management Service. Minerals Management Service, Anchorage, Alaska.

Delvigne, G.A.L., J.A.Van der Stel and C.E. Sweeney. 1987. *Measurements of vertical turbulent dispersion and diffusion of oil droplets and oiled particles.* OCS Study MMS 87-111. Minerals Management Service, Anchorage, Alaska.

Deneer, J.W., T. Sinnige, W. Seinen and J.L.M. Hermens. 1988. The joint acute toxicity to Daphnia magna of industrial organic chemicals at low concentrations. *Aquatic Toxicology* 12:33–38.

DiToro, D.M., C.S. Zarba, D.J. Hansen, W.J. Berry, R.C. Swartz, C.E. Cowan, S.P. Pavlou, H.E. Allen, N.A. Thomas and P.R. Paquin. 1991. Technical basis for establishing sediment quality criteria for nonionic organic chemicals using equilibrium partitioning. *Journal of Environmental Toxicology and Chemistry* 10:1541–1583.

DiToro, D.M., J.A. McGrath and D.J. Hansen. 2000. Technical basis for narcotic chemicals and polycyclic aromatic hydrocarbon criteria. I. Water and tissue. *Journal of Environmental Toxicology and Chemistry* 19:1951–1970.

Dodge, R.E., S.C. Wyers, H.R. Frith, A.H. Knap, S.R. Smith and T.D. Sleeter. 1984. Effects of oil and oil dispersants on the skeletal growth of the hermatypic coral *Diplora strigosa*. *Coral Reefs* 3(4):191–198.

Dodge, R.E., B.J. Baca, A.H. Knap, S.C. Snedaker and T.D. Sleeter. 1995. *The Effects of Oil and Chemically Dispersed Oil in Tropical Ecosystems: 10 Years of Monitoring Experimental Sites.* Technical Report Series 95-014. Marine Spill Response Corporation, Washington, D.C.

Doron, P., L. Bertuccioli, J. Katz and T.R. Osborn. 2001. Turbulence characteristics and dissipation estimates in the coastal ocean bottom boundary layer from PIV data. *Journal of Physical Oceanography* 31(8):2108–2134.

Duesterloh, W., J.W. Short and M.G. Barron. 2002. Photoenhanced toxicity of weathered Alaska North Slope crude oil to the calanoid copepods *Calanus marshallae* and *Metridia okhotensis*. *Environmental Science and Technology* 36:3953–3959.

Duke, N.C. and K.A. Burns. 1999. *Fate and effects of oil and dispersed oil on mangrove ecosystems in Australia.* Report to the Australian Petroleum Production Exploration Association. Australian Institute of Marine Science, Townsville, Queensland, Australia and CRC Reef Research Centre, James Cook University, Townsville, Queensland, Australia. [Online] Available at: http://www.aims.gov.au/pages/research/mangroves/fae/fae01.html [February 14, 2005].

Duke, N.C., Z.S. Pinzon and M.C. Prada T. 1997. Large-scale damage to mangrove forests following two large oil spills in Panama. *Biotropica* 29:2–14.

Duke, N.C., K.A. Burns and O. Dalhaus. 1998a. Effects of oils and dispersed-oils on mangrove seedlings in planthouse experiments: a preliminary assessment of results two months after oil treatments. *Australian Petroleum Production and Exploration Association Limited Journal* 38:631–636.

Duke, N.C., K.A. Burns, J.C. Ellison, R.J. Rupp and O. Dalhaus. 1998b. Effects of oil and dispersed-oil mixtures on mature mangroves in field trials at Gladstone. *Australian Petroleum Production and Exploration Association Limited Journal* 38:637–645.

Duke, N.C., J.C. Ellison and K.A. Burns. 1998c. Surveys of oil spill incidents affecting mangrove habitat in Australia: a preliminary assessment of incidents, impacts on mangroves, and recovery of deforested areas. *Australian Petroleum Production and Exploration Association Limited Journal* 38:646–654.

Duke, N.C., K.A. Burns and R.P.J. Swannell. 1999. *Research into the bioremediation of oil spills in tropical Australia: with particular emphasis on oiled mangrove and salt marsh habitat.* Final Report to the Australian Maritime Safety Authority. Australian Institute of Marine Science, Townsville, Queensland, Australia and AEA Technology, Oxfordshire, United Kingdom.

Duke, N.C., K.A. Burns, R.P.J. Swannell, O. Dalhaus and R.J. Rupp. 2000. Dispersant use and a bioremediation strategy as alternate means of reducing impacts of large oil spills on Mangroves: The Gladstone Field Trials. *Marine Pollution Bulletin* 41:7–12, 403–412.

Dutka, B.J. and K.K. Kwan. 1984. Study of long term effects of oil and oil-dispersant mixtures on freshwater microbial populations in man made ponds. *Science of the Total Environment* 35:135–148.

Duval, W.S., L.A. Harwood and R.P. Fink. 1982. *The sublethal effects of dispersed oil on an estuarine isopod.* Technology Development Report, EPS-4-EC-82-1. Environment Canada, Ottawa, Ontario, Canada.

Ecological Steering Group on the Oil Spill in Shetland. 1994. *The Environmental Impact of the Wreck of the Braer.* Scottish Office, Edinburgh, United Kingdom.

Eganhouse, R.P. and J.A. Calder. 1976. The solubility of medium molecular weight aromatic hydrocarbons and the effects of hydrocarbon co-solutes and salinity. *Geochimica et Cosmochimica Acta* 40:555–561.

Elliott, A.J., N. Hurford and C.J. Penn. 1986. Shear diffusion and the spreading of oil slicks. *Marine Pollution Bulletin* 17(7):308–313.

Environment Canada. 2005. *Oil Properties Database.* Environment Canada, Ottawa, Ontario, Canada. [Online] Available at: http://www.etcentre.org/databases/spills_e.html [April 18, 2005].

Environmental Protection Agency (EPA). 1992. *Peer Review Workshop Report on a Framework for Ecological Risk Assessment.* U.S. Environmental Protection Agency, Risk Assessment Forum, Washington, D.C.

Environmental Protection Agency (EPA). 2001. *EPA requirements for quality assurance project plans.* EPA QA/R-5, EPA/240/B-01/003. U.S. Environmental Protection Agency, Office of Environmental Information, Washington D.C.

Environmental Protection Agency (EPA). 2002a. *Short-Term Methods for Estimating the Chronic Toxicity of Effluents and Receiving Waters to Freshwater Organisms.* EPA-821-R-02-013. U.S. Environmental Protection Agency, Washington D.C. [Online] Available at: http://www.epa.gov/waterscience/WET/disk3/ [April 27, 2005].

Environmental Protection Agency (EPA). 2002b. *Methods for Measuring the Acute Toxicity of Effluents and Receiving Waters to Freshwater and Marine Organisms.* EPA-821-R-02-012. U.S. Environmental Protection Agency, Washington D.C. [Online] Available at: http://www.epa.gov/waterscience/WET/disk2/ [April 27, 2005].

Environmental Protection Agency (EPA). 2003. Swirling flask dispersant effectiveness test. Pp. 224–229 in *Code of Federal Regulations Title 40: Protection of the Environment* (40 CFR). Pt. 300, Appendix C. Federal Register, Washington, D.C. [Online] Available at: http://www.epa.gov/epahome/cfr40.htm [March 23, 2005].

Environmental Protection Agency (EPA). 2005. *Terms of Environment.* Environmental Protection Agency, Washington, D.C. [Online] Available at: http://www.epa.gov/OCEPA terms/ [April 27, 2005].

Epstein, N., R.P.M. Bak and B. Rinkevich. 2000. Toxicity of third generation dispersants and dispersed Egyptian crude oil on Red Sea coral larvae. *Marine Pollution Bulletin* 40:497–503.

Etkin, D.S. 1999. *Oil spill dispersants: From technology to policy.* Cutter Information Corporation, Arlington, Massachusetts.

Exxon. 1992. *Exxon Oil Spill Response Manual.* Exxon Production Research Company, Houston, Texas.

Exxon Biomedical Sciences Incorporated. 1992. *Microtox® toxicity tests. Test material: Corexit 9527.* Technical report. ExxonMobile, East Millstone, New Jersey.

Exxon Biomedical Sciences Incorporated 1993a. *Mysid acute toxicity test. Flowthrough continuous exposure with* Mysidopsis bahia. *Test material: Corexit 9527.* Technical report. ExxonMobile, East Millstone, New Jersey.

Exxon Biomedical Sciences Incorporated. 1993b. *Mysid acute toxicity test. Continuous exposure with* Holmesimysis costata. *Test material: Corexit® 9527.* Technical report. ExxonMobile, East Millstone, New Jersey.

Exxon Biomedical Sciences Incorporated. 1993c. *Mysid acute toxicity test. Flowthrough continuous exposure with* Holmesimysis costata. *Test material: Corexit 9527.* Technical report. ExxonMobile, East Millstone, New Jersey.

Exxon Biomedical Sciences Incorporated. 1993d. *Fish acute toxicity test flow-through continuous exposure with* Menidia beryllina. *Test material: Corexit 9527.* Technical report. ExxonMobile, East Millstone, New Jersey.

ExxonMobil. 2000. *Dispersant Guidelines.* ExxonMobil Research and Engineering Company, Fairfax, Virginia.

Fannelop, T.K. and K. Sjoen. 1980. Hydrodynamics of underwater blowouts. *Norwegian Maritime Researh* 4:17–33.

Fay, J.A. 1969. The spread of oil slicks on a calm sea. Pp. 53–63 in *Oil on the Sea,* D.P. Hoult, ed. Plenum Press, New York, New York.

Fiest, D.L. and P.D. Boehm. 1980. Subsurface distributions of petroleum from an offshore well blowout, Bay of Campeche. Pp. 169–185 in *Symposium on Preliminary Results from the September 1979 Researcher/Pierce IXTOC I Cruise.* Key Biscayne, Florida. National Oceanic and Atmospheric Administration Publications Office, Boulder, Colorado.

Fingas, M. 1985. The effectiveness of oil spill dispersants. *Spill Technology Newsletter* 10(4-6):47–64.

Fingas, M. 1996. The evaporation of crude oil and petroleum products. Ph.D. Thesis, McGill University, Montreal, Canada.

Fingas, M. 1997. The evaporation of oil spills: Prediction of equations using distillation data. Pp. in 1–20 *Proceedings of the Twentieth Arctic and Marine Oil Spill Program (AMOP) Technical Seminar,* Vancouver, British Columbia, Canada. Environment Canada, Ottawa, Ontario, Canada.

Fingas, M. 1999a. The evaporation of oil spills: Development and implementation of new prediction methodology. Pp. 281–287 in *Proceedings of the 1999 International Oil Spill Conference,* Seattle, Washington. American Petroleum Institute, Washington, D.C.

Fingas, M. 1999b. *In-Situ* Burning of Oil Spills: A Historical Perspective. Pp. 55–65 in *Workshop Proceedings* In-Situ *Burning of Oil Spills,* New Orleans, Louisiana. NIST Special Publication 935. National Institute of Standards and Technology, Gaithersburg, Maryland.

Fingas, M. 2002a. *A review of literature related to oil spill dispersants especially relevant to Alaska.* Report prepared for the Prince William Sound Regional Citizens' Advisory Council. Prince William Sound Regional Citizens' Advisory Council, Anchorage, Alaska.

Fingas, M. 2002b. *A white paper on oil spill dispersant field testing.* Report prepared for the Prince William Sound Regional Citizens' Advisory Council. Prince William Sound Regional Citizens' Advisory Council, Anchorage, Alaska.

Fingas, M. 2003. *Review of monitoring protocols for dispersant effectiveness.* Report prepared for the Prince William Sound Regional Citizens' Advisory Council. Prince William Sound Regional Citizens' Advisory Council, Anchorage, Alaska.

Fingas, M. 2004a. Dispersant tank testing—A review of procedures and considerations. Pp. 1003–1016 in *Proceedings of the Twenty-Seventh Arctic Marine Oilspill Program (AMOP) Technical Seminar,* Edmonton, Alberta, Canada. Environment Canada, Ottawa, Ontario, Canada.

Fingas, M. 2004b. Energy and work input in laboratory vessels. Pp. 1–18 in *Proceedings of the Twenty-Seventh Arctic Marine Oilspill Program (AMOP) Technical Seminar,* Edmonton, Alberta, Canada. Environment Canada, Ottawa, Ontario, Canada.

Fingas, M. and B. Fieldhouse. 1994. Studies of water-in-oil emulsions and techniques to measure emulsion treating agents. Pp. 233–244 in *Proceedings of the Seventeenth Arctic and Marine Oilspill Program (AMOP) Technical Seminar,* Vancouver, British Columbia, Canada. Environment Canada, Ottawa, Ontario, Canada.

Fingas, M. and B. Fieldhouse. 2003. Studies of the formation process of water-in-oil emulsions. *Marine Pollution Bulletin* 47:369–396.

Fingas, M. and B. Fieldhouse. 2004a. Modeling of water-in-oil emulsions. Pp. 335–350 in *Proceedings of the Twenty-Seventh Arctic Marine Oilspill Program (AMOP) Technical Seminar,* Edmonton, Alberta, Canada. Environment Canada, Ottawa, Ontario, Canada.

Fingas, M. and B. Fieldhouse. 2004b. Formation of water-in-oil emulsions and application to oil spill modeling. *Journal of Hazardous Materials* 107:37–50.

Fingas, M. and L. Ka'aihue. 2004a. Weather windows for oil spill countermeasures. Pp. 881–955 in *Proceedings of the Twenty-Seventh Arctic Marine Oilspill Program (AMOP) Technical Seminar*, Edmonton, Alberta, Canada. Environment Canada, Ottawa, Ontario, Canada.

Fingas, M. and L. Ka'aihue. 2004b. Dispersant field testing—a review of procedures and considerations. Pp. 1017–1046 in *Proceedings of the Twenty-Seventh Arctic Marine Oilspill Program (AMOP) Technical Seminar*, Edmonton, Alberta, Canada. Environment Canada, Ottawa, Ontario, Canada.

Fingas, M. and L. Ka'aihue. 2004c. Dispersant tank testing—a review of procedures and considerations. Pp. 1003–1016 in *Proceedings of the Twenty-Seventh Arctic Marine Oilspill Program (AMOP) Technical Seminar*, Edmonton, Alberta, Canada. Environment Canada, Ottawa, Ontario, Canada.

Fingas, M. and N. Laroche. 1991. *An Introduction to* In-Situ *Burning of Oil Spills.* Paper presented at *In-Situ* Burning Workshop, Sacramento, California, May 21–22, 1991, unpublished.

Fingas, M. and M. Punt. 2000. In-situ *burning: a cleanup technique for oil spills on water.* Environment Canada Special Publication, Ottawa, Ontario, Canada.

Fingas, M. and M. Sydor. 1980. *Development of an Oil Spill model for the St. Lawrence River.* Technical Bulletin No. 116. Environment Canada, Inland Waters Directorate, Water Planning and Management Branch, Ottawa, Ontario, Canada.

Fingas, M.F., D.L. Munn, B. White, R.G. Stoodley and I.D. Crerar. 1989. Laboratory testing of dispersant effectiveness: the importance of oil-to-water ratio and settling time. Pp. 365–373 in *Proceedings of the 1989 International Oil Spill Conference*, San Antonio, Texas. American Petroleum Institute, Washington, D.C.

Fingas, M.F., B. Kolokowski and E.J. Tennyson. 1990. Study of oil spill dispersants effectiveness and physical studies. Pp. 265–287 in *Proceedings of the Thirteenth Arctic Marine Oil Spill Program (AMOP) Technical Seminar*. Edmonton, Alberta, Canada. Environment Canada, Ottawa, Ontario, Canada.

Fingas, M., I. Bier, M. Bobra and S. Callaghan. 1991. Studies on the physical and chemical behavior of oil and dispersant mixtures. Pp. 411–414 in *Proceedings of the 1991 Oil Spill Conference*, San Diego, California. American Petroleum Institute, Washington, D.C.

Fingas, M.F., D.A. Kyle, Z. Wang and F. Ackerman. 1994. Testing of oil spill dispersant effectiveness in the laboratory. Pp. 905–933 in *Proceedings of the Seventeenth Arctic and Marine Oilspill Program (AMOP) Technical Seminar*, Vancouver, British Columbia, Canada. Environment Canada, Ottawa, Ontario, Canada.

Fingas, M., B. Fieldhouse and J.V. Mullin. 1995a. Water-in-oil emulsions: How they are formed and how they are broken. Pp. 829–830 in *Proceedings of the 1995 International Oil Spill Conference*, Long Beach, California. American Petroleum Institute, Washington, D.C.

Fingas, M., B. Fieldhouse, L. Gamble and J.V. Mullin. 1995b. Studies of water-in-oil emulsions: Stability, classes, and measurement. Pp. 21–42 in *Proceedings of the Seventeenth Arctic and Marine Oilspill Program (AMOP) Technical Seminar*, Vancouver, British Columbia, Canada. Environment Canada, Ottawa, Ontario, Canada.

Fingas, M.F., D.A. Kyle, P. Lambert, Z. Wang and J.V. Mullin. 1995c. Analytical procedures for measuring oil spill dispersant effectiveness in the laboratory. Pp. 339–354 in *Proceedings of the Eighteenth Arctic and Marine Oilspill Program (AMOP) Technical Seminar*, Edmonton, Alberta, Canada. Environment Canada, Ottawa, Ontario, Canada.

Fingas, M.F., D. Kyle and E. Tennyson. 1995d. Dispersant effectiveness: Studies into the causes of effectiveness variations. Pp. 92–132 in *The Use of Chemicals in Oil Spill Re-*

sponse, ASTM STP 1252, P. Lane, Ed. American Society for Testing and Materials, Philadelphia, Pennsylvania.

Fingas, M.F., E. Huang, B. Fieldhouse, L. Wang and J.V. Mullin. 1996a. The effect of energy, settling time and shaking time on the swirling flask dispersant apparatus. *Spill Science and Technology Bulletin* 3(4):193–194.

Fingas, M., B. Fieldhouse and J.V. Mullin. 1996b. Studies of water-in-oil emulsions: The role of asphaltenes and resins. Pp. 73–88 in *Proceedings of the Nineteenth Arctic and Marine Oilspill Program (AMOP) Technical Seminar*, Calgary, Alberta, Canada. Environment Canada, Ottawa, Ontario, Canada.

Fingas, M., B. Fieldhouse and J.V. Mullin. 1998. Studies of water-in-oil emulsions: Stability and oil properties. Pp. 1–25 in *Proceedings of the Twenty-First Arctic and Marine Oil Spill Program (AMOP) Technical Seminar*, Edmonton, Alberta, Canada. Environment Canada, Ottawa, Ontario, Canada.

Fingas, M., B. Fieldhouse and J. Mullin. 1999. Studies of water-in-oil emulsions: Energy threshold on emulsion formation. Pp. 57–68 in *Proceedings of the Twenty-Second Arctic and Marine Oil Spill Program (AMOP) Technical Seminar*, Calgary, Alberta, Canada. Environment Canada, Ottawa, Ontario, Canada.

Fingas, M., B. Fieldhouse, J. Lane, and J. Mullin. 2000a. Studies of water-in-oil emulsions: Long-term stability, oil properties, and emulsions formed at sea. Pp. 145–160 in *Proceedings of the Twenty-Third Arctic and Marine Oil Spill Program (AMOP) Technical Seminar*, Vancouver, British Columbia, Canada. Environment Canada, Ottawa, Ontario, Canada.

Fingas, M., B. Fieldhouse, J. Lane and J. Mullin. 2000b. Studies of water-in-oil emulsions: Energy and work threshold for emulsion formation. Pp. 19–36 in *Proceedings of the Twenty-Third Arctic and Marine Oil Spill Program (AMOP)* Technical Seminar, Vancouver, British Columbia, Canada. Environment Canada, Ottawa, Ontario, Canada.

Fingas, M.F., B. Fieldhouse, P. Lambert, Z. Wang, J. Noonan, J. Lane and J.V. Mullin. 2002a. Water-in-oil emulsions formed at sea, in test tanks, and in the laboratory. Environment Canada Manuscript Report EE-169. Environment Canada, Ottawa, Ontario, Canada.

Fingas, M.F., B. Fieldhouse, J. Noonan, P. Lambert, J. Lane and J. Mullin. 2002b. Studies of water-in-oil emulsions: testing of emulsion formation in OHMSETT, year II. Pp. 29–44 in *Proceedings of Twenty-Fifth Arctic and Marine Oilspill Technical Seminar*, Calgary, Alberta, Canada. Environment Canada. Ottawa, Ontario, Canada.

Fingas, M.F., B. Fieldhouse and Z. Wang. 2003a. The long term weathering of water-in-oil emulsions. *Spill Science and Technology Bulletin* 8(2) 137–143.

Fingas, M., Z. Wang, B. Fieldhouse and P. Smith. 2003b. The correlation of chemical characteristics of an oil to dispersant effectiveness. Pp. 679–730 in *Proceedings of the Twenty-Sixth Arctic and Marine Oilspill Program (AMOP) Technical Seminar*, Victoria, British Columbia, Canada. Environment Canada, Ottawa, Ontario, Canada.

Fiocco, R., P.S. Daling, G. DeMarco, R.R. Lessard and G.P. Canevari. 1999. Chemical dispersibility of heavy Bunker Fuel oil. Pp. 173–186 in *Proceedings of the Twenty-Second Arctic and Marine Oil Spill Program (AMOP) Technical Seminar*, Calgary, Alberta, Canada. Environment Canada, Ottawa, Ontario, Canada.

Fischer, H.B., E.J. List, R.C.Y. Koh, J. Imberger and N.H. Brooks. 1979. *Mixing in Inland and Coastal Waters*, Academic Press, New York, New York.

Foght, J.M. and D.W.S. Westlake. 1982. Effect of the dispersant Corexit 9527 on the microbial degradation of Prudhoe Bay oil. *Canadian Journal of Microbiology* 28:117–122.

Foght, J.M., P.M. Fedorak and D.W.S. Westlake. 1983. Effect of the dispersant Corexit 9527 on the microbial degradation of sulfur heterocycles in Prudhoe Bay oil. *Canadian Journal of Microbiology* 61:623–627.

Foght, J.M., N.J. Fairbairn and D.W.S. Westlake. 1987. Effect of oil dispersants on microbially-mediated processes in freshwater systems. Pp. 252–263 in *Oil in Freshwater: Chemistry,*

Biology, Countermeasure Technology, J.H. Vandermeulen and S.E. Hrudey. Pergamon Press, Oxford, England.

Foght, J.M., D.L. Gutnick and D.W.S. Westlake. 1989. Effect of emulsan on biodegradation of crude oil by pure and mixed bacterial cultures. *Applied and Environmental Microbiology* 55(1):36–42.

Foy, M.G. 1982. *Acute lethal toxicity of Prudhoe Bay Crude oil and Corexit 9527 to Arctic marine fish and invertebrates.* Technology Development Report, EPS 4-EC-82-3. Environment Canada, Ottawa, Ontario, Canada.

Franklin, F.L. and R. Lloyd. 1986. The relationship between oil droplet size and the toxicity of dispersant/oil mixtures in the Standard MAFF "Sea" test. *Oil & Chemical Pollution* 3:37–52.

Franks, F. 1966. Solute-water interactions and the solubility behavior of long-chain paraffin hydrocarbons. *Nature* 210:87–88.

French-McCay, D.P. 1998. Modeling the Impacts of the North Cape Oil Spill. Pp. 378–430 in *Proceedings of the Twenty-first Arctic and Marine Oilspill Program (AMOP) Technical Seminar*, Edmonton, Alberta, Canada. Environment Canada, Ottawa, Ontario, Canada.

French-McCay, D.P. 2001. *Development and application of an oil toxicity and exposure model, OilToxEx.* Report prepared for the National Oceanic and Atmospheric Administration, Damage Assessment Center. National Oceanic and Atmospheric Administration, Silver Spring, Maryland.

French-McCay, D.P. 2002. Development and application of an oil toxicity and exposure model, OilToxEx. *Journal of Environmental Toxicology and Chemistry* 21:2080–2094.

French-McCay, D.P. 2003. Development and Application of Damage Assessment Modeling: Example Assessment for the North Cape Oil Spill. *Marine Pollution Bulletin* 47(9-12):341–359.

French-McCay, D.P. 2004. Oil spill impact modeling: Development and validation. *Journal of Environmental Toxicology and Chemistry* 23(10):2441–2456.

French-McCay, D.P. and J.R. Payne. 2001. Model of oil fate and water concentrations with and without application of dispersants. Pp. 611–645 in *Proceedings of the Twenty-Fourth Arctic and Marine Oilspill (AMOP) Technical Seminar*, Edmonton, Alberta, Canada. Environment Canada, Ottawa, Ontario, Canada.

French-McCay, D.P. and H. Rines. 1997. Validation and use of spill impact modeling for impact assessment. Pp. 829–834 in *Proceedings of the 1997 International Oil Spill Conference*, Fort Lauderdale, Florida. American Petroleum Institute, Washington, D.C.

French-McCay, D.P., M. Reed, K. Jayko, S. Feng, H. Rines, S. Pavignano, T. Isaji, S. Puckett, A. Keller, F.W. French III, D. Gifford, J. McCue, G. Brown, E. MacDonald, J. Quirk, S. Natzke, R. Bishop, M. Welsh, M. Phillips and B.S. Ingram. 1996. *The CERCLA Type A Natural Resource Damage Assessment Model for Coastal and Marine Environments (NRDAM/CME), Technical Documentation, Volume I–VI.* Contract No. 14-0001-91-C-11. Final Report, submitted to the Office of Environmental Policy and Compliance, U.S. Department of the Interior, Washington, D.C.

Fucik, K.W., T.J. Bright and K.S. Goodman. 1984. Measurements of damage, recovery, and rehabilitation of coral reefs exposed to oil. Pp. 115–133 in *Restoration of Habitats Impacted by Oil Spills*, J. Cairns and A.L. Buikema, eds. Butterworth Press, London, England.

Fucik, K.W., K.A. Carr and B.J. Balcom. 1994. *Dispersed oil toxicity tests with biological species indigenous to the Gulf of Mexico.* OCS Study, MMS 94-0021, Report submitted to the Minerals Management Service, Gulf of Mexico Outer Continental Shelf (OCS) Region, New Orleans, Louisiana.

Fucik, K.W., K.A. Carr and B.J. Balcom. 1995. Toxicity of oil and dispersed oil to the eggs and larvae of seven marine fish and invertebrates from the Gulf of Mexico. Pp. 135–171 in *The Use of Chemicals in Oil Spill Response*, P. Lane, Ed. American Society for Testing and Materials, Philadelphia, Pennsylvania.

Fuller, C. and J.S. Bonner. 2001. Comparative Toxicity of Oil, Dispersant and Dispersed Oil to Texas Marine Species. Pp. 1243–1248 in *Proceedings of the 2001 International Oil Spill Conference*, Tampa, Florida. American Petroleum Institute, Washington, D.C.

Fuller, C., J. Bonner, T. McDonald, C. Page, G. Bragin, J. Clark, D. Aurand, A. Hernandez and A. Ernest. 1999. Comparative toxicity of simulated beach sediments impacted with both whole and chemical dispersions of weathered Arabian Medium crude oil. Pp. 659–670 in *Proceedings of the Twenty-Second Arctic and Marine Oil Spill Program (AMOP) Technical Seminar*, Calgary, Alberta, Canada. Environment Canada, Ottawa, Ontario, Canada.

Fuller, C., J.S. Bonner, M.C. Sterling, T.O. Ojo and C.A. Page. 2003. Field instruments for real time *in-situ* crude oil concentration measurements. Pp. 755–764 in *Proceedings of the Twenty-Sixth Arctic and Marine Oilspill Program (AMOP) Technical Seminar*, Victoria, British Columbia, Canada. Environment Canada, Ottawa, Ontario, Canada.

Gagnon, R.M. and D.A. Holdaway. 2000. EROD induction and biliary metabolite excretion following exposure to the water accommodated fraction of crude oil and to chemically dispersed crude oil. *Archives of Environmental Contamination and Toxicology* 38(1):70–77.

Galt, J.A. 1995. The Integration of Trajectory Models and Analysis into Spill Response Systems: The Need for a Standard. Pp. 499–507 in *Proceedings of the Second International Oil Spill Research and Development Forum*, London England. International Maritime Organization (IMO), London, England.

Garcia, J.M., L.Y. Wick and H. Harms. 2001. Influence of the nonionic surfactant Brij 35 on the bioavailability of solid and sorbed dibenzofuran. *Environmental Science & Technology* 35(10):2033–2039.

Garrett, R.M., I.J. Pickering, E.H. Copper and R.C. Prince. 1998. Photooxidation of polycyclic aromatic hydrocarbons in crude oils. Pp. 99–114 in *Proceedings of the Twenty-First Arctic and Marine Oilspill Program (AMOP) Technical Seminar*, Edmonton, Alberta, Canada. Environment Canada, Ottawa, Ontario, Canada.

Gearing, P.J. and J.N. Gearing. 1982a. Behavior of No. 2 fuel oil in the water column of controlled ecosystems. *Marine Environmental Research* 6:115–132.

Gearing, P.J. and J.N. Gearing. 1982b. Transport of No. 2 fuel oil between water column, surface microlayer, and atmosphere in controlled ecosystems. *Marine Environmental Research* 6:133–143.

Gearing, P.J., J.N. Gearing, R.J. Pruell, T.L. Wade and J.G. Quinn. 1980. Partitioning of No. 2 fuel oil in controlled ecosystems: Sediments and suspended particulate matter. *Environmental Science and Technology* 14:1129–1136.

George, S.E., G.M. Nelson, M.J. Kohan, S.H. Warren, B.T. Eischen and L.R. Brooks. 2001. Oral treatment of Fischer 344 rats with weathered crude oil and a dispersant influences intestinal metabolism and microbiota. *Journal of Toxicology and Environmental Health, Part A* 63:297–316.

George-Ares, A. and J.R. Clark. 2000. Aquatic toxicity of two Corexit® dispersants. *Chemosphere* 40:897–906.

George-Ares, A., J.R. Clark, G.R. Biddinger and M.L. Hinman. 1999. Comparison of Test Methods and Early Toxicity Characterization for Five Dispersants. *Ecotoxicology and Environmental Safety* 42:138–142.

Georgiades, E.T., D.A. Holdaway, S.E. Brennan, J.S. Butty and A. Temara. 2003. The impact of oil-derived products on the behaviour and biochemistry of the eleven-armed asteroid *Coscinasterias muricata* (Echinodermata). *Marine Environmental Research* 55:257–276.

Getter, C.D. and T.G. Ballou. 1985. Field experiments on the effects of oil and dispersant on mangroves. Pp. 577–582 in *Proceedings of the 1985 International Oil Spill Conference*, Los Angeles, California. American Petroleum Institute, Washington, D.C.

Gilfillan, E.S., D.S. Page and J.C. Foster. 1986. *Tidal area dispersant project: Fate and Effects of Chemically Dispersed Oil in the Nearshore Benthic Environment.* Final Report, Publication Number 4440. American Petroleum Institute, Washington, D.C.

Goodman, R.H. 2003. Is SMART Really that Smart. Pp. 779–786 in *Proceedings of the Twenty-Sixth Arctic and Marine Oilspill Program (AMOP) Technical Seminar,* Victoria, British Columbia, Canada. Environment Canada, Ottawa, Ontario, Canada.

Goodman, R.H. and M.F. Fingas. 1988. The use of remote sensing in the determination of dispersant effectiveness. *Spill Technology Newsletter* 13(3):55–58.

Goodman, R.H. and M.R. MacNeill. 1984. The use of remote sensing in the determination of dispersant effectiveness. Pp. 143–160 in *Oil Spill Chemical Dispersants: Research, Experience, and Recommendations,* T.E. Allen, Ed. American Society for Testing and Materials, Philadelphia, Pennsylvania.

Gordon, D.C., Jr., P.D. Keizer and N.J. Prouse. 1973. Laboratory studies of the accommodation of some crude and residual fuel oils in sea water. *Journal of the Fisheries Research Board of Canada* 30:1611–1618.

Gould, J.R. and J. Lindstedt-Siva. 1991. Santa Barbara to Mega Borg and beyond: A review of API's spill program and priorities. Pp. 341–352 in *Proceedings of the 1991 International Oil Spill Conference,* San Diego, California. American Petroleum Institute, Washington, D.C.

Green, D.R., J. Buckley and B. Humphrey. 1982. *Fate of chemically dispersed oil in the sea: A report on two field experiments.* Environment Canada Report EPS 4-EC-82-5. Environment Canada, Ottawa, Ontario, Canada.

Gregory, C.L., A.A. Allen and D.H. Dale. 1999. Assessment of potential oil spill recovery capabilities. Pp. 527–534 in *Proceedings of the 1999 International Oil Spill Conference,* Seattle, Washington. American Petroleum Institute, Washington, D.C.

Grothe, D.R., K.L. Dickson, D.K. Reed-Judkins, Eds. 1996. *Whole Effluent Toxicity Testing: An Evaluation of Methods and Prediction of Receiving System Impacts.* Society of Environmental Toxicology and Chemistry Press, Pensacola, Florida.

Gugg, P.M., C.B. Henry and S.P. Glenn. 1999. Proving dispersants work. Pp. 1007–1010 in *Proceedings of the 1999 International Oil Spill Conference,* Seattle, Washington. American Petroleum Institute, Washington, D.C.

Gulec, I. and D.A. Holdaway. 2000. Toxicity of crude oil and dispersed crude oil to ghost shrimp *Palaemon serenus* and larvae of Australian Bass *Macquaria novemaculeata. Environmental Toxicology* 15:91–98.

Gulec, I., B. Leonard and D.A. Holdaway. 1997. Oil and Dispersed Oil Toxicity to Amphipods and Snails. *Spill Science and Technology Bulletin* 4:1–6.

Guyomarch, J., F.X. Merlin and S. Colin. 1999a. Study of the feasibility chemical dispersion of viscous oils and water-in-oil emulsions. Pp. 219–230 in *Proceedings of the Twenty-Second Arctic and Marine Oil Spill Program (AMOP) Technical Seminar,* Calgary, Alberta, Canada. Environment Canada, Ottawa, Ontario, Canada.

Guyomarch, J., F.-X. Merlin and P. Bernanose. 1999b. Oil interaction with mineral fines and chemical dispersion: behaviour of the dispersed oil in coastal or estuarine conditions. Pp. 137–149 in *Proceedings of the Twenty-Second Arctic and Marine Oil Spill Program (AMOP) Technical Seminar,* Calgary, Alberta, Canada. Environment Canada, Ottawa, Ontario, Canada.

Guyomarch, J., O. Kerfourn and F.X. Merlin. 1999c. Dispersants and demulsifiers: studies in the laboratory, harbor and polludrome. Pp. 195–202 in *Proceedings of the 1999 International Oil Spill Conference,* Seattle, Washington. American Petroleum Institute, Washington, D.C.

Guyomarch, J., S. Le Floch and F.-X. Merlin. 2002. Effect of suspended mineral load, water salinity and oil type on the size of oil–mineral aggregates in the presence of chemical dispersant. *Spill Science & Technology Bulletin* 8(1):95–100

Hammond, T.M., C. Pattiaratchi, D. Eccles, M. Osborne, L. Nash and M. Collins. 1987. Ocean Surface Current Radar (OSCR) Vector Measurements on Inner Continental Shelf. *Continental Shelf Research* 7:411–431.

Harris, C. 1997. The Sea Empress incident: overview and response at sea. Pp. 177–184 in *Proceedings of the 1997 International Oil Spill Conference*, Fort Lauderdale, Florida. American Petroleum Institute, Washington, D.C.

Hartwick, E.B., R.S.S. Wu and D.B. Parker. 1982. Effects of a crude oil and an oil dispersant Corexit 9527 on populations of the littleneck clam *Protothaca staminea*. *Marine Environmental Research* 6:291–306.

Henrichs, S., M. Luoma and S. Smith. 1997. *A study of the adsorption of aromatic hydrocarbons by marine sediments.* Final Report OCS Study MMS 97-0002. Report submitted to the Minerals Management Service, Gulf of Mexico Outer Continental Shelf (OCS) Region, New Orleans, Louisiana.

Henry, C. 2004. *Response goals and the role of dispersants.* Presentation to the National Research Council Committee on Understanding Oil Spill Dispersants: Efficacy and Effects, March 15–16, 2004, Washington, D.C., unpublished. National Research Council, Washington, D.C.

Henry, C. and P.O. Roberts. 2001. Background fluorescence values and matrix effects observed using SMART protocols in the Atlantic Ocean and Gulf of Mexico. Pp. 1203–1207 in *Proceedings of the 2001 International Oil Spill Conference*, Tampa, Florida. American Petroleum Institute, Washington, D.C.

Henry, C.B., P.O. Roberts and E.B. Owens. 1999. A primer on *in-situ* fluorimetry to monitor dispersed oil. Pp. 225–228 in *Proceedings of the 1999 International Oil Spill Conference*, Seattle, Washington. American Petroleum Institute, Washington, D.C.

Herbes, S.E. 1977. Partitioning of polycyclic aromatic hydrocarbons between dissolved and particulate phases in natural waters. *Water Research* 11:493–496.

Hill, P.S., A. Khelifa and K. Lee. 2003. Time scale for oil droplet stabilization by mineral particles in turbulent suspensions. *Spill Science and Technology Bulletin* 8(1):73–82.

Hillman, S.O., S.D. Hood, M.T. Bronson and G. Shufelt. 1997. Dispersant field monitoring procedures. Pp. 521–539 in *Proceedings of the Twentieth Arctic and Marine Oil Spill Program (AMOP) Technical Seminar*, Vancouver, British Columbia, Canada. Environment Canada, Ottawa, Ontario, Canada.

Ho, K., L. Patton, J.S. Latimer, R.J. Pruell, M. Pelletier, R. McKinney and S. Jayaraman. 1999. The chemistry and toxicity of sediment affected by oil from the North Cape spilled into Rhode Island Sound. *Marine Pollution Bulletin* 38:314–323.

Hodgins, D.O., R.H. Goodman and M.F. Fingas. 1993. Forcasting Surface Current Measured with HF Radar. Pp. 1083–1094 in *Proceedings of the Sixteenth Arctic and Marine Oil Spill Program (AMOP) Technical Seminar*, Calgary, Canada. Environment Canada. Ottawa, Ontario, Canada.

Hodson, P.V., D.G. Dixon and K.L.E. Kaiser. 1988. Estimating the Acute Toxicity of Waterborne Chemicals in Trout from Measurements of Median Lethal Dose and the Octanol-Water Partition Coefficient. *Journal of Environmental Toxicology and Chemistry* 7:443–454.

Hoffman, E.J. and J.G. Quinn. 1978. A comparison of Argo Merchant oil and sediment hydrocarbons from Nantucket Shoals. Pp. 80–88 in *In the wake of the Argo Merchant—Proceedings of a Symposium*. University of Rhode Island, Center for Ocean Management Studies, Kingston, Rhode Island.

Hoffman, E.J. and J.G. Quinn. 1979. Gas chromatographic analysis of Argo Merchant oil and sediment hydrocarbons at the wreck site. *Marine Pollution Bulletin* 10:20–24.

Hokstad, J.N., B. Knudsen and P.S. Daling. 1996. *Oil-surfactant interaction and mechanism studies—Part 1: Leaching of surfactants form oil to water. Chemical composition of dispersed oil.* IKU No. 22.2043.00/21/95, ESCOST report No. 21. SINTEF Group report to Esso Norge A.S. SINTEF Group, Trondheim, Norway.

Hoult, D.P. 1972. Oil spreading on the sea. *Annual Review of Fluid Mechanics* 4:341–367.

Howlett, E., K. Jayko and M. Spaulding. 1993. Interfacing Real-Time Information with OILMAP. Pp. 539–548 in *Proceedings of the Sixteenth Arctic and Marine Oil Spill Program (AMOP) Technical Seminar*, Calgary, Canada. Environment Canada. Ottawa, Ontario, Canada.

Huang, J.C. and F.C. Monastero. 1982. *Review of the state-of-the-art of oil spill simulation models.* Final report submitted by Raytheon Ocean Systems Company, East Providence, Rhode Island to the American Petroleum Institute, Washington, D.C. American Petroleum Institute, Washington, D.C.

Huang, W. and M. Spaulding. 1995. 3D model of estuarine circulation and water quality induced by surface discharges. *Journal of Hydraulic Engineering*, 121(4):300–311.

Humphrey, B., P.D. Boehm, M.C. Hamilton and R.J. Nordstrom. 1987. The fate of chemically dispersed and untreated crude oil in Arctic oil spill. *Arctic* 40 (Supplement 1):149–161.

Hurlbert, S.H. 1984. Pseudoreplication and the design of ecological field experiments. *Ecological Monographs* 54(2):187–211.

Ichiye, T. 1967. Upper ocean boundary-layer flow determined by dye diffusion. *Physics of Fluids* 10:270–277.

Inchcape Testing Services. 1995. *Laboratory test data for Corexit 9500 and Corexit 9527.* Inchcape Testing Services, Houston, Texas.

International Petroleum Industry Environmental Conservation Association (IPIECA). 2000. *Choosing Spill Response Options to Minimize Damage—Net Environmental Benefit Analysis.* IPIECA Report Series, Volume 10. International Petroleum Industry Environmental Conservation Association, London, England.

Jahns, H.O., J.R. Bragg, L.C. Dash and E.H. Owens. 1991. Natural Cleaning of Shorelines Following the Exxon Valdez Spill. Pp. 167–176 in *Proceedings of the 1991 International Oil Spill Conference*, San Diego, California. American Petroleum Institute, Washington, D.C.

Jasper, W.L., T.J. Kim and M.P. Wilson. 1978. Drop Size Distributions in a Treated Oil-Water System. Pp. 203–216 in *Chemical Dispersants for the Control of Oil Spills*, L.T. McCarthy, G.P. Lindblom and H.F. Walter, Eds. American Society for Testing and Materials, Philadelphia, Pennsylvania.

Jenssen, B.M. 1994. Review article: effects of oil pollution, chemically treated oil, and cleaning on the thermal balance of birds. *Environmental Pollution* 86:207–215.

Jenssen, B.M. and M. Ekker. 1991. Effects of plumage contamination with crude oil dispersant mixtures on thermoregulation in common eiders and mallards. *Archives of Environmental Contamination and Toxicology* 20:398–403.

Jezequel, R., S. Lefloch, F.-X. Merlin, J. Drewes and K. Lee. 1998. The influence of microorganisms on oil-mineral fine interactions in low energy coastal environment: Preliminary results. Pp. 957–962 in *Proceedings of the Twenty-first Arctic and Marine Oilspill Program (AMOP) Technical Seminar*, Edmonton, Alberta, Canada. Environment Canada, Ottawa, Ontario, Canada.

Jezequel, R., F.-X. Merlin and K. Lee. 1999. The influence of microorganisms on oil-mineral fine interactions in low energy coastal environment. Pp. 771–775 in *Proceedings of the 1999 International Oil Spill Conference*, Seattle, Washington. American Petroleum Institute, Washington, D.C.

Johannes, R.E. 1975. Pollution and degradation of coral reef communities. Pp. 13–51 in *Tropical Marine Pollution*, E.J. Ferguson Wood and R.E. Johannes, Eds. Elsevier, New York, New York.

Johansen, O. 1984. The Halten Bank experiment—observations and model studies of drift and fate of oil in the marine environment, Pp. 18–36 in *Proceedings of the Eleventh Arctic and Marine Oilspill Program (AMOP) Technical Seminar*, Edmonton, Alberta, Canada. Environment Canada, Ottawa, Ontario, Canada.

Johansen, O. 2000. DeepBlow—a Lagrangian plume model for deep water blowouts. *Spill Science and Technology Bulletin* 6(2):103–111.

Johansen, O. and I.M. Carlsen. 2002. *Assessment of methods for dispensing dispersant into subsea blowouts.* SINTEF Applied Chemistry, Trondheim, Norway.

Johansson, S.U., U. Larsson and P.D. Boehm. 1980. The Tsesis oil spill impact on the pelagic ecosystem. *Marine Pollution Bulletin* 11:284–293.

Jones, R.K. 1996. Method for estimating boiling temperatures of crude oils. *Journal of Environmental Engineering* 122(8):761–763.

Jones, R.K. 1997. A Simplified Pseudo-Component of Oil Evaporation Model. Pp. 43–61 in *Proceedings of the Twentieth Arctic and Marine Oil Spill Program (AMOP) Technical Seminar*, Vancouver, British Columbia, Canada. Environment Canada, Ottawa, Ontario, Canada.

Jordan, R.E. and J.R. Payne. 1980. *Fate and Weathering of Petroleum Spills in the Marine Environment: A Literature Review and Synopsis.* Ann Arbor Science Publishers, Incorporated, Ann Arbor, Michigan.

Kaku, V.J., M.C. Boufadel and A.D. Venosa. 2002. Evaluation of the mixing energy in the EPA flask tests for dispersants effectiveness. Pp. 211–218 in *Oil Spills 2002*. Wessex Institute of Technology, Ashurst, Southampton, United Kingdom. [Online] Available at: http://www.temple.edu/environment/Papers/kakugreece.pdf [December 20, 2004].

Karickhoff, S.W. 1981. Semi-empirical estimation of sorption of hydrophobic pollutants on natural sediments and soils. *Chemosphere* 10:833–846.

Kaser, R.M., J. Gahn and C. Henry. 2001. Blue Master: Use of Corexit 9500 to Disperse IFO 180 Spill. Pp. 815–819 in *Proceedings of the 2001 International Oil Spill Conference*, Tampa, Florida. American Petroleum Institute, Washington, D.C.

Kennedy, C.J., N.J. Gassman and P.J. Walsh. 1992. The fate of benzo[a]pyrene in the scleractinian corals *Favia fragrum* and *Montastrea annularis*. *Marine Biology* 112:313–318

Kennicutt, M.C. II, T.L. Wade, N.L. Guinasso, Jr. and J.M. Brooks. 1991. The Mega Borg Incident: A comparison of response, mitigation and impact. Pp. 275–280 in *Proceedings of the 1991 Offshore Technology Conference*, Houston, Texas. Offshore Technology Conference, Houston, Texas.

Kerr, C.L. and C.S. Barrientos, Eds. 1979. *Workshop on the physical behavior of oil in the marine environment.* Princeton University Press, Princeton, New Jersey.

Khelifa, A., P. Stoffyn-Egli, P.S. Hill and K. Lee. 2002. Characteristics of oil droplets stabilized by mineral particles: effect of oil types and temperature. *Spill Science and Technology Bulletin* 8(1):19–30.

Khelifa, A., P.S. Hill, L.O. Ajijolaiya and K. Lee. 2004. Modeling the effect of sediment size on OMA formation. Pp. 383–395 in *Proceedings of the Twenty-Seventh Arctic Marine Oilspill Program (AMOP) Technical Seminar*, Edmonton, Alberta, Canada. Environment Canada, Ottawa, Ontario, Canada.

Kiesling, R.W., S.K. Alexander and J.W. Webb. 1988. Evaluation of alternative oil-spill cleanup techniques in a *Spartina-alterniflora* salt-marsh. *Environmental Pollution* 55(3): 221–238.

Kirstein, B.E. 1992. *Adaptation of the Minerals Management Service's Oil-Weathering Model for Use in the Gulf of Mexico Region.* Contract No. 14-35-0001-30537. Minerals Management Service, Gulf of Mexico Outer Continental Shelf (OCS) Region, New Orleans, Louisiana.

Knap, A.H. 1987. Effects of chemically dispersed oil on the brain coral *Diploria strigosa*. *Marine Pollution Bulletin* 18(3):119–122.

Knap, A.H., T.D. Sleeter, R.E. Dodge, S.C. Wyers, H.R. Frith and S.R. Smith. 1983. The effects of oil spills and dispersants use on corals: A review and multidisciplinary experimental approach. *Oil and Petrochemical Pollution* 1:157–169.

Knap, A.H., S.C. Wyers, R.E. Dodge, T.D. Sleeter, H.R. Frith, S.R. Smith and C.B. Cook. 1985. The effects of chemically and physically dispersed oil on the brain coral, *Diploria strigosa* (Dana). Pp. 547–551 in *Proceedings of the 1985 International Oil Spill Conference*, Los Angeles, California. American Petroleum Institute, Washington, D.C.

Knudsen, O.O., P.J. Brandvik and A. Lewis. 1994. Treating oil spills with W/O emulsion inhibitors—a laboratory study of surfactant leaching from the oil to the water phase. Pp. 1023–1034 in *Proceedings of the Seventeenth Arctic and Marine Oilspill Program (AMOP) Technical Seminar*, Vancouver, British Columbia, Canada. Environment Canada. Ottawa, Ontario, Canada.

Kochany, J. and R.J. Maguire. 1994. Abiotic Transformations of Polynuclear Aromatic Hydrocarbons and Polynuclear Aromatic Nitrogen Heterocycles in Aquatic Environments. *Science of the Total Environment* 144:17–31.

Koh, R.C.Y. and Fan, L-N. 1970. *Mathematical models for the prediction of temperature distributions resulting from the discharge of heated water into large bodies of water.* Report prepared under Contract 14-12-570 by Tetra Tech, Inc., Pasadena CA, for U.S. Environmental Protection Agency's Water Quality Office. U.S. Environmental Protection Agency, Office of Water, Washington, D.C.

Kolluru, V.S., M.L. Spaulding and E.L. Spaulding. 1993. Application and Verification of Worldwide Oil Spill Model (WOSM) to Selected Spill Events. Pp. 573–585 in *Proceedings of the Sixteenth Arctic and Marine Oil Spill Program (AMOP) Technical Seminar*, Calgary, Canada. Environment Canada. Ottawa, Ontario, Canada.

Kovats, von E. 1958. Gas-Chromatographische Charakterisierung Organischer Verbindungen, Teil 1: Retentionsindices Aliphatischer Halogenide, Alkohole, Aldehyde und Ketone. *Helvetica Chimica Acta* 10:1915–1932.

Kraly, J., R.G. Pond, D.V. Aurand, G.M. Coelho, A.H. Walker, B. Martin, J. Caplis and M. Snowby. 2001. Ecological risk assessment principles applied to oil spill response planning. Pp. 177–183 in *Proceedings of the 2001 International Oil Spill Conference*, Tampa, Florida. American Petroleum Institute, Washington, D.C.

Kucklick, J.H. and D. Aurand. 1997. Historical dispersant and *in-situ* burning opportunities in the United States. Pp. 205–210 in *Proceedings of the 1997 International Oil Spill Conference*, Fort Lauderdale, Florida. American Petroleum Institute, Washington, D.C.

LaBelle, R.P. and W.R. Johnson, 1993. Stochastic Oil Spill Analysis for Cook Inlet/Shelikof Strait. Pp. 573–585 in *Proceedings of the Sixteenth Arctic and Marine Oil Spill Program (AMOP) Technical Seminar*, Calgary, Alberta, Canada. Environment Canada. Ottawa, Ontario, Canada.

Lai, H.C. and C.P. Lim. 1984. Comparative toxicities of various crude oils to mangroves. Pp. 12–138 in *Fate and Effects of Oil in the Mangrove Environment*, H.C. Lai and M.C. Feng, Eds. Universiti Sains Malaysia, Pulau Pinang, Malaysia.

Lambert, G., D.B. Peakall, B.J.R. Philogene and F.R. Engelhardt. 1982. Effect of oil and oil dispersant mixtures on the basal metabolic rate of ducks. *Bulletin of Environmental Contamination and Toxicology* 29:520–524.

Lambert, P., B. Fieldhouse, Z. Wang, M. Fingas, M. Goldthorp, L. Pearson and E. Collazzi. 2001a. A laboratory study of a flow-through fluorometer for measuring oil-in-water levels. Pp. 23–45 in *Twenty-Fourth Arctic and Marine Oilspill (AMOP) Technical Seminar*, Edmonton, Alberta, Canada. Environment Canada, Ottawa, Ontario, Canada.

Lambert, P., M. Goldthorp, B. Fieldhouse, Z. Wang, M.F. Fingas, L. Pearson and E. Collazzi. 2001b. A review of oil-in-water monitoring techniques. Pp. 1375–1380 in *Proceedings of the 2001 International Oil Spill Conference*, Tampa, Florida. American Petroleum Institute, Washington, D.C.

Landrum, P.F., J.P. Giesy, J.T. Oris and P.M. Allred. 1987. Photoinduced toxicity of polycyclic aromatic hydrocarbons to aquatic organisms. Pp. 304–318 in *Oil in Freshwater*, J.H. Vandermeulen and S.E. Hrudey, Eds. Pergamon Press, New York, New York.

Lane, P.A., J.H. Vandermeulen, M.J. Crowell and D.G. Patriquin. 1987. Impact of experimentally dispersed crude oil on vegetation in a northwestern Atlantic salt marsh-preliminary observations. Pp. 509–514 in *Proceedings of the 1987 International Oil Spill Conference*, Baltimore, Maryland. American Petroleum Institute, Washington, D.C.

LaRiviere, D.J., R.L. Autenrieth and J.S. Bonner. 2003. Redox dynamics during recovery of an oil-impacted estuarine wetland. *Water Research* 37:3307–3318.

Law, A.T. 1995. Toxicity study of the oil dispersant Corexit 9527 on *Macrobrachium rosenbergii* (de Man) egg hatchability by using a flow-through bioassay technique. *Environmental Pollution* 88:341–343.

Law, R., C.A. Kelly, K.L. Graham, R.J. Woodhead, P.E.J. Dyrynda and E.A. Dyrynda. 1997. Hydrocarbon and PAH in fish and shellfish from Southwest Wales following the Sea Empress oilspill in 1996. Pp. 205–211 in *Proceedings of the 1997 International Oil Spill Conference*, Fort Lauderdale, Florida. American Petroleum Institute, Washington, D.C.

Lawrence, A.S.C. and W. Killner. 1948. Emulsions of seawater in Admiralty fuel oil with special reference to their demulsification. *Journal of the Institute of Petroleum* 34(299):821–857.

Ledwell, J.R., A.J. Watson and C.S. Law. 1998. Mixing of a tracer in the pycncline, *Journal of Geophysical Research* 103(C10):21499–21529.

Lee, H. 1992. Models, muddles and mud: Predicting bioaccumulation of sediment associated pollutants. Pp. 267–293 in *Sediment Toxicity Assessment*, G.A. Burton, Ed. CRC Press, Boca Raton, Florida.

Lee, H.W., N. Kobayashi and C.-R. Ryu. 1990. *Review on oil spills and their effects.* Report Number CACR-90-03. Center for Applied Coastal Research, Department of Civil Engineering, University of Delaware, Newark, Delaware.

Lee, K. 2002. Oil-particle interactions in aquatic environments: influence on the transport, fate, effect and remediation of oil spills. *Spill Science & Technology Bulletin* 8(1):3–8.

Lee, K., C.S. Wong, W.J. Cretney, F.A. Whitney, T.R. Parsons, C.M. Lalli and J. Wu. 1985. Microbial response to crude oil and Corexit 9527: SEAFLUXES enclosure study. *Microbial Ecology* 11:337–351.

Lee, K., T. Lunel, P. Wood, R. Swannell and P. Stoffyn-Egli. 1997a. Shoreline cleanup by acceleration of clay-oil flocculation processes. Pp. 235–240 in *Proceedings of the 1997 International Oil Spill Conference*, Fort Lauderdale, Florida. American Petroleum Institute, Washington, D.C.

Lee, K., S. St-Pierre and A.M. Weise. 1997b. Enhanced oil biodegradation with mineral fine interaction. Pp. 715–722 in *Proceedings of the Twentieth Arctic and Marine Oilspill Program (AMOP) Technical Seminar*, Vancouver, British Columbia. Environment Canada, Ottawa, Ontario, Canada.

Lee, K., P. Stoffyn-Egli, P.A. Wood and T. Lunel. 1998. Formation and structure of oil-mineral fines aggregates in coastal environments. Pp. 911–921 in *Proceedings of the Twenty-First Arctic and Marine Oilspill Program (AMOP) Technical Seminar*, Edmonton, Alberta, Canada. Environment Canada, Ottawa, Ontario, Canada.

Lee, K., P. Stoffyn-Egli and E.H. Owens. 2001. Natural dispersion of oil in a freshwater ecosystem: Desaguadero pipeline spill, Bolivia. Pp. 1445–1448 in *Proceedings of the 2001 International Oil Spill Conference*, Tampa, Florida. American Petroleum Institute, Washington, D.C.

Lee, K., P. Stoffyn-Egli and E.H. Owens. 2002. The OSSA II pipeline oil spill: natural mitigation of a riverine oil spill by oil–mineral aggregate formation. *Spill Science and Technology Bulletin* 7(3–4):149–154.

Leech, M., M. Walker, M. Wiltshire and A. Tyler. 1993. OSIS: A Windows 3 Oil Spill Information System. Pp. 549–572 in *Proceedings of the Sixteenth Arctic and Marine Oil Spill Program (AMOP) Technical Seminar*, Calgary, Alberta, Canada. Environment Canada. Ottawa, Ontario, Canada.

Le Floch, S., J. Guyomarch, F.-X. Merlin, P. Stoffyn-Egli, J. Dixon and K. Lee. 2002. The influence of salinity on oil-mineral aggregate formation. *Spill Science & Technology Bulletin* 8(1):65–71.

Legore, R.S., D.S. Marzalek, J.E. Hoffman and J.E. Cuddeback. 1983. A field experiment to assess impact of chemically dispersed oil on Arabian Gulf corals. Pp. 51–60 in *Proceedings of the Middle East Oil Technical Conference*, Manama, Bahrain. Society of Petroleum Engineers, Richardson, Texas.

Legore, S., D.S. Marszalek, L.J. Danek, M.S. Tomlinson, J.E. Hoffman and J.E. Cuddeback. 1989. Effect of chemically dispersed oil on Arabian Gulf corals: A field experiment. Pp. 375–380 in *Proceedings of the 1989 International Oil Spill Conference*, San Antonio, Texas. American Petroleum Institute, Washington, D.C.

Lehr, W. 1996. Progress in Oil Spread Modeling. Pp. 889–894 in *Proceedings of the Nineteenth Arctic and Marine Oilspill Program (AMOP) Technical Seminar*, Calgary, Alberta, Canada. Environment Canada, Ottawa, Ontario, Canada.

Lehr, W.J., H.M. Cekirge, R.J. Fraga and M.S. Belen. 1984. Empirical studies of the spreading of oil spills. *Oil and Petrochemical Pollution* 2:7–12.

Lehr, W.J., R. Overstreet, R. Jones and G. Watabayashi. 1992. ADIOS-Automatic Data Inquiry for Oil Spills. Pp. 31–45 in *Proceedings of the Fifteenth Arctic and Marine Oilspill Program (AMOP) Technical Seminar*, Edmonton, Alberta, Canada. Environment Canada, Ottawa, Ontario, Canada.

Lehr, W.J., D. Wesley, D. Simecek-Beatty, R. Jones, G. Kachook and J. Lankford. 2000. Algorithm and interface modifications of the NOAA oil spill behavior model. Pp. 525–539 in *Proceedings of the Twenty-Third Arctic and Marine Oil Spill Program (AMOP)* Technical Seminar, Vancouver, British Columbia, Canada. Environment Canada, Ottawa, Ontario, Canada.

Lehr, W., R. Jones, M. Evans, D. Simecek-Beatty and R. Overtreet. 2002. Revisions of the ADIOS oil spill model. *Environmental Modeling and Software* 17:191–199.

Leibovich, S. and J.L. Lumley. 1982. Interaction of turbulence and Langmuir cells in vertical transport of oil droplets. Pp. 271–276 in *Proceedings of the First International Conference on Meteorology and Air/Sea Interaction of the Coastal Zone*, The Hague, Netherlands. American Meteorological Society, Boston, Massachusetts.

Lessard, R.R., D.V. Aurand, G. Coelho, J.C. Clark, G. Bragin, C.M. Fuller, T.J. McDonald, R. Jamail and A. Steen. 1999. Design and implementation of a mesocosm experiment on the environmental consequences of nearshore dispersant use. Pp. 1027–1030 in *Proceedings of the 1999 International Oil Spill Conference*, Seattle, Washington. American Petroleum Institute, Washington, D.C.

Levine, E. 1999. Development and implementation of the dispersant observation job aid. Pp. 1015–1018 in *Proceedings of the 1999 International Oil Spill Conference*, Seattle, Washington. American Petroleum Institute, Washington, D.C.

Lewis, A. 2004. *Experimental and field case studies in dispersant effectiveness*. Presentation to the National Research Council Committee on Understanding Oil Spill Dispersants: Efficacy and Effects, March 15–16, 2004, Washington, D.C., unpublished. National Research Council, Washington, D.C.

Lewis, A. and D. Aurand. 1997. Putting Dispersants to work: Overcoming obstacles. Pp. 157–164 in *Proceedings of the 1997 International Oil Spill Conference*, Fort Lauderdale, Florida. American Petroleum Institute, Washington, D.C.

Lewis, A., P.S. Daling, R. Fiocco and A.B. Nordvik. 1994. Chemical dispersion of oil and water-in-oil emulsions—A comparison of bench-scale test methods and dispersant treatment in meso-scale flume. Pp. 979–1010 in *Proceedings of the Seventeenth Arctic and Marine Oilspill Program (AMOP) Technical Seminar*, Vancouver, British Columbia, Canada. Environment Canada. Ottawa, Ontario, Canada.

Lewis, A., P.S. Daling, T. Strom-Kristiansen, A.B. Nordvik and R. Fiocco. 1995a. Weathering and chemical dispersion of oil at sea. Pp. 157–164 in *Proceedings of the 1995 International Oil Spill Conference*, Long Beach, California. American Petroleum Institute, Washington, D.C.

Lewis, A., P.S. Daling, T. Strom-Kristiansen and P.J. Brandvik. 1995b. The behavior of Sture blend crude oil spilled at sea and treated with dispersant. Pp. 453–469 in *Proceedings of the Eighteenth Arctic and Marine Oilspill Program (AMOP) Technical Seminar*, Edmonton, Alberta, Canada. Environment Canada. Ottawa, Ontario, Canada.

Lewis, A., A. Crosbie, L. Davies and T. Lunel. 1998a. Large scale field experiments into oil weathering at sea and aerial application of dispersants. Pp. 319–343 in *Proceedings of the Twenty-first Arctic and Marine Oilspill Program (AMOP) Technical Seminar*, Edmonton, Alberta, Canada. Environment Canada, Ottawa, Ontario, Canada.

Lewis A., A. Crosbie, L. Davies and T. Lunel. 1998b. Dispersion of emulsified oil at sea. AEA Technology report. AEAT-3475. AEA Technology, National Environmental Technology Centre (NETCEN, Didcot, Oxfordshire, England.

Li, M. and C. Garrett. 1998. The relationship between oil droplet size and upper ocean turbulence. *Marine Pollution Bulletin* 36(12):961–970.

Linden, O., A. Rosemarin, A. Lindskog, C. Hoglund and S. Johansson. 1987. Effects of oil and oil dispersant on an enclosed marine ecosystem. *Environmental Science and Technology* 2:374–382.

Lindstedt-Siva, J. 1987. Advance planning for dispersant use. Pp. 329–333 in *Proceedings of the 1987 International Oil Spill Conference*, Baltimore, Maryland. American Petroleum Institute, Washington, D.C.

Lindstedt-Siva J. 1991. U.S. oil spill policy hampers response and hurts science. Pp. 349–352 in *Proceedings of the 1991 International Oil Spill Conference*, San Diego, California. American Petroleum Institute, Washington, D.C.

Lindstrom, J.E. and J.F. Braddock. 2002. Biodegradation of petroleum hydrocarbons at low temperature in the presence of the dispersant Corexit 9500. *Marine Pollution Bulletin* 44(8):739–747.

Lindstrom, J.E., D.M. White and J.F. Braddock. 1999. *Biodegradation of dispersed oil using Corexit 9500*. Report to the Alaska Department of Environmental Conservation, Division of Spill Prevention and Response. Alaska Department of Environmental Conservation, Juneau, Alaska.

Literathy, P., S. Haider, O. Samhan and G. Morel. 1989. Experimental studies on biological and chemical oxidation of dispersed oil in seawater. *Water Science and Technology* 21:845–856.

Little, D.I. and D.L. Scales. 1987a. Effectiveness of a type III dispersant on low-energy shorelines. *Proceedings of the 1987 International Oil Spill Conference*, Baltimore, Maryland. American Petroleum Institute, Washington, D.C.

Little, D.I. and D.L. Scales. 1987b. The persistence of oil stranded on sediment shorelines. *Proceedings of the 1987 International Oil Spill Conference*, Baltimore, Maryland. American Petroleum Institute, Washington, D.C.

Little, D.I. and D.L. Scales. 1987c. The effectiveness of a new type III dispersant in the treatment of weathered crude and emulsified fuel oils on saltmarshes and sandflats. Pp. 217–219 in *Fate and Effects of Oil in Marine Ecosystems*, J. Kuiper and W.J. Van Den Brink, Eds. Martinus Nijhoff Publishers, Dordrecht, The Netherlands.

Long, S.M. and D.A. Holdaway. 2002. Acute toxicity of crude and dispersed oil to *Octopus pallidus* (Hoyle, 1885) hatchlings. *Water Research* 36:2769–2776.

Louchouarn, P., J.S. Bonner, P. Tissot, T.J. McDonald, C.B. Fuller and C.A. Page. 2000. Quantitative determination of oil films/slicks from water surfaces using a modified solid-phase extraction (SPE) sampling method. Pp. 59–68 in *Proceedings of the Twenty-Third Arctic and Marine Oil Spill Program (AMOP)* Technical Seminar, Vancouver, British Columbia, Canada. Environment Canada, Ottawa, Ontario, Canada.

Lunel, T. 1993. Dispersion: Oil droplet size measurement at sea. Pp. 1023–1056 in *Proceedings of the Sixteenth Arctic and Marine Oil Spill Program (AMOP) Technical Seminar*, Calgary, Alberta, Canada. Environment Canada. Ottawa, Ontario, Canada.

Lunel, T. 1994a. Dispersion of a large experimental slick by aerial application of dispersant. Pp. 951–979 in *Proceedings of the Seventeenth Arctic and Marine Oilspill Program (AMOP) Technical Seminar*, Vancouver, British Columbia, Canada. Environment Canada. Ottawa, Ontario, Canada.

Lunel, T. 1994b. Field trials to determine quantitative estimates of dispersant efficiency at sea. Pp. 1011–1022 in *Proceedings of the Seventeenth Arctic and Marine Oilspill Program (AMOP) Technical Seminar*, Vancouver, British Columbia, Canada. Environment Canada. Ottawa, Ontario, Canada.

Lunel, T. 1995a. Dispersant effectiveness at sea. Pp. 147–155 in *Proceedings of the 1995 International Oil Spill Conference*, Long Beach, California. American Petroleum Institute, Washington, D.C.

Lunel, T. 1995b. Understanding the mechanism of dispersion through oil droplet size measurements at sea. Pp. 240–285 in *The Use of Chemicals in Oil Spill Response*, ASTM STP 1252, P. Lane, Ed. American Society for Testing and Materials, Philadelphia, Pennsylvania.

Lunel, T. 1998. *Sea Empress* spill: Dispersant operations, effectiveness, and effectiveness monitoring. Pp. 59–78 in *Proceedings of the Dispersant Use in Alaska: A Technical Update*, Anchorage, Alaska. Prince William Sound Oil Spill Recovery Institute, Cordova, Alaska.

Lunel, T. and L. Davies. 1996. Dispersant effectiveness in the field on fresh oils and emulsions. Pp. 1355–1394 in *Proceedings of the Nineteenth Arctic and Marine Oilspill Program (AMOP) Technical Seminar*, Calgary, Alberta, Canada. Environment Canada, Ottawa, Ontario, Canada.

Lunel, T. and A. Lewis. 1993a. Effectiveness of demulsifiers in sea trials: The use of fluorometry, surface sampling, and remote sensing to determine effectiveness. Pp. 179–202 in *Formation and Breaking of Water-in-Oil Emulsions: Workshop Proceedings*, A.H. Walker, D.L. Ducey, Jr, J.R. Gould and A.B. Nordvik, Eds. Marine Spill Response Corporation, Washington, D.C.

Lunel, T. and A. Lewis. 1993b. Oil concentrations below a demulsifier-treated slick. Pp. 955–972 in *Proceedings of the Sixteenth Arctic and Marine Oil Spill Program (AMOP) Technical Seminar*, Calgary, Alberta, Canada. Environment Canada. Ottawa, Ontario, Canada.

Lunel, T., G. Baldwin and F. Merlin. 1995a. Comparison of meso-scale and laboratory dispersant tests with dispersant effectiveness measured at sea. Pp. 629–651 in *Proceedings of the Eighteenth Arctic and Marine Oilspill Program (AMOP) Technical Seminar*, Edmonton, Alberta, Canada. Environment Canada. Ottawa, Ontario, Canada.

Lunel, T., L. Davies and P.J. Brandvik. 1995b. Field trials to determine dispersant effectiveness at sea. Pp. 603–627 in *Proceedings of the Eighteenth Arctic and Marine Oilspill Program (AMOP) Technical Seminar*, Edmonton, Alberta, Canada. Environment Canada. Ottawa, Ontario, Canada.

Lunel, T., L. Davies, A.C.T. Chen and R.A. Major. 1995c. Field test of dispersant application by fire monitor. Pp. 559–574 in *Proceedings of the Eighteenth Arctic and Marine Oilspill Program (AMOP) Technical Seminar*, Edmonton, Alberta, Canada. Environment Canada. Ottawa, Ontario, Canada.

Lunel, T.J., J. Rusin, N. Bailey, C. Halliwell and L. Davies. 1996. A successful at-sea response to the Sea Empress spill. Pp. 1499–1520 in *Proceedings of the Nineteenth Arctic and Marine Oilspill Program (AMOP) Technical Seminar*, Calgary, Alberta, Canada. Environment Canada, Ottawa, Ontario, Canada.

Lunel, T.J. Rusin, N. Bailey, C. Halliwell and L. Davies. 1997a. The net environmental benefit of a successful dispersant application at the *Sea Empress* incident. Pp. 185–194 in *Proceedings of the 1997 International Oil Spill Conference*, Fort Lauderdale, Florida. American Petroleum Institute, Washington, D.C.

Lunel, T., P. Wood and L. Davies, 1997b. Dispersant effectiveness in field trials and in operational response. Pp. 923–926 in *Proceedings of the 1997 International Oil Spill Conference*, Fort Lauderdale, Florida. American Petroleum Institute, Washington, D.C.

Lyklema, J. 2000. *Fundamentals of Interface and Colloid Science, Volume III: Liquid-Fluid Interfaces.* Academic Press, Incorporated, San Diego, California.

Mackay, D. 1987. *Formation and stability of water-in-oil emulsions.* DIWO Report No. 1. IKU. SINTEF Group, Trondheim, Norway.

Mackay, D. 1993. *Effectiveness of dispersants applied following the Exxon Valdez spill.* Report from the University of Toronto for Exxon Company. ExxonMobil Research and Engineering Company, Fairfax, Virginia.

Mackay, D. 1995. *Effectiveness of chemical dispersants under breaking wave conditions.* Pp. 310–340 in *Use of Chemicals in Oil Spill Response.* ASTM special technical publication 1252. American Society of Testing Materials, Philadelphia, Pennsylvania.

Mackay, D. and A. Chau. 1986. The effectiveness of chemical dispersants: a discussion of laboratory and field tests results. *Oil and Chemical Pollution* 3:405–415.

Mackay, D. and K. Hossain. 1982. *An exploratory study of naturally and chemically dispersed oil.* EE-35. Report to Environment Canada. Environment Canada, Ottawa, Ontario, Canada.

Mackay, D. and P.J. Leinonen. 1977. *Mathematical model of the behavior of oil spills on water with natural and chemical dispersion.* Economic and Technical Review Report EPS-3-EC-77-19. Prepared Environment Canada. Environment Canada, Ottawa, Ontario, Canada.

Mackay, D. and R.M. Matsugu. 1973. Evaporation rates of liquid hydrocarbon spills on land and water. *Canadian Journal of Chemical Engineering* 8:434–439.

Mackay, D. and W.Y. Shiu. 1976. Aqueous solubilities of weathered northern crude oils. *Bulletin of Environmental Contamination and Toxicology* 15(1):101–109.

Mackay, D. and P.G. Wells. 1983. Effectiveness, behavior, and toxicity of dispersants. Pp. 65–71 in *Proceedings of the 1983 International Oil Spill Conference*, San Antonio, Texas. American Petroleum Institute, Washington, D.C.

Mackay, D., J.S. Nadeau and C. Ng. 1978. A small-scale laboratory dispersant effectiveness test. Pp. 35–49 in *Chemical Dispersants for the Control of Oil Spills*, L.T. McCarthy, G.P. Lindblom and H.F. Walter, Eds. American Society for Testing and Materials, Philadelphia, Pennsylvania.

Mackay, D., S. Paterson and K. Trudel. 1980a. *A mathematical model of oil spill behavior.* Publication EE-7. Report for Environment Canada. Department of Chemical and Applied Chemistry, University of Toronto, Canada.

Mackay, D., I. Buist, R. Mascarenhas and S. Paterson. 1980b. *Oil spill processes and models.* Publication EE-8. Report for Environment Canada. Environment Canada, Ottawa, Ontario, Canada.

Mackay, D, W.Y. Shiu, K. Hossain, W. Stiver, D. McCurdy and S. Peterson. 1982. *Development and Calibration of an Oil Spill Behavior Model.* Report No. CG-D-27-83. U.S. Coast Guard, Research and Development Center, Groton, Connecticut.

Mackay, D.A., A. Chau, K. Hossain and M. Bobra. 1984. Measurement and prediction of the effectiveness of oil spill chemical dispersants. *Oil Spill Chemical Dispersants, Research Experience and Recommendations*, T.E. Allen, Ed. ASTM special technical publication 840. American Society of Testing Materials, Philadelphia, Pennsylvania.

Mackay, D., H. Puig and L.S. McCarty. 1992. An equation describing the time course and variability in uptake and toxicity of narcotic chemicals to fish. *Journal of Journal of Environmental Toxicology and Chemistry* 11:941–951.

MacKinnon, D.S. and P.A. Lane. 1993. *Saltmarsh revisited—The long-term effects of oil and dispersant on saltmarsh vegetation.* Environmental Studies Research Fund, Report No. 122. National Energy Board, Calgary, Alberta, Canada.

MacNaughton, S.J., R.P.J. Swannell, F. Daniel and L. Bristow. 2003. Biodegradation of dispersed Forties crude and Alaskan North Slope oils in microcosms under simulated marine conditions. *Spill Science and Technology Bulletin* 8(2):179–186.

Major, R.A., A.C.T. Chen and P. Nicholson. 1994. Wave basin tests of boat dispersant application systems. Pp 1035–1051 in *Proceedings of the Seventeenth Arctic and Marine Oilspill Program (AMOP) Technical Seminar*, Vancouver, British Columbia, Canada. Environment Canada. Ottawa, Ontario, Canada.

Malinky, G., and D.G. Shaw. 1979. Modeling the association of petroleum hydrocarbons and sub-arctic sediments. Pp. 621–623 in *Proceedings of the 1979 International Oil Spill Conference*, Los Angeles, California. American Petroleum Institute, Washington, D.C.

Marine and Freshwater Resources Institute. 1998. *Toxicity and effectiveness of the oil spill dispersant Corexit 9500.* Laboratory report. Marine and Freshwater Resources Institute, Queenscliff, Australia.

Masutani, S. and E. Adams. 2004. Liquid droplet contaminant plumes in the deep ocean. *Japanese Journal of Multiphase Flow* 18(2):135–152.

McAuliffe, C.D. 1963. Solubility in water of C1-C9 hydrocarbons. *Nature* 200:1092–1093.

McAuliffe, C.D. 1966. Solubility in water of paraffin, cycloparaffin, olefin, acetylene, cycloolefin, and aromatic hydrocarbons. *Journal of Physical Chemistry* 70:1267–1275.

McAuliffe, C.D. 1989. The weathering of volatile hydrocarbons from crude oil slicks on water. Pp. 357–364 in *Proceedings of the 1989 International Oil Spill Conference*, San Antonio, Texas. American Petroleum Institute, Washington, D.C.

McAuliffe, C.D., B.L. Steelman, W.R. Leek, D.E. Fitzgerald, J.P. Ray and C.D. Barker. 1981. The 1979 Southern California Dispersant Treated Research Oil Spills. Pp. 269–282 in *Proceedings of the 1981 International Oil Spill Conference*, Atlanta, Georgia. American Petroleum Institute, Washington, D.C.

McCarty, L.S. 1986. The relationship between aquatic toxicity QSARs and bioconcentration for some organic chemicals. *Journal of Environmental Toxicology and Chemistry* 5:1071–1080.

McCarty, L.S. and D. Mackay. 1993. Enhancing ecotoxicological modeling and assessment. *Environmental Science and Technology* 27(9):1719–1728.

McCarty, L.S., D. Mackay, A.D. Smith, G.W. Ozburn and D.G. Dixon. 1992. Residue-based Interpretation of Toxicity and Bioconcentration QSARs from Aquatic Bioassays: Neutral Narcotic Organics. *Journal of Environmental Toxicology and Chemistry* 11:917–930.

McDonagh, M. and K. Colcomb-Heiliger. 1992. Aerial spraying of demulsifiers to enhance the natural dispersion of oil slicks. Pp. 107–122 in *Proceedings of the Fifteenth Arctic and Marine Oilspill Program (AMOP) Technical Seminar*, Edmonton, Alberta, Canada. Environment Canada, Ottawa, Ontario, Canada.

McGrath, J.A., F.L. Hellweger, T.F. Parkerton and D.M. Di Toro. 2003. *Application of the Narcosis Target Lipid Model to Complex Mixtures Using Gasolines as a Case Study.* Paper presented at the Annual Meeting of the Society of Environmental Toxicology and Chemistry (SETAC), Austin, Texas. Society of Environmental Toxicology and Chemistry, Pensacola, Florida.

McGrath, J.A., T.F. Parkerton and D.M. DiToro. 2004. Application of the narcosis target lipid model to algal toxicity and deriving predicted no-effect concentrations. *Journal of Environmental Toxicology and Chemistry* 23:2503–2517.

McLean, J.D. and P.K. Kilpatrick. 1997a. Effects of asphaltene solvency on stability of water-in-crude-oil emulsions. *Journal of Colloidal and Interfacial Science* 189:242–255.

McLean, J.D. and P.K. Kilpatrick. 1997b. Effects of asphaltene aggregation in model heptane-toluene mixtures on stability of water-in-oil emulsions. *Journal of Colloid and Interface Science* 196:23–34.

McLean, J.D., P.M. Spiecker, A.P. Sullivan and P.K. Kilpatrick. 1998. The role of petroleum asphaltenes in the stabilization of water-in-oil emulsions. Pp. 377–422 in *Structure and Dynamics of Asphaltenes*, O.C. Mullins and E.Y. Sheu, Eds., Plenum Press, New York, New York.

McNaughton, S.J., R. Swannell, F. Daniel and L. Bristow. 2003. Biodegradation of dispersed Forties crude and Alaskan North Slope oils in microcosms under simulated marine conditions. *Spill Science and Technology Bulletin* 8:179–186.

Meeks, D.G. 1981. A view on the laboratory testing and assessment of oil spill dispersant efficiency. Pp. 19–29 in *Proceedings of the 1981 International Oil Spill Conference*, Atlanta, Georgia. American Petroleum Institute, Washington, D.C.

Menon, N.N. and N.R. Menon. 1999. Uptake of polycyclic aromatic hydrocarbons from suspended oil borne sediments by the marine bivalve *Sunetta scripta*. *Aquatic Toxicology* 45:63–69.

Michel, J. and C.B. Henry. 1997. Oil uptake and depuration in oysters after use of dispersants in shallow water in El Salvador. *Spill Science and Technology Bulletin* 4:57–70.

Michel, J., D. Scholz, S.R. Warren Jr. and A.H. Walker. 2004. *A Decision-Maker's Guide to* In-situ *Burning*. American Petroleum Institute, Washington, D.C.

Middaugh, D.P. and D.D. Whiting. 1995. Responses of embryonic and larval inland silversides, *Menidia beryllina*, to No.-2 Fuel-oil and oil dispersants in seawater. *Archives of Environmental Contamination and Toxicology* 29(4):535–539.

Miller, R.M. and R. Bartha. 1989. Evidence from liposome encapsulation for transport-limited microbial metabolism of solid alkanes. *Applied and Environmental Microbiology* 55(2):269–274.

Mitchell, F.M. and D.A. Holdaway. 2000. The acute and chronic toxicity of the dispersants Corexit 9527 and 9500, water accommodated fraction (WAF) of crude oil and dispersant enhanced WAF (DEWAF) to *Hydra viridissima* (green hydra). *Water Research* 34:343–348.

Miura, Y., M. Okazaki, S.-I. Hamada, S.-I. Murakawa and R. Yugen. 1977. Assimilation of liquid hydrocarbons by microorganisms. I. Mechanism of hydrocarbon uptake. *Biotechnology and Bioengineering* 19:701–714.

Moles, A., L. Holland and J. Short. 2002. Effectiveness in the laboratory of Corexit 9527 and 9500 in dispersing fresh, weathered, and emulsion of Alaska North Slope crude oil under subarctic conditions. *Spill Science and Technology Bulletin* 7(5-6):241–247.

Montgomery, D.C. 1997. *Design and Analysis of Experiments, Fourth Edition.* John Wiley & Sons, Incorporated. New York, New York.

Morales, R.A., A.J. Elliot and T. Lunel. 1997. The Influence of Tidal Currents and Wind on Mixing in the Surface Layers on the Sea. *Marine Pollution Bulletin* 34(1):15–25.

Mulkins-Phillips, G.J. and J.E. Stewart. 1974. Effect of four dispersants on biodegradation and growth of bacteria on crude oil. *Applied Microbiology* 28(4):547–552.

Mulyono, M., E. Jasjfi and M. Maloringan. 1994. Oil dispersants: do they do any good? Pp. 539–549 in *Proceeding of the Second International Conference on Health, Safety and Environment in Oil and Gas Exploration and Production*, Jakarta, Indonesia. Society of Petroleum Engineers, Richardson, Texas.

Muschenheim, D.K. and K. Lee. 2002. Removal of oil from the sea surface through particulate interactions: review and prospectus. *Spill Science and Technology Bulletin* 8, 9–18.

National Oceanic and Atmospheric Administration (NOAA). 1994. *ADIOS, Automated Data Injury for Oil Spills, User's Manual.* NOAA/Hazardous Materials Response and Assessment Division, Seattle, Washington.

National Oceanic and Atmospheric Administration (NOAA). 1999. *Dispersant application observer job aid.* NOAA Hazardous Materials Response Division, Seattle, WA.

National Research Council (NRC). 1985. *Oil in the Sea: Inputs, Fates, and Effects.* National Academy Press, Washington, D.C.

National Research Council (NRC). 1989. *Using Oil Spill Dispersants on the Sea.* National Academy Press, Washington, D.C.

National Research Council (NRC) 1999. *Spills of Nonfloating Oils, Risks and Response.* National Academy Press, Washington, D.C.

National Research Council (NRC). 2003. *Oil in the Sea III: Inputs, Fates and Effects.* National Academies Press, Washington, D.C.

National Response Team Response Committee. 2002. *NRT-RRT Fact Sheet.* U.S. National Response Team, Washington, D.C. [Online] Available at: http://www.nrt.org/Production/NRT/NRTWeb.nsf/AllAttachmentsByTitle/A-58Riskcomm1/$File/riskcomm1.pdf?OpenElement [January 14, 2005].

Neff, J.M. 2002. *Bioaccumulation in Marine Organisms: Effect of Contaminants from Oil Well Produced Water.* Elsevier Science Publishers, Amsterdam, The Netherlands.

Neff, J.M. and Burns, W.A. 1996. Estimation of Polycyclic Aromatic Hydrocarbon Concentrations in the Water Column Based on Tissue Residues in Mussels and Salmon: An Equilibrium Partitioning Approach. *Journal of Environmental Toxicology and Chemistry* 15(12):2240–2253.

Neff, J.M. and W.E. Haensly. 1982. Long-term impact of the Amoco Cadiz crude oil spill on oysters *Crassostrea gigas* and plaice *Pleuronectes platessa* from Aber Benoit and Aber Wrac'h, Brittany, France. Pp. 269–327 in *Report of the NOAA-CNEXO Joint Scientific Commission, Ecological Study of Amoco Cadiz Oil Spill.* U.S. Department of Commerce, National Oceanic and Atmospheric Administration, Boulder, Colorado.

Negri, A.P. and A.J. Heyward. 2000. Inhibition of fertilization and larval metamorphosis of the coral *Acropora millepora* (Ehrenberg, 1834) by petroleum products. *Marine Pollution Bulletin* 41:420–427.

New York Department of Health. 2005. *Glossary of Environmental Health Terms.* New York Department of State, New York, New York. [Online]. Available at: http://www.health.state.ny.us/nysdoh/consumer/environ/toxglos.htm [April 27, 2005].

Newsted, J.L. and J.P. Giesy. 1987. Predictive models for photoinduced acute toxicity of polycyclic aromatic hydrocarbons to *Daphnia magna. Journal of Environmental Toxicology and Chemistry* 6:445–461.

Nichols, J.A. and H.D. Parker. 1985. Dispersants: Comparison of laboratory tests and field trials with practical experience at spills. Pp. 421–427 in *Proceedings of the 1985 International Oil Spill Conference,* Los Angeles, California. American Petroleum Institute, Washington, D.C.

Nilsen, J., A. Naess and Z. Volent. 1985. *Measurements of oil concentrations in the water column, under breaking waves.* Report STF 60 A 85079, Norwegian Hydrotechnical Laboratory. SINTEF Trondheim, Norway.

Nirmalakhandan, N. and R.E. Speece. 1988. Structure-Activity Relationships, Quantitative Techniques for Predicting the Behavior of Chemicals in the Ecosystem. *Journal of Environmental Science Technology* 22(6):606–615.

Norwegian Institute for Water Research. 1994. *Marine algal growth inhibition test.* Laboratory Report. Norwegian Institute for Water Research, Oslo, Norway.

Nyman, J.A. 1999. Effect of crude oil and chemical additives on metabolic activity of mixed microbial populations in fresh marsh soils. *Microbial Ecology* 37:152–162.

Ohwada, K., M. Nishimura, M. Wada, H. Nomura, A. Shibata, K. Okamoto, K. Toyoda, A. Yoshida, H. Takada and M. Yamada. 2003. Study of the effect of water-soluble fractions of heavy-oil on coastal marine organisms using enclosed ecosystems, mesocosms. *Marine Pollution Bulletin* 47:78–84.

Ojo, T.O. and J.S. Bonner. 2002. Three-Dimensional Self-Calibrating Coastal Oil Spill Trajectory Tracking and Contaminant Transport using HF Radar. Pp. 215–226 in *Proceedings of Twenty-Fifth Arctic and Marine Oilspill Technical Seminar*, Calgary, Alberta, Canada. Environment Canada. Ottawa, Ontario, Canada

Ojo, T.O., J.S. Bonner, C.A. Page, M. Sterling, C. Fuller and F.J. Kelly. 2003. Field simulation experiment of aerial dispersant application for spill of opportunity. Pp. 813–824 in *Proceedings of the Twenty-Sixth Arctic and Marine Oilspill Program (AMOP) Technical Seminar*, Victoria, British Columbia, Canada. Environment Canada, Ottawa, Ontario, Canada.

Okubo, A. 1971. Oceanic diffusion diagrams. *Deep Sea Research* 18:789–802.

Omotoso, O.E., V.A. Munoz, R.J. Mikula. 2002. Mechanisms of crude oil–mineral interactions. *Spill Science and Technology Bulletin* 8(1):45–54.

Onishi, Y., D.S. Trent, T.E. Michener, J.E. Van Beek and C.A. Rieck. 1999. Simulation of Radioactive Tank Waste Mixing with Chemical Reactions, FEDSM99-7786. In *Proceedings of Third American Society of Mechanical Engineers/Japan Society of Mechanical Engineers (ASME/JSME) Joint Fluids Engineering Conference*, San Francisco, California. American Society of Mechanical Engineers, New York, New York.

Ordsie, C.J. and G.C. Garofalo. 1981. Lethal and sublethal effects of short term acute doses of Kuwait crude oil and a dispersant Corexit 9527 on bay scallops *Argopecten irradians* and two predators at different temperatures. *Marine Environmental Research* 5:195–210.

Osborn, T.R. 1974. Vertical profiling of velocity microstructure. *Journal of Physical Oceanography* 4:109–115.

Owens, E.H. and K. Lee. 2003. Interaction of oil and mineral fines on shorelines: Review and assessment. *Spill Science and Technology Bulletin* 47(9-12):397–405.

Pace, C.B. and J.R. Clark. 1993. *Evaluation of a toxicity test method used for dispersant screening in California.* MSRC Technical Report Series 93-028. Marine Spill Response Corporation, Washington, D.C.

Pace, C.B., J.R. Clark and G.E. Bragin. 1995. Comparing crude oil toxicity under standard and environmentally realistic exposures. Pp. 1003–1004 in *Proceedings of the 1995 International Oil Spill Conference*, Long Beach, California. American Petroleum Institute, Washington, D.C.

Page, C., P. Sumner, R. Autenrieth, J. Bonner and T. McDonald. 1999. Materials balance on a chemically dispersed oil and a whole oil exposed to an experimental beach front. Pp. 645–658 in *Proceedings of the Twenty-Second Arctic and Marine Oil Spill Program (AMOP) Technical Seminar*, Calgary, Alberta, Canada. Environment Canada, Ottawa, Ontario, Canada.

Page, C.A., J.S. Bonner, P.L. Sumner, T.J. McDonald, R.L. Autenrieth and C.B. Fuller. 2000a. Behaviour of a chemically-dispersed oil and a whole oil on a near-shore environment. *Water Research* 34:2507–2516.

Page, C.A., J.S. Bonner, P.L. Sumner and R.L. Autenrieth. 2000b. Solubility of petroleum hydrocarbons in oil/water systems. *Marine Chemistry* 70:79–87.

Page, C.A., R.L. Autenrieth, J.S. Bonner and T. McDonald. 2001. Behaviour of chemically dispersed oil in a wetland environment. Pp. 821–823 in *Proceedings of the 2001 International Oil Spill Conference*, Tampa, Florida. American Petroleum Institute, Washington, D.C.

Page, C.A., J.S. Bonner, T.J. McDonald and R.L. Autenrieth. 2002. Behavior of a chemically dispersed oil in a wetland environment. *Water Research* 36:3821–3833.

Parker, P.L. and S. Macko. 1978. An intensive study of the heavy hydrocarbons in the suspended particulate matter of seawater. Chapter 11 in *Environmental studies, south Texas outer continental shelf, biology and chemistry*. BLM Contract AA550-CT7-11. U.S. Department of Interior, Bureau of Land Management (BLM), Washington, D.C.

Parker, C.A., M. Freegarde and C.G. Hatchard. 1971. The effect of some chemical and biological factors on the degradation of crude oil at sea. Pp. 237–244 in *Water Pollution by Oil*, P. Hepple, Ed. Institute of Petroleum, London, United Kingdom.

Payne, J.R. and A.A. Allen. 2004. *Use of natural oil seeps for evaluation of dispersant application and monitoring techniques.* Final report for Cooperative Institute for Coastal and Estuarine Environmental Technology /University of New Hampshire (UNH) Subcontract number 03-690. National Oceanic and Atmospheric Administration/UNH Cooperative Institute for Coastal and Estuarine Environmental Technology, Durham, New Hampshire. [Online] Available at: www.crrc.unh.edu. [April 22, 2005].

Payne, J.R. and A.A. Allen. In press. Use of natural oil seeps for evaluation of dispersant application and monitoring techniques. *Proceedings of the 2005 International Oil Spill Conference*, Miami Beach, Florida. American Petroleum Institute, Washington, D.C.

Payne, J.R. and W.B. Driskell. 2001. Source characterization and identification of New Carissa oil in NRDA environmental samples using a combined statistical and fingerprinting approach. Pp. 1403–1409 in *Proceedings of the 2001 International Oil Spill Conference*, Tampa, Florida. American Petroleum Institute, Washington, D.C.

Payne, J.R. and W.B. Driskell. 2003. The importance of distinguishing dissolved- versus oil-droplet phases in assessing the fate, transport, and toxic effects of marine oil pollution. Pp. 771–778 in *Proceedings of the 2003 International Oil Spill Conference*, Vancouver, Canada. American Petroleum Institute. Washington, D.C.

Payne, J.R. and French-McCay, D. 2001. *Development of a conceptual model for predicting pollutant movement from an oil spill with and without dispersant treatment: background information/literature review and oil spill modeling conceptualization.* Draft prepared for Environmental Protection Agency, Office of Research and Development, National Exposure Research Laboratory, Ecosystems Research Division. Environmental Protection Agency, Athens, Georgia.

Payne, J.R. and McNabb, G.D., Jr. 1984. Weathering of petroleum in the marine environment. *Marine Technology Society Journal* 18(3):24–40.

Payne, J.R. and C.R. Phillips. 1985a. Photochemistry of petroleum in water. *Environmental Science and Technology* 19(7):569–579.

Payne, J.R. and C.R. Phillips. 1985b. *Petroleum Spills in the Marine Environment: The Chemistry and Formation of Water-in-Oil Emulsions and Tar Balls.* Lewis Publishers, Incorporated, Chelsea, Michigan.

Payne, J.R., N.W. Flynn, P.J. Mankiewicz and G.S. Smith. 1980a. Surface evaporation/dissolution partitioning of lower-molecular-weight aromatic hydrocarbons in a down-plume transect from the IXTOC I wellhead. Pp. 119–166 in *Proceedings of the Symposium on Preliminary Results from the September 1979 Researcher/Pierce IXTOC-I Cruise*, Key Biscayne, Florida. National Oceanic and Atmospheric Administration Publications Office, Boulder, Colorado.

Payne, J.R., G.S. Smith, P.J. Mankiewicz, R.F. Shokes, N.W. Flynn, W. Moreno and J. Altamirano. 1980b. Horizontal and vertical transport of dissolved and particulate-bound higher-molecular-weight hydrocarbons from the IXTOC I blowout. Pp. 239–263 in *Proceedings of the Symposium on Preliminary Results from the September 1979 Researcher/Pierce IXTOC-I Cruise*, Key Biscayne, Florida. National Oceanic and Atmospheric Administration Publications Office, Boulder, Colorado.

Payne, J.R., B.E. Kirstein, G.D. McNabb Jr., J.L. Lambach, C. de Oliveira, R.E. Jordan and W. Hom. 1983. Multivariate analysis of petroleum hydrocarbon weathering in the subarctic marine environment. Pp. 423–434 in *Proceedings of the 1983 International Oil Spill Conference*, San Antonio, Texas. American Petroleum Institute, Washington, D.C.,

Payne, J.R., B.E. Kirstein, G.D. McNabb, Jr., J.L. Lambach, R. Redding, R.E. Jordan, W. Hom, C. de Oliveira, G.S. Smith, D.M. Baxter and R. Geagel. 1984. Multivariate analysis of petroleum weathering in the marine environment—subarctic. Volume I, Technical Results; Volume II, Appendices in *Outer Continental Shelf Environmental Assessment Program, Final Reports of Principal Investigators, Volume 21 and 22*. Volume 21 NTIS Accession Number PB85-215796; Volume 22 NTIS Accession Number PB85-215739. Report to the U.S. Department of Commerce. National Oceanic and Atmospheric Administration, Ocean Assessment Division, Juneau, Alaska.

Payne, J.R., B.E. Kirstein, J.R. Clayton, Jr., C. Clary, R. Redding, Jr., G.D. McNabb and G.H. Farmer. 1987a. *Integration of suspended particulate matter and oil transportation study*. Final Report submitted to Minerals Management Service. Minerals Management Service, Environmental Studies Branch, Anchorage, Alaska.

Payne, J.R., C.R. Phillips and W. Hom. 1987b. Transport and Transformations: Water Column Processes. Pp. 175–232 in *Long Term Environmental Effects of Offshore Oil and Gas Development*, D.F. Boesch and N.N. Rabelais, Eds. Elsevier Applied Science, New York, New York.

Payne, J.R., G.D. McNabb, Jr., L.E. Hachmeister, B.E. Kirstein, J.R. Clayton, Jr., C.R. Phillips, R.T. Redding, C.L. Clary, G.S. Smith and G.H. Farmer. 1987c. Development of a Predictive Model for the Weathering of Oil in the Presence of Sea Ice. Pp. 147–465 in *Outer Continental Shelf Environmental Assessment Program, Final Reports of Principal Investigators*. NTIS Accession Number PB-89-159776. Report to the U.S. Department of Commerce. National Oceanic and Atmospheric Administration, Ocean Assessment Division, Juneau, Alaska.

Payne, J.R., J.R. Clayton, Jr., G.D. McNabb, Jr., B.E. Kirstein, C.L. Clary, R.T. Redding, J.S. Evans, E. Reimnitz and E.W. Kempema. 1989. Oil-Ice-Sediment Interactions During Freezeup and Breakup. Pp. 1–382 in *Outer Continental Shelf Environmental Assessment Program, Final Reports of Principal Investigators*. NTIS Accession Number PB-90-156217. Report to the U.S. Department of Commerce. National Oceanic and Atmospheric Administration, Ocean Assessment Division, Juneau, Alaska.

Payne, J.R., J.R. Clayton, Jr., G.D. McNabb, Jr. and B.E. Kirstein. 1991a. Exxon Valdez oil weathering fate and behavior: Model predictions and field observations. Pp. 641–654 in *Proceedings of the 1991 International Oil Spill Conference*, San Diego, California. American Petroleum Institute, Washington, D.C.

Payne, J.R., G.D. McNabb, Jr. and J.R. Clayton, Jr. 1991b. Oil-weathering behavior in arctic environments. *Polar Research* 10(2):631–662.

Payne, J.R., J.R. Clayton, Jr., C.R. Phillips, J. Robinson, D. Kennedy, J. Talbot, G. Petrae, J. Michel, T. Ballou and S. Onstad. 1991c. Dispersant trials using the Pac Baroness, a spill of opportunity. Pp. 427–433 in *Proceedings of the 1991 International Oil Spill Conference*, San Diego, California. American Petroleum Institute, Washington, D.C.

Payne, J.R., L.E. Hachmeister, G.D. McNabb, Jr., H.E. Sharpe, G.S. Smith and C.A. Manen. 1991d. Brine-induced advection of dissolved aromatic hydrocarbons to arctic bottom waters. *Environmental Science and Technology* 25(5):940–951.

Payne, J.R., T.J. Reilly, R.J. Martrano, G.P. Lindblom, M.C. Kennicutt II and J.M. Brooks. 1993. Spill-of-opportunity testing of dispersant effectiveness at the Mega Borg oil spill. Pp. 791–793 in *Proceedings of the 1993 International Oil Spill Conference*, Tampa, Florida. American Petroleum Institute, Washington, D.C.

Payne, J.R., T.J. Reilly and D.P. French-McCay. 1999. Fabrication of a portable large-volume water sampling system to support oil spill NRDA efforts. Pp. 1179–1184 in *Proceedings of the 1999 International Oil Spill Conference*, Seattle, Washington. American Petroleum Institute, Washington, D.C.

Payne, J.R., J.R. Clayton, Jr. and B.E. Kirstein. 2003. Oil/suspended particulate material interactions and sedimentation. *Spill Science & Technology* 8(2):201–221.

Peakall, D.B., P.G. Wells and D. Mackay. 1987. A hazard assessment of chemically dispersed oil spills and seabirds. *Marine Environmental Research* 22:91–106.

Pelletier, M.C., R.M. Burgess, K.T. Ho, A. Kuhn, R.A. McKinney and S.A. Ryba. 1997. Phototoxicity of individual polycyclic aromatic hydrocarbons and petroleum to marine invertebrate larvae and juveniles. *Journal of Environmental Toxicology and Chemistry* 16:2190–2199.

Peters, E.C., P.A. Meyers, P.P. Yevich and N.J. Blake. 1981. Bioaccumulation and histopathological effects of oil on a stony coral. *Marine Pollution Bulletin* 12(10):333–339.

Peters, E.C., N.J. Gassman, J.C. Firman, R.H. Richmond and E.A. Power. 1997. Ecotoxicology of tropical marine ecosystems. *Environmental Toxicology and Chemistry* 16(1):12–40.

Pezeshki, S.R., M.W. Hester, Q. Lin and J.A. Nyman. 2000. The effects of oil spill and cleanup on dominant US Gulf coast marsh macrophytes: a review. *Environmental Pollution* 108(2):129–139.

Pittinger, C.A., R. Bachman, A.L. Barton, J.R. Clark, P.L. deFur, S.J. Ellis, M.W. Slimak, R.G. Stahl and R.S. Wentzel. 1998. A multi-stakeholder framework for ecological risk assessment—Summary of a SETAC technical workshop. Pp. 23–25 in *Summary of SETAC Workshop on Framework for Ecological Risk Management*, Williamsburg, Virginia. Society of Environmental Toxiciology and Chemistry, Pensacola, Florida.

Pollino, C.A. and D.A. Holdaway. 2002a. Reproductive potential of crimson-spotted rainbowfish (*Melanotaenia fluviatilis*) following short-term exposure to bass strait crude oil and dispersed crude oil. *Environmental Toxicology* 17:138–145.

Pollino, C.A. and D.A. Holdaway. 2002b. Toxicity testing of crude oil and related compounds using early life stages of the crimson-spotted rainbowfish (*Melanotaenia fluviatilis*). *Ecotoxicology and Environmental Safety* 52:180–189.

Pollino, C.A. and D.A. Holdaway. 2003. Hydrocarbon-induced changes to metabolic and detoxification enzymes of the Australian Crimson-Spotted rainbowfish (*Melanotaenia fluviatilis*). *Environmental Toxicology* 18:21–28.

Pond, R., J.H. Kucklick and A.H. Walker. 1997. *Dispersant use: Real-time operational monitoring and long-term data gathering.* Prepared by Scientific and Environmental Associates, Incorporated for the Marine Preservation Association. Marine Preservation Association, Scottsdale, Arizona.

Pond, R.G., D.V. Aurand and J.A. Kraly (compilers). 2000. *Ecological Risk Assessment Principles Applied to Oil Spill Response Planning in the Galveston Bay Area.* Texas General Land Office, Austin, Texas.

Poremba, K. 1993. Influence of synthetic and biogenic surfactants on the toxicity of water-soluble fractions of hydrocarbons in sea water determined with the bioluminescence inhibition test. *Environmental Pollution* 80:25–29.

Poremba, K. and W. Gunkel. 1990. Marine Biosurfactants, III. Toxicity Testing with Marine Microorganisms and Comparison with Synthetic Surfactants. *Verlag der Zeitschrift fur Naturforschung* 46c:210–216.

Porter, M.R. 1991. *Handbook of Surfactants.* Blackie and Sons, Ltd., Glasgow, Scotland.

Prince William Sound Regional Citizens' Advisory Council (PWSRCAC). 2004. *Heated oil and under-reported dispersant volumes MAR MMS/Exxon cold water dispersant tests at OHMSETT.* Report prepared by the PWSRCAC. Prince William Sound Regional Citizens' Advisory Council, Anchorage, Alaska.

Proctor, R., A. Elliot and R. Flather. 1994. Forecast and hindcast simulations of the Braer oil-spill. *Marine Pollution Bulletin* 28(4):219–229.

Proctor, R., R. Flather, A. Roger and A.J. Elliott. 1994. Modeling tides and surface drift in the Arabian Gulf—application to the Gulf oil spill. *Continental Shelf Research* 14(5):531–545.

Ramachandran, S.D., P.V. Hodson, C.W. Khan and K. Lee. 2003. PAH uptake by juvenile rainbow trout exposed to dispersed crude oil. Pp. 743–754 in *Proceedings of the Twenty-*

Sixth Arctic and Marine Oilspill Program (AMOP) Technical Seminar, Victoria, British Columbia, Canada. Environment Canada, Ottawa, Ontario, Canada.

Ramachandran, S.D., C.W. Khan, P.V. Hodson, K. Lee and T. King. 2004. Role of droplets in promoting uptake of PAHs by fish exposed to chemically dispersed crude oil. Pp. 765–772 in *Proceedings of the Twenty-Seventh Arctic Marine Oilspill Program (AMOP) Technical Seminar*, Edmonton, Alberta, Canada. Environment Canada, Ottawa, Ontario, Canada.

Rand, G.M., Ed. 1995. *Fundamentals of aquatic toxicology: Effects, environmental fate, and risk assessment*. CRC Press, Incorporated, Boca Raton, Florida.

Reddy, P.G., H.D. Singh, M.G. Pathak, S.D. Bhagat and J.N. Baruah. 1983. Isolation and functional characterization of hydrocarbon emulsifying and solubilizing factors produced by a *Pseudomonas* species. *Biotechnology and Bioengineering* 25:387–401.

Reed, M., D. French-McCay, S. Feng and W. Knauss. 1991. A Three-Dimensional Natural Resource Damage Assessment and Coupled Geographical Information System. Pp. 631–637 in *Proceedings of 1991 National Conference of the American Society of Civil Engineers (ASCE) Hydraulic Engineering*, Nashville, Tennessee. American Society of Civil Engineers, Reston, Virginia.

Reed, M., O. Johansen, P.J. Brandvik, P. Daling, A. Lewis, R. Fiocco, D. Mackay and R. Prentki. 1999. Review: Oil Spill Modeling Toward the Close of the 20th Century: Overview of the State of the Art. *Spill Sciences and Technology Bulletin* 5(1):3–16.

Reed, M., P.S. Daling, A. Lewis, M.K. Ditlevsen, B. Brørs, J. Clark and D. Aurand. 2004. Modeling of Dispersant Application to Oil Spills in Shallow Coastal Waters. *Environmental Modeling and Software* 19(7-8):681–690.

Rewick, R.T., K.A. Sabo, J. Gates, J.H. Smith and L.T. McCarthy. 1981. An evaluation of oil spill dispersant testing requirements. Pp. 5–10 in *Proceedings of the 1981 International Oil Spill Conference*, Atlanta, Georgia. American Petroleum Institute, Washington, D.C.

Rhoton, S.L., R.A. Perkins, J.F. Braddock and C. Behr-Andres. 2001. A cold-weather species' response to chemically dispersed fresh and weathered Alaska North Slope crude oil. Pp. 1231–1236 in *Proceedings of the 2001 International Oil Spill Conference*, Tampa, Florida. American Petroleum Institute, Washington, D.C.

Ritchie, W. and M. O'Sullivan. 1994. *The environmental impact of the wreck of the Braer*. The Scottish Office, Edinburgh, Scotland.

Rogers, R.D., J.C. McFarlane and A.J. Cross. 1980. Adsorption and desorption of benzene in two soils and montmorillonite clay. *Environmental Science and Technology* 14:457–461.

Rosenberg, M. and E. Rosenberg. 1981. Role of adherence in growth of *Acinetobacter calcoaceticus* RAG-1 on hexadecane. *Journal of Bacteriology* 148(1):51–57.

Ross, S. and R. Belore. 1993. Effectiveness of dispersants on thick oil slicks. Pp. 1011–1022 in *Proceedings of the Sixteenth Arctic and Marine Oil Spill Program (AMOP) Technical Seminar*, Calgary, Alberta, Canada. Environment Canada. Ottawa, Ontario, Canada.

Ross, S. and I. Buist. 1995. Preliminary laboratory study to determine the effect of emulsification on oil spill evaporation. Pp. 91–312 in *Proceedings of the Eighteenth Arctic and Marine Oilspill Program (AMOP) Technical Seminar*, Edmonton, Alberta, Canada. Environment Canada. Ottawa, Ontario, Canada.

Ruxton, G.D. and N. Colegrave. 2003. *Experimental Design for the Life Sciences*. Oxford University Press, Incorporated, New York, New York.

Schippers, C., K. Gebner, T. Muller and T. Scheper. 2000. Microbial degradation of phenanthrene by addition of a sophorolipid mixture. *Journal of Biotechnology* 83:189–198.

Schlautman, M.A. and J.J. Morgan. 1993. Effects of aqueous chemistry on the binding of polycyclic aromatic hydrocarbons by dissolved humic materials. *Environmental Science and Technology* 27(5):961–969.

Schroh, K. 1995. Advanced aerial surveillance system for detection of marine pollution and international aerial surveillance cooperation in the North and Baltic Seas. Pp. 21–26 in

Proceedings of the 1995 International Oil Spill Conference, Long Beach, California. American Petroleum Institute, Washington, D.C.

Schwarzenbach, R.P., P.M. Gschwend and D.M. Imaboden. 1993. Pp. 436–484 in *Environmental Organic Chemistry*. Wiley Interscience, New York, New York.

Sea Empress Environmental Evaluation Committee (SEEEC). 1996. *Sea Empress Environmental Evaluation Committee (SEEEC) Initial Report*. SEEEC Secretariat, Cardiff, Wales, United Kingdom.

Shafir, S., J.V. Rijnb and B. Rinkevicha. 2003. The use of coral nubbins in coral reef ecotoxicology testing. *Biomolecular Engineering* 20(4-6):401–406.

Shigenaka, G. 2001. *Toxicity of Oil to Reef-Building Corals: A Spill Response Perspective*. NOAA Technical Memorandum NOA OR&R 8. National Oceanic and Atmospheric Administration, Hazardous Materials Response Division, Seattle, Washington.

Short, J.W. and P.M. Harris. 1996. Chemical sampling and analysis of petroleum hydrocarbons in near-surface seawater of Prince William Sound after the Exxon Valdez oil spill. *American Fisheries Society Symposium* 18:17–28.

Sigman, M.E., P.F. Schuler, M.M. Gosh and R.T. Dabestani. 1998. Mechanism of Pyrene Photochemical Oxidation in Aqueous and Surfactant Solutions. *Environmental Science and Technology* 32:3980–3985.

Simecek-Beatty, D., C. O'Conner and W.J. Lehr. 2002. 3-D Modeling of Chemically Dispersed Oil. Pp. 1149–1159 in *Proceedings of Twenty-Fifth Arctic and Marine Oilspill Technical Seminar*, Calgary, Alberta, Canada. Environment Canada. Ottawa, Ontario, Canada.

Singer, M.E. and W.R. Finnerty. 1984. Microbial metabolism of straight-chain and branched alkanes. Pp. 1–59 in *Petroleum Microbiology*, R.M. Atlas, Ed. Macmillan Publishing Company, New York, New York.

Singer, M.M., D.L. Smalheer, R.S. Tjeerdema and M. Martin. 1990. Toxicity of an oil dispersant to the early life stages of four California marine species. *Journal of Environmental Toxicology and Chemistry* 9:1389–1397.

Singer, M., D.L. Smalheer, R.S. Tjeerdema and M. Martin. 1991. Effects of spiked exposure to an oil dispersant on the early life stages of four marine species. *Journal of Environmental Toxicology and Chemistry* 10:1367–1374.

Singer, M.M., S. George, D. Benner, S. Jacobson, R.S. Tjeerdema and M.L. Sowby. 1993. Comparative toxicity of 2 oil dispersants to the early-life stages of 2 marine species. *Journal of Environmental Toxicology and Chemistry* 12(10):1855–1863.

Singer, M.M., S. George, S. Jacobson, I. Lee, R.S. Tjeerdema and M.L. Sowby. 1994a. Comparative effects of oil dispersants to the early-life stages of topsmelt (*Atherinops-affinis*) and kelp (*Macrocystis-pyrifera*). *Journal of Environmental Toxicology and Chemistry* 13(4): 649–655.

Singer, M.M., S. George, S. Jacobson, I. Lee, R.S. Tjeerdema and M.L. Sowby. 1994b. Comparative toxicity of Corexit(R)-7664 to the early-life stages of 4 marine species. *Archives of Environmental Contamination and Toxicology* 27(1):130–136.

Singer, M.M., S. George and R.S. Tjeerdema. 1995. Relationship of some physical-properties of oil dispersants and their toxicity to marine organisms. *Archives of Environmental Contamination and Toxicology* 29(1):33–38.

Singer, M.M., S. George, S. Jacobson, I. Lee, L.L. Weetman, R.S. Tjeerdema and M.L. Sowby. 1996. Comparison of acute aquatic effects of the oil dispersant Corexit 9500 with those of other Corexit series dispersants. *Ecotoxicology and Environmental Safety* 35(2):183–189.

Singer, M.M., S.George, I. Lee, S. Jacobson, L.L. Weetman, G. Blondina, R.S. Tjerdeema, D. Aurand and M.L. Sowby. 1998. Effects of dispersant treatment on the acute toxicity of petroleum hydrocarbons. *Archives of Environmental Contamination and Toxicology* 34(2): 177–87.

Singer, M.M., D. Aurand, G.E. Bragins, J.R. Clark, G.M. Coelho, M.L. Sowby and R.S. Tjeerdema. 2000. Standardization of the preparation and quantitation of water-accommodated fractions of petroleum for toxicity testing. *Marine Pollution Bulletin* 40(11):1007–1016.

Singer, M.M., D. Aurand, G. Coelho, G.E. Bragin, J.R. Clark, S. Jacobson, M.L. Sowby and R.S. Tjeerdema. 2001a. Making, measuring, and using water-accommodated fractions of petroleum for toxicity testing. Pp. 1269–1274 in *Proceedings of the 2001 International Oil Spill Conference*, Tampa, Florida. American Petroleum Institute, Washington, D.C.

Singer, M.M., S. Jacobson, R.S. Tjeerdema and M.L. Sowby. 2001b. Acute effects of fresh versus weathered oil to marine organisms: California findings. Pp. 1263–1268 in *Proceedings of the 2001 International Oil Spill Conference*, Tampa, Florida. American Petroleum Institute, Washington, D.C.

Sjoblom, J., N. Aske, I.H. Auflem, O. Brandal, T.E. Harve, O. Saether, A. Westvik, E.E. Johnsen and H. Kallevik. 2003. Our current understanding of water-in-crude oil emulsions. Recent characterization techniques and high pressure performance. *Advances in Colloid and Interface Science* 100-102:399–473.

Slade, G.J. 1982. Effect of Ixtoc I crude oil and Corexit® 9527 on spot (*Leiostomus xanthurus*) egg mortality. *Bulletin of Environmental Contamination and Toxicology* 29:525–530.

S.L. Ross. 1997. *A review of dispersant use on spills of North Slope crude oil in Prince William Sound and the Gulf of Alaska.* Report No. C\634.96.1. Report by S.L. Ross Environmental Research Ltd. for Prince William Sound Regional Citizens' Advisory Council, Anchorage, Alaska.

S.L. Ross. 2000. *Feasibility of using OHMSETT for dispersant testing.* Report to the MAR, Incorporated. MAR, Incorporated, Atlantic Highlands, New Jersey.

S.L. Ross. 2002. *Dispersant Effectiveness Testing in Cold Water.* Report by S.L. Ross Environmental Research Ltd. for the Minerals Management Service and ExxonMobil Research and Engineering Company. S.L. Ross, Ottawa, Ontario Canada.

S.L. Ross and MAR Incorporated. 2003. *Dispersant Effectiveness Testing on Alaskan Oils in Cold Water.* Report by S.L. Ross Environmental Research Ltd. and MAR Incorporated for the Minerals Management Service. S.L. Ross, Ottawa, Ontario, Canada; MAR Incorporated, Leonardo, New Jersey.

Smith, C.J., R.D. Delaune, W.H. Patrick, Jr. and J.W. Fleeger. 1984. Impact of dispersed and undispersed oil entering a Gulf Coast salt marsh. *Journal of Environmental Toxicology and Chemistry* 3(4):609–616.

Smith, J.E. 1968. *Torrey Canyon Pollution and Marine Life.* Cambridge University Press, New York, New York.

Society of Environmental Toxicology and Chemistry (SETAC). 2003. *Population-Level Ecological Risk Assessment.* Workshop held in Roskilde, Denmark, 23–27 August 2003. SETAC, Pensacola, Florida.

Socolofsky, S.A. and E.E. Adams. 2002. Multi-phase plumes in uniform and stratified crossflow. *Journal of Hydraulic Research*, 40(6):661–672.

Sorial, G.A., K.M. Koran, E. Holder, A.D. Venosa and D.W. King. 2001. Development of a rational oil spill dispersant effectiveness protocol. Pp. 471–478 in *Proceedings of the 2001 International Oil Spill Conference*, Tampa, Florida. American Petroleum Institute, Washington, D.C.

Sorial, G.A., A.D. Venosa, K.M. Koran, E. Holder and D.W. King. 2004a. Oil spill dispersant effectiveness protocol. I: Impact of operational variables. *Journal of Environmental Engineering* 130:1073–1084.

Sorial, G.A., A.D. Venosa, K.M. Koran, E. Holder and D.W. King. 2004b. Oil spill dispersant effectiveness protocol. II: Performance of revised protocol. *Journal of Environmental Engineering* 130:1085–1093.

Spaulding, M. 1988. A state-of-the-art review of oil spill trajectory and fate modeling. *Oil and Chemical Pollution* 4:39–55.

Spaulding, M.L., E. Howlett, E.L. Anderson and K. Jayko. 1992. OILMAP: A Global Approach to Spill Modeling. Pp. 15–21 in *Proceedings of the Fifteenth Arctic and Marine Oilspill Program (AMOP) Technical Seminar*, Edmonton, Alberta, Canada. Environment Canada, Ottawa, Ontario, Canada.

Spaulding, M., V. Kolluru, E. Anderson and E. Howlett. 1994. Application of a 3-dimensional oil spill model to hindcast the Braer spill. *Spill Science and Technology Bulletin* 1(1):23–35.

Speight, J.G. 1991. *The Chemistry and Technology of Petroleum.* Marcel Dekker, Incorporated, New York, New York.

Stephens, F.L., J.S. Bonner, R.L. Autenrieth and T.J. McDonald. 1999. TLC/FID analysis of compositional hydrocarbon changes associated with bioremediation. Pp. 219–224 in *Proceedings of the 1999 International Oil Spill Conference*, Seattle, Washington. American Petroleum Institute, Washington, D.C.

Stephenson, R. 1997. Effects of oil and other surface-active organic pollutants on aquatic birds. *Environmental Conservation* 24:121–129.

Stephenson, R. and C.A. Andrews. 1997. The effect of water surface tension on feather wettability in aquatic birds. *Canadian Journal of Zoology* 74:288–294.

Sterling, M.C., J.S. Bonner, C.A. Page, C.B. Fuller, A.N.S. Ernest and R.L Autenrieth. 2003. Partitioning of crude oil polycyclic aromatic hydrocarbons in aquatic systems. *Environmental Science and Technology* 37:4429–4434.

Sterling, M.C., Jr., J.S. Bonner, A.N.S. Ernest, C.A. Page and R.L. Autenrieth. 2004a. Characterizing aquatic sediment-oil aggregates using *in-situ* instruments. *Marine Pollution Bulletin* 48:533–542.

Sterling, M.C., Jr., J.S. Bonner, C.A. Page, C.B. Fuller, A.N.S. Ernest and R.L. Autenrieth. 2004b. Modeling crude oil droplet-sediment aggregation in nearshore waters. *Environmental Science and Technology* 38:4627–4634.

Sterling, M.C., J.S. Bonner, A.N.S. Ernest, C.A. Page and R.L. Autenrieth. 2004c. Chemical dispersant effectiveness testing: influence of droplet coalescence. *Marine Pollution Bulletin* 48:969–977.

Stevens, L. and J. Roberts. 2003. Dispersant effectiveness on heavy fuel oil and crude oil in New Zealand. Pp. 1–5 in *Proceedings of the 2003 International Oil Spill Conference*, Vancouver, Canada. American Petroleum Institute. Washington, D.C.

Stiver, W. and D. Mackay. 1984. Evaporation rate of oil spills of hydrocarbons and petroleum mixtures. *Environmental Science and Technology* 18:834–840.

Stoermer, S., G. Butler, C. Henry. 2001. Application of Dispersants to Mitigate Oil Spills in the Gulf of Mexico: The Poseidon Pipeline Spill Case Study. Pp. 1227–1299 in *Proceedings of the 2001 International Oil Spill Conference*, Tampa, Florida. American Petroleum Institute, Washington, D.C.

Stoffyn-Egli, P. and K. Lee. 2002. Observation and characterization of oil–mineral aggregates. *Spill Science and Technology* 8(1):31–44.

Stolzenbach, K.D., O.S. Madsen, E.E. Adams, A.M. Pollack and C.K. Cooper. 1977. *A Review and Evaluation of Basic Techniques Predicting the Behavior of Surface Oil Slicks.* Report No. 222 of the Ralph M. Parsons Laboratory for Water Resources and Hydrodynamics. Massachusetts Institute of Technology, Cambridge, Massachusetts.

Strom-Kristiansen, T., P.S. Daling, A. Lewis and A.B. Nordvik. 1994. *Weathering properties and chemical dispersibility of crude oils transported in U.S. waters.* Technical Report Series 93-032. Marine Spill Response Corporation, Washington, D.C.

Strom-Kristiansen, T., P.S. Daling, P.J. Brandvik and H. Jensen. 1995. Mechanical recovery of chemically-treated oil slicks. Pp. 407–421 in *Proceedings of the Nineteenth Arctic and Ma-*

rine Oilspill Program (AMOP) Technical Seminar, Calgary, Alberta, Canada. Environment Canada, Ottawa, Ontario, Canada.

Subba-Rao and M. Alexander. 1982. Effects of sorption on mineralization of low concentrations of aromatic hydrocarbons in lake water. *Applied Environmental Microbiology* 44:659–668.

Suter, G.W., II. 1993. *Ecological Risk Assessment*. Lewis Publishers, Boca Raton, Florida.

Sutton, C. and J.A. Calder. 1974. Solubility of higher-molecular-weight n-paraffins in distilled water and seawater. *Environmental Science and Technology* 8:654–657.

Sutton, C. and J.A. Calder. 1975a. Reply to correspondence to the editor. *Environmental Science and Technology* 9:354–6.

Sutton, C. and J.A. Calder. 1975b. Solubility of alkylbenzenes in distilled water and seawater at 25.0° C. *Journal of Chemical and Engineering Data* 20(3):320–322.

Swannell, R.P.J. and F. Daniel. 1999. Effect of dispersants on oil biodegradation under simulated marine conditions. Pp. 166–176 in *Proceedings of the 1999 International Oil Spill Conference*, Seattle, Washington. American Petroleum Institute, Washington, D.C.

Swannell, R.J.P., F. Daniel, B.C. Croft, M.A. Englehardt, S. Wilson, D.J. Mitchell and T. Lunel. 1997. Influence of physical and chemical dispersion on the biodegradation of oil under simulated marine conditions. Pp. 617–641 in *Proceedings of the Twentieth Arctic and Marine Oilspill Program (AMOP) Technical Seminar*, Vancouver, British Columbia. Environment Canada, Ottawa, Ontario, Canada.

Swartz, R.C., D.W. Schults, R.O. Ozretich, J.O. Lamberson, F.A. Cole, T.H. DeWitt, M.S. Redmond and S.P. Ferraro. 1995. S PAH: A model to predict the toxicity of polynuclear aromatic hydrocarbon mixtures in field collected sediments. *Journal of Environmental Toxicology and Chemistry* 14:1977–1987.

Tang, L. 2004. Cylindrical liquid-liquid jet instability. Ph.D. Dissertation, University of Hawaii, Manoa.

Tasaki, R. and A. Ogawa. 1999. Emulsification of crude oil: A new equation and governing parameters. Pp. 1011–1014 in *Proceedings of the 1999 International Oil Spill Conference*, Seattle, Washington. American Petroleum Institute, Washington, D.C.

Tennekes, H. and J.L. Lumley. 1972. *A First Course in Turbulence*. Massachusetts Institute of Technology Press, Cambridge, Massachusetts.

Terray, E.A., M.A. Donelan, Y.C. Agrawal, W.M. Drennan, K.K. Kahma, A.J. Williams, III, P.A. Hwang and S.A. Kitaigorodskii. 1996. Estimates of kinetic energy dissipation under breaking waves. *Journal of Physical Oceanography* 26:792–807.

Thomas, D. and T. Lunel. 1993. The *Braer* incident: Dispersion in action. Pp. 843-859 in *Proceedings of the Sixteenth Arctic and Marine Oil Spill Program (AMOP) Technical Seminar*, Calgary, Alberta, Canada. Environment Canada. Ottawa, Ontario, Canada.

Thorhaug, A., K. Aiken, M. Anderson, B. Carby, V. Gordon, F. McDonald, J. McFarlane, B. Miller, R. Reese, M. Rodriquez, G. Sidrak, H. Teas and W. Walker. 1989. Dispersant Use for Tropical Nearshore Waters: Jamaica. Pp. 415-418 in *Proceedings of the 1989 International Oil Spill Conference*, San Antonio, Texas. American Petroleum Institute, Washington, D.C.

Transtronics. 2000. *Viscosity*. Transtronics, Lawrence, Kansas. [Online] Available at: http://xtronics.com/reference/viscosity.htm [January 18, 2005].

Traxler, R.W. and L.S. Bhattacharya. 1978. Effect of a chemical dispersant on microbial utilization of petroleum hydrocarbons. Pp. 181–187 in *Chemical Dispersants for the Control of Oil Spills*, L.T. McCarthy, G.P. Lindblom and H.F. Walter, Eds. American Society for Testing and Materials. Philadelphia, Pennsylvania.

Trudel, B.K, Ed. 1998. *Dispersant Use in Alaska: A Technical Update*, Conference Proceedings. Prince William Sound Oil Spill Recovery Institute, Cordova, Alaska.

Trudel, K. 2002. *Lecture Notes, Oil Spill Dispersant Course for Supervisors*. S.L. Ross Environmental Research, Ottawa, Ontario, Canada, Unpublished.

Tsahalis, D.T. 1979. Contingency Planning for Oil Spills: RiverSpill—A River Simulation Model. Pp. 27–36 in *Proceedings of the 1979 International Oil Spill Conference*, Los Angeles, California. American Petroleum Institute, Washington, D.C.

Twardus, E.M. 1980. *A Study to Evaluate the Combustibility and Other Physical and Chemical Properties of Aged Oils and Emulsions.* Report of Environment Canada, Research and Development Division, Environmental Emergency Branch, Environmental Impact Control Directorate, Environmental Protection Service, Ottawa, Ontario, Canada.

Underwood, A.J. 1994. On beyond BACI: Sampling designs that might reliably detect environmental disturbances. *Ecological Applications* 4(1):3–15.

Unsal, M. 1991. Comparative toxicity of crude-oil, dispersant and oil-dispersant mixture to prawn *Palaemon-elegans*. *Toxicology and Environmental Chemistry* 31-2:451–459.

U.S. Coast Guard, National Oceanic and Atmospheric Administration, Environmental Protection Agency, Centers for Disease Control and Prevention and Minerals Management Service. 2001. *Special monitoring of applied response technologies (SMART).* U.S. Coast Guard, Washington, D.C. [Online] Available at: http://response.restoration.noaa.gov/oilaids/SMART/SMART.html [March 17, 2005].

Valiela I. 1984. Marine Ecological Processes. Springer-Verlag, New York, Inc.

Van Hamme, J.D. and O.P. Ward. 1999. Influence of chemical surfactants on the biodegradation of crude oil by a mixed bacterial culture. *Canadian Journal of Microbiology* 45:130–137.

Van Hamme, J.D. and O.P. Ward. 2001. Physical and metabolic interactions of Pseudomonas sp strain JA5-B45 and Rhodococcus sp strain F9-D79 during growth on crude oil and effect of a chemical surfactant on them. *Applied and Environmental Microbiology* 67(10): 4874–4879.

van Loosdrecht, M.C.M., J. Lyklema, W. Norde and A.J.B. Zehnder. 1990. Influence of interfaces on microbial activity. *Microbiolological Reviews* 54:75–87.

van Wezel, A., D. de Vries, D. Sijm and A. Opperhuizen. 1996. Use of the lethal body burden in the evaluation of mixture toxicity. *Ecotoxicology and Environmental Safety* 35:236–241.

Vandermeulen, J.H. 1980. Chemical dispersion of oil in coastal low-energy systems: saltmarshes and Tidal Rivers. Pp. 27–29 in *Chemical Dispersion of Oil Spills: An International Research Symposium*, D. Mackay, P.G. Wells and S. Paterson, Eds. Publication number EE-17. University of Toronto, Toronto, Ontario, Canada.

Varadaraj, R., M.L. Robbins, J. Bock, S. Pace and D. MacDonald. 1995. Dispersion and biodegradation of oil spills on water. Pp. 101–106 in *Proceedings of the 1995 International Oil Spill Conference*, Long Beach, California. American Petroleum Institute, Washington, D.C.

Varhaar, H.J.M, C.J. VanLeeuwen and J.L.M. Hermens. 1992. Classifying Environmental Pollutants, 1: Structure-activity Relationships for Prediction of Aquatic Toxicity. *Chemosphere* 25:471–491.

Venosa, A.D., G.A. Sorial, T.L. Richardson, F. Uraizee and M.T. Suidan. 1999. Research leading to revisions in EPA's dispersant effectiveness protocol. Pp. 1019–1022 in *Proceedings of the 1999 International Oil Spill Conference*, Seattle, Washington. American Petroleum Institute, Washington, D.C.

Venosa, A.D., G.A. Sorial and D.W. King. 2001. Round-robin testing of a new EPA dispersant effectiveness protocol. Pp. 467–470 in *Proceedings of the 2001 International Oil Spill Conference*, Tampa, Florida. American Petroleum Institute, Washington, D.C.

Venosa, A.D., D.W. King and G.A. Sorial. 2002. The baffled flask test for dispersant effectiveness: a round robin evaluation of reproducibility and repeatability. *Spill Science and Technology Bulletin* 7(5-6):299–308.

Veron, F. and W.K. Melville. 2001. Pulse-to-pulse coherent Doppler measurements of waves and turbulence. *Journal of Atmospheric and Oceanic Technology* 16(11):1580–1597.

Volkman, J.K., G.J. Miller, A.T. Revill and D.W. Connell. 1994. Oil spills, Part 6. Pp. 509–695 in *Environmental Implications of Offshore Oil and Gas Development in Australia—The Findings of an Independent Scientific Review*, J.M. Swan, J.M. Neff and P.C. Young, Eds. Australian Petroleum Exploration Association (APEA) and Energy Research and Development Corporation (ERDC), Sydney, Australia.

Wade, T.L. and J.G. Quinn. 1980. Incorporation, distribution and fate of saturated petroleum hydrocarbons in sediments from a controlled marine ecosystem. *Marine Environmental Research* 3:15–33.

Walker, A.H. and D.R. Henne. 1991. The Region III Regional Response Team technical symposium on dispersants: An interactive, educational approach to enlightened decision making. Pp. 405–410 in *Proceedings of the 1991 International Oil Spill Conference*, San Diego, California. American Petroleum Institute, Washington, D.C.

Walker, M.I. and T. Lunel. 1995. Response to oil spills at sea using both demulsifiers and dispersants. Pp. 537–558 in *Proceedings of the Eighteenth Arctic and Marine Oilspill Program (AMOP) Technical Seminar*, Edmonton, Alberta, Canada. Environment Canada. Ottawa, Ontario, Canada.

Walker, A.H., D.L. Ducey, Jr., J.R. Gould and A.B. Nordvik, Eds. 1993a. *Formation and Breaking of Water-in-Oil Emulsions: Workshop Proceedings.* Marine Spill Response Corporation, Washington, D.C.

Walker, M., M. McDonaugh, D. Albone, S. Grigson, A. Wilkinson and G. Baron. 1993b. Comparison of observed and predicted changes to oil after spills. Pp. 389–393 in *Proceedings of the 1993 International Oil Spill Conference*, Tampa, Florida. American Petroleum Institute, Washington, D.C.

Walker, A.H., T. Lunel, P.J. Brandvik and A. Lewis. 1995. Emulsification processes at sea—Forties crude oil. Pp. 471–491 in *Proceedings of the Eighteenth Arctic and Marine Oilspill Program (AMOP) Technical Seminar*, Edmonton, Alberta, Canada. Environment Canada. Ottawa, Ontario, Canada.

Ward, G.A., B. Baca, W. Cyriacks, R.E. Dodge, A. Knap. 2003. Continuing long-term studies of the TROPICS Panama oil and dispersed oil spill sites. Pp.1–9 in *Proceedings of the 2003 International Oil Spill Conference*, Vancouver, Canada. American Petroleum Institute. Washington, D.C.

Wardrup, J.A. 1987. *The effects of oils and dispersants on mangroves: a review and bibliography.* Occasional paper no. 2. University of Adelaide, Environmental Studies, Adelaide, Australia.

Watkinson, R.J. and P. Morgan. 1990. Physiology of aliphatic hydrocarbon-degrading microorganisms. *Biodegradation* 1:79–92.

Weise, A.M., C. Nalewajko and K. Lee. 1999. Oil-mineral fine interactions facilitate oil biodegradation in seawater. *Environmental Technology* 20:811–824.

Wells, P. and K.G. Doe. 1976. Results of the E.P.S. oil dispersant testing program: concentrates, effectiveness testing, and toxicity to marine organisms. *Spill Technology Newsletter* 1:9–16.

Wells, P.G., S. Abernethy and D. Mackay. 1982. Study of oil water partitioning of a chemical dispersant using an acute bioassay with marine crustaceans. *Chemosphere* 11:1071–1086.

Wells, P.G., S. Abernathy and D. Mackay. 1985. Acute toxicity of solvents and surfactants of dispersants in two planktonic crustaceans. Pp. 228–240 in *Proceedings of the Eighth Arctic and Marine Oilspill Program (AMOP) Technical Seminar*, Edmonton, Alberta, Canada. Environment Canada, Ottawa, Ontario, Canada.

Wetzel, D.L. and E.S. Van Fleet. 2001. Cooperative studies on the toxicity of dispersants and dispersed oil to marine organisms: A 3-year Florida study. Pp. 1237–1241 in *Proceedings of the 2001 International Oil Spill Conference*, Tampa, Florida. American Petroleum Institute, Washington, D.C.

Wheelock, C.E., T.A. Baumgartner, J.W. Newman, M.F. Wolfe and R.S. Tjeerdema. 2002. Effect of nutritional state on Hsp60 levels in the rotifer *Brachionus plicatilis* following toxicant exposure. *Aquatic Toxicology* 61:89–93.

White, D.M., I. Ask and C. Behr-Andres. 2002. Laboratory study on dispersant effectiveness in Alaskan seawater. *Journal of Cold Regions Engineering* 16(1):17–27.

Wiechart, J., M.L. Rideout, D.I. Little, M. McCormick, E.H. Owens and B.K. Trudel. 1991. Development of dispersant pre-approval for Washington and Oregon coastal waters. Pp. 435–438 in *Proceedings of the 1991 International Oil Spill Conference*, San Diego, California. American Petroleum Institute, Washington, D.C.

Winters, J.K. 1978. Fate of petroleum-derived aromatic compounds in seawater held in outdoor tanks. Chapter 12 in *Environmental studies, south Texas outer continental shelf, biology and chemistry*. BLM Contract AA550-CT7-11. U.S. Department of Interior, Bureau of Land Management (BLM), Washington, D.C.

Wodzinski, R.S. and D. Bertolini. 1972. Physical state in which naphthalene and bibenzyl are utilized by bacteria. *Applied Microbiology* 23(6):1077–1081.

Wodzinski, R.S. and J.E. Coyle. 1974. Physical state of phenanthrene for utilization by bacteria. *Applied Microbiology* 27(6):1081–1084.

Wolfe, M.F., J.A. Schlosser, G.J.B. Schwartz, S. Singaram, E.E. Mielbrecht, R.S. Tjeerdema and M.L. Sowby. 1998a. Influence of dispersants on the bioavailability and trophic transfer of petroleum hydrocarbons to primary levels of a marine food chain. *Aquatic Toxicology* 42:211–227.

Wolfe, M.F., G.J.B. Schwartz, S. Singaram, E.E. Mielbrecht, R.S. Tjeerdema and M.L. Sowby. 1998b. Effects of salinity and temperature on the bioavailability of dispersed petroleum hydrocarbons to the Golden-Brown Algae, *Isochrysis galbana. Archives of Environmental Contamination and Toxicology* 35:268–273.

Wolfe, M.F., G.J.B. Schwartz, S. Singaram, E.E. Mielbrecht, R.S. Tjeerdema and M.L. Sowby. 1998c. Influence of dispersants on the bioavailability of napthalene from the water-accommodated fraction of crude oil to the Golden-Brown Algae, *Isochrysis galbana. Archives of Environmental Contamination and Toxicology* 35:274–280.

Wolfe, M.F., G.J.B. Schwartz, S. Singaram, E.E. Mielbrecht, R.S. Tjeerdema and M.L. Sowby. 1999a. Influence of dispersants on the bioavailability and trophic transfer of phenanthrene to algae and rotifers. *Aquatic Toxicology* 48:13–24.

Wolfe, M.F., H.E. Olsen, K.A. Gasuad, R.S. Tjeerdema and M.L. Sowby. 1999b. Induction of heat shock protein (hsp)60 in *Isochrysis galbana* exposed to sublethal preparations of dispersant and Prudhoe Bay crude oil. *Marine Environmental Research* 47:473–489.

Wolfe, M.F., G.J.B. Schwartz, S. Singaram, E.E. Mielbrecht, R.S. Tjeerdema and M.L. Sowby. 2001. Influence of dispersants on the bioavailability and trophic transfer of petroleum hydrocarbons to larval topsmelt (*Atherinops affinis*). *Aquatic Toxicology* 52:49–60.

Wood, P.A., T. Lunel, F. Daniel, R. Swannell, K. Lee and P. Stoffyn-Egli. 1998. Influence of oil and mineral characteristics on oil-mineral interaction. Pp. 51–77 in *Proceedings of the Twenty-first Arctic and Marine Oilspill Program (AMOP) Technical Seminar*, Edmonton, Alberta, Canada. Environment Canada, Ottawa, Ontario, Canada.

Wright, A.L., R.W. Weaver and J.W. Webb. 1997. Oil bioremediation in salt marsh mesocosms as influenced by N and P fertilization, flooding, and season. *Water Air and Soil Pollution* 95(1-4):179–191.

Wright, D.A. and G.M. Coehlo. 1996. *Dispersed oil and dispersant fate and effects research: MD program results for 1995.* MSRC Technical Report Series 95-013, Draft report. Marine Spill Response Corporation, Washington, D.C.

Wu, R.S.S., P.K.S. Lam and B.S. Zhou. 1997. Effects of two oil dispersants on phototaxis and swimming behaviour of barnacle larvae. *Hydrobiologia* 352:9–16.

Wyers, S.C. 1985. Sexual reproduction of the coral *Diploria strigosa* (Scleratinia, Faviidae) in Bermuda: Research in progress. *Proceedings of the Fifth International Coral Reef Congress, Tahiti* 5:301–502.

Wyers, S.C., H.R. Frith, R.E. Dodge, S.R. Smith, A.H. Knap and T.D. Sleeter. 1986. Behavioral effects of chemically dispersed oil and subsequent recovery in *Diploria strigosa* (DANA). *Marine Ecology* 7:23–42.

Yamada, M., H. Takada, K. Toyoda, A. Yoshida, A. Shibata, H. Nomura, M. Wada, M. Nishimura, K. Okamoto and K. Ohwada. 2003. Study on the fate of petroleum-derived polycyclic aromatic hydrocarbons (PAHs) and the effect of chemical dispersant using an enclosed ecosystem mesocosm. *Marine Pollution Bulletin* 47:105–113.

Yapa, P.D. and H.T. Shen. 1994. Modeling of river oil spills: a review. *Journal of Hydraulic Research* 32(5):765–782.

Yapa, P.D. and L. Zheng. 1997. Simulation of oil spills from underwater accidents I: model development. *Journal of Hydraulic Research* 35(5):673–687.

Yapa, P.D. and L. Zheng. 1999. Modeling underwater oil/gas jets and plumes. *Journal of Hydraulic Engineering* 125(5):481–491.

Yapa, P.D., H.T. Shen, D.S. Wang and K. Angammana. 1992. An Integrated Computer Model for Simulating Oil Spills in the St. Lawrence River. *Journal of Great Lakes Research* 18(2):34–51.

Yapa, P.D., H.T. Shen and K.S. Angammana. 1994. Modeling Oil Spills in a River-Lake System. *Journal of Marine Systems* 4:453–471.

Youssef, M. and M. Spaulding. 1993. Drift current under the action of wind and waves, Pp. 587–615 in *Proceedings of the Sixteenth Arctic and Marine Oil Spill Program (AMOP) Technical Seminar*, Calgary, Alberta, Canada. Environment Canada. Ottawa, Ontario, Canada.

Zachleder, V. and Z. Tukaj. 1993. Effect of fuel-oil and dispersant on cell-cycle and macromolecular-synthesis in the chlorococcal alga *Scenedesmus-armatus*. *Marine Biology* 117(2): 347–353.

Zagorski, W. and D. Mackay. 1982. *Studies of water-in-oil emulsions*. EPS Report EE-34. Environment Canada. Ottawa, Ontario, Canada.

Zhang, D.F., A.K. Easton and J.M. Steiner. 1997. Simulation of Coastal Oil Spills Using the Random Walk Particle Method with Gaussian Kernel Weighting. *Spill Sciences and Technology Bulletin* 4(2):71–88.

Zhang, Y. and R.M. Miller. 1992. Enhanced octadecane dispersion and biodegradation by a *Pseudomonas* rhamnolipid surfactant (biosurfactant). *Applied and Environmental Microbiology* 58(10):3276–3282.

Zhang, Y. and R.M. Miller. 1994. Effect of a *Pseudomonas* rhamnolipid biosurfactant on cell hydrophobicity and biodegradation of octadecane. *Applied and Environmental Microbiology* 60(6):2101–2106.

Zurcher, F. and M. Thuer. 1978. Rapid weathering processes of fuel oil in natural waters—analysis and interpretations. *Environmental Science and Technology* 12:838–843.

Appendixes

A

Committee and Staff Biographies

COMMITTEE

Jacqueline Michel (Chair) received her Ph.D. in geology from the University of South Carolina in 1980. Currently, she is the President of Research Planning, Inc. She is an expert in oil and chemical response and contingency planning. Dr. Michel has been providing scientific support to the National Oceanic and Atmospheric Administration's Hazardous Materials Response Division since 1978. She has served on several NRC committees including the Committee on Oil in the Sea: Inputs, Fates, and Effects, the Committee on Marine Transportation of Heavy Oil, and chaired the Committee on Spills of Emulsified Fuels. Dr. Michel is currently a member of the NRC's Ocean Studies Board.

E. Eric Adams gained his Ph.D. in Hydrodynamics from Massachusetts Institution of Technology (MIT) in 1975. He is a Senior Research Engineer and Lecturer for the Department of Civil and Environmental Engineering at MIT. Dr. Adams is also the Director of the Civil and Environmental Engineering Department Master of Engineering Program and serves as the Associate Director for Research for the MIT Sea Grant College Program. His research specializes in environmental fluid mechanics, physical and mathematical modeling of pollutant transport and mixing, and hydrologic tracer studies. His articles span topics such as modeling descending carbon dioxide injections in the ocean, as well as the role of slip velocity in controlling the behavior of stratified multi-phase plumes. Dr. Adams is a member of the Massachusetts Bays Circulation and Water Quality

Modeling Model Evaluation Group and serves on the Technical Advisory Committee for the Boston Harbor Navigation Improvement Project.

Yvonne Addassi received her Master of Science in Ecology with emphasis in Environmental Policy from the University of California, Davis in 1997. Currently, she is a staff environmental scientist for the Office of Spill Prevention and Response for the California Department of Fish and Game. Her primary responsibilities include program coordination for the statewide licensing, approval and use of oil spill cleanup agents as well as the use of applied response technologies (ART), such as *in-situ* burning and dispersants; primary research and policy development for preparation and updating the legislative report on the feasibility of requiring alternative oil spill clean-up technologies; serving as state liaison for applied response technologies including Western States Task Force ART subcommittee, Regional IX Regional Response Team ART subcommittee; development and implementation of state-policies for the use of ARTs, specifically *in-situ* burning and dispersants, coordination of three dispersant area subcommittees utilizing net environmental analysis as a means of trade-off quantification. Ms. Addassi has published several papers on *in-situ* burning and NEBA including: Utilizing Net Environmental Benefit Analysis (NEBA), the Use of *In-Situ* Burning as a Mechanism to Minimize Environmental Impacts of a Marine Oil Spill, and Case Study: SS JACOB LUCKENBACH: Adaptation of Traditional Incident Command Structure to Meet the Unique Needs of Long-Term Wildlife Operations. She serves as an advisory board member for the Oiled Wildlife Care Network and also serves on the board of directors for the California Association of Professional Scientists.

Tom Copeland received B.A. degrees in English and Economics from Whitman College in 1971. He has been a commercial fisherman in Alaska from 1963 until his recent retirement. Currently, Mr. Copeland farms bamboo in Everson WA. In 1989, he responded to the *Exxon Valdez* oil spill, worked for passage of OPA 90, and eventually was a founding member of the Oil Spill Prevention and Response Committee of the Prince William Sound Regional Citizen's Advisory Council. One of the committee's duties is to advise the council on questions concerning dispersants. Before his recent retirement, Mr. Copeland served 12 years on the committee along with three terms on the council representing both aquiculture and environmental groups.

Mark S. Greeley holds a Ph.D. in biology from the University of Alabama at Birmingham. Dr. Greeley is a research scientist in the Environmental Sciences Division of Oak Ridge National Laboratory (ORNL). He manages the Aquatic Toxicology Laboratory at ORNL and the Biological Monitoring and Abatement Program for the National Nuclear Security Administration's Y-12 National Security Complex. Dr. Greeley's research interests focus on reproductive and developmental toxicology, aquatic ecotoxicology, biomarkers of contaminant exposure and effects, and methods of environmental assessment. He has been involved in a number of projects assessing the effects of environmental pollution on aquatic and terrestrial communities. His research projects include the use of zebrafish as a model for studying the functional relationship between gene and protein expression in response to toxicant exposure.

Bela James graduated in 1972 from Texas A&M University with a Ph.D. in Biological Oceanography. He presently is a spill response, environmental specialist for Shell Global Solutions (US) Inc. in Houston and has been Shell's leading spills technology expert for the past 13 years. Dr. James has been involved for many years with National Oceanic and Atmospheric Administration Scientific Support Coordination staff, Regional Response Teams, and American Society for Testing and Materials (ASTM) spill response subcommittees. He worked on a recent deepwater discharge study off Norway and developed deepwater spill modeling and response guidelines.

Beth McGee has a B.A. in Biology from the University of Virginia, a M.S. in Ecology from the University of Delaware, and a Ph.D. in Environmental Science from the University of Maryland. Dr. McGee is currently the Maryland Senior Scientist with the Chesapeake Bay Foundation. Her background and expertise is in environmental toxicology and benthic ecology, particularly the fate and effects of contaminants on aquatic organisms. For over 15 years, Dr. McGee has been very active in Chesapeake Bay water quality issues, conducting research and serving on several technical subcommittees and advisory groups. In addition, she has worked for a variety of state and federal agencies, including the U.S. Fish and Wildlife Service, EPA, and the Maryland Department of the Environment, giving her extensive knowledge of the Clean Water Act, the Endangered Species Act, CERCLA, and the Oil Pollution Act. She is currently a Board Member of the Society of Environmental Toxicology and Chemistry—North America.

Carys Mitchelmore gained her Ph.D. from the University of Birmingham in 1997 investigating toxicity processes in aquatic organisms. Dr. Mitchelmore is an Assistant Professor for the University of Maryland, Center for Environmental Science, Chesapeake Biological Laboratory, in Solomons, Maryland. Her expertise lies in aquatic toxicology and her research experience includes investigating the basic processes involved in contaminant driven toxicity in a variety of aquatic organisms including coral reefs, and in developing novel tools (biomarkers) to assess contaminant impacts. Dr. Mitchelmore has authored and coauthored several journal articles in the areas of aquatic biochemistry, genetic toxicology and endocrine disruption in both vertebrate and invertebrate species.

Yasuo Onishi received a Ph.D. in Mechanics and Hydraulics from the University of Iowa in 1972. He is a Chief Scientist in the Environmental Technology Directorate of the Pacific Northwest National Laboratory. Dr. Onishi is also a member of the Graduate Adjunct Faculty for the Department of Civil and Environmental Engineering at Washington State University. His principal discipline is fluid mechanics/hydrology, environmental risk assessment, and reactive fluid dynamics; specifically, transport and fate of sediment/contaminants (e.g., toxic chemicals, heavy metals, radionuclides, oil) in natural environment, aquatic biota/human health assessment, and chemical reactions/transport of multi-component, multi-phase, Newtonian/non-Newtonian fluids with chemically active solids, liquid, and gaseous chemicals. Dr. Onishi served as the U.S. Coordinator for "Radionuclide Behavior in Soil-Water" of the bilateral "U.S./ (former) U.S.S.R. Joint Coordinating Committee on Civilian Nuclear Power Safety", and has been working for over ten years with scientists from the former Soviet Union to assess impacts on aquatic environment and human health caused by the Chernobyl plant nuclear accident.

James R. Payne received his Ph.D. in Chemistry from the University of Wisconsin—Madison in 1974, and he was a Woods Hole Oceanographic Institution Postdoctoral Scholar from 1974 to 1975. Currently, he is the President of Payne Environmental Consultants, Inc., which specializes in oil and chemical pollution studies for government and industry. Over the 30 years of his professional career, Dr. Payne has been involved in numerous projects dealing with marine- and water-pollution issues, including laboratory-scale and outdoor flow-through wave-tank studies of oil weathering behavior in arctic and subarctic environments. He has also supported NOAA natural resource damage assessment efforts after the *Exxon Valdez, American Trader, Kuroshima, New Carissa,* and *Westchester* oil spills. As a result of his environmental studies and field investigations, Dr. Payne has authored or co-authored three books and chapters in four others. He has published over 30 peer-reviewed articles and/or papers in various conference proceedings, and he has prepared over 45 environmental reports for use by various governmental agencies and private clients. In addition to his other publications, Dr. Payne contributed background chapters for the 1985 NRC publication *Oil in the Sea—Inputs, Fates,* *and Effects*, and he was a member of NRC Ocean Studies Board Committees dealing with the Effectiveness of Oil Spill Dispersants (1985–1988) and Spills of Emulsified Fuels (2001). He also served on the NRC Polar Research Board Committee to review the Oil Spill Recovery Institute's arctic and subarctic research programs (2002).

David Salt has an Ordinary National Diploma (OND) in Mechanical Engineering. He currently serves as Technical Director for the Oil Spill Response Limited, Global Alliance with responsibility for the technical response preparedness of the Alliance and all technical issues related to the response activities of the organization. Previously, Mr. Salt joined Oil Spill Response Limited-Southampton in 1981, where he served for several years as Operations Manager and two years as General Manager. Formerly, Mr. Salt worked as an engineer officer for BP on tankers for five years. After his time at sea, he transferred to Sullom Voe terminal, where he became Pollution Officer and was responsible for the maintenance and operation of the response stockpile and leading the response to terminal based incidents. In 1992, Mr. Salt was posted to East Asia Response Limited to establish the center in Singapore. He currently serves as Secretary to the International Technical Advisory Committee, a pan-industry and response community technical group looking at best practice and response issues. Mr. Salt has been particularly involved in the use of dispersants and the introduction of a number of aerial dispersant platforms to satisfy particular response needs in both the UK and West Africa. He has been involved in a huge range of international spills during his time with OSRL including these major incidents: *Haven, Nagasaki Spirit, Evoikas, Natuna Sea, Exxon Valdez, Patmos, Sea Empress, Katina P.*, and *Toledo.*

Brian Wrenn earned his Ph.D. in environmental science in civil engineering from the University of Illinois at Urbana-Champaign in 1992, and M.S. in biological oceanography from the University of Miami in 1984. He is currently an assistant professor of civil engineering and environmental biotechnology at Washington University. Dr. Wrenn's research interests include: bioremediation, biological treatment of industrial and hazardous wastes, water and wastewater treatment, environmental microbiology and biodegradation, environmental and analytical chemistry, biodegradation kinetics, and analysis of biological treatability. He is currently a member of the American Chemical Society, the American Society for Microbiology, the Water Environment Federation, and the Association of Environmental Engineering and Science Professors.

STAFF

Dan Walker (Study Director) obtained his Ph.D. in geology from the University of Tennessee in 1990. A Scholar at the Ocean Studies Board, Dr. Walker also holds a joint appointment as a Guest Investigator at the Marine Policy Center of the Woods Hole Oceanographic Institution. Since joining the Ocean Studies Board in 1995, he has directed a number of studies including *Future Needs in Deep Submergence Science: Occupied and Unoccupied Vehicles in Basic Ocean Research* (2004), *Environmental Information for Naval Warfare* (2003), *Oil in the Sea III: Inputs, Fates and Effects* (2002), *Spills of Emulsified Fuels: Risks and Response* (2002), *Clean Coastal Waters: Understanding and Reducing the Effects of Nutrient Pollution* (2000), *Science for Decisionmaking: Coastal and Marine Geology at the U.S. Geological Survey* (1999), *Global Ocean Sciences: Toward an Integrated Approach* (1998), and *The Global Ocean Observing System: Users, Benefits, and Priorities* (1997). A member of the American Geophysical Union, the Geological Society of America, and the Oceanography Society, Dr. Walker was recently named Editor of the *Marine Technology Society Journal*. A former member of both the Kentucky and the North Carolina state geologic surveys, Dr. Walker's interests focus on the value of environmental information for policymaking at local, state, and national levels.

Sarah Capote gained her B.A. in history from the University of Wisconsin-Madison in the winter of 2001. She is a senior program assistant with the Ocean Studies Board. During her tenure with the Board, Ms. Capote worked on the following reports: *Exploration of the Seas: Voyage into the Unknown* (2003), *Nonnative Oysters in the Chesapeake Bay* (2004), *Future Needs in Deep Submergence Science: Occupied and Unoccupied Vehicles in Basic Ocean Research* (2004), the interim report for *Elements of a Science Plan for the North Pacific Research Board* (2004), *A Vision for the International Polar Year 2007–2008* (2004), *Marine Mammal Populations and Ocean Noise: Determining When Noise Causes Biologically Significant Effects* (2005), and *Final Comments on the Science Plan for the North Pacific Research Board* (2005).

B

Dispersant Authorizations

Caribbean	Pre-authorization	Case-by-Case Authorization	Restricted/Exclusion Areas	Special Considerations
Puerto Rico	1) Waters at least 1/2 nautical mile (nm) seaward of any shoreline extending to the Exclusive Economic Zone (EEZ); 2) Water depth at least 60 feet.	1) Waters within 1/2 nm of the shoreline; 2) Water depth less than 60 feet; 3) Waters designated as marine reserves, National Marine Sanctuaries, National or State Wildlife Refuges, or proposed or designated Critical Habitats; 4) Waters in mangroves or coastal wetland ecosystems (including submerged algal beds and submerged sea grass beds), or directly over coral communities which are in less than 60 feet of water.	N/A	N/A
U.S. Virgin Islands	1) Waters at least 1 nm seaward of any shoreline extending to the EEZ; 2) Water depth at least 60 feet.	1) Waters within 1 nm of the shoreline; 2) Water depth less than 60 feet; 3) Waters designated as marine reserves, National Marine Sanctuaries, National or State Wildlife Refuges, or proposed or designated Critical Habitats; 4) Waters in mangroves or coastal wetland ecosystems (including submerged algal beds and submerged sea grass beds), or directly over coral communities which are in less than 60 feet of water.	1) Waters of the Virgin Islands National Park including waters 1 nm seaward of the park boundary;** 2) Waters of the Buck Island Reef National Monument including waters 1 nm seaward from the park boundary.**	**Dispersants may be applied to a Restricted Area in the event that: 1) Dispersant application is necessary to prevent or substantially reduce a hazard to human life; and/or 2) An emergency modification of this Agreement is made on an incident-specific basis.

Region 1	Pre-authorization	Expedited Authorization	Trial Application Zones	Restricted/Exclusion Areas	Special Considerations
Maine & New Hampshire	1) Waters at least 1/2 nm seaward of shoreline extending to the EEZ.	1) Waters within 1/2 nm seaward of shoreline; 2) Concurrence of Federal On-Scene Coordinator (FOSC) and State OSC(s).	N/A	N/A	1) Isle of Shoals (New Hampshire Fish & Game) 2) 1/2 nm to 2 nm (Dept. of the Interior 1-hour consultation) 3) 1/2 to 2 nm from DOI-owned or -managed islands between January 1st and March 1st & May 1st & August 1st (DOI concurrence); 4) Jeffrey's Ledge between April 1st & September 30th (NMFS).
Massachusetts & Rhode Island	1) Waters under the jurisdiction of the COTP-Boston and COTP-Providence seaward of 2 nm of the mainland or of designated islands extending to the EEZ; 2) Water depth is greater than 40 feet.	N/A	N/A	N/A	1) Jeffrey's Ledge between May 1st and September 30th (NMFS); 2) Stellwagen Bank between May 1st and November 15th (SBNMS Sanctuary Manager); 3) Great South

continues

Region 1	Pre-authorization	Expedited Authorization	Trial Application Zones	Restricted/Exclusion Areas	Special Considerations
Massachusetts & Rhode Island (cont'd.)					Channel between May 1st and June 30th and October 1st and November 15th (NFMS); 4) Cape Cod Bay between February 1st and May 15th.
Long Island Sound, CT	1) Waters under the jurisdiction of the COTP-NY and/or COTP-Long Island Sound at least 3 nm seaward of the Territorial Sea Baseline;[1] 2) Waters along the coastline of NJ and/or the south shore of Long Island Sound, NY (north of the demarcation of the jurisdiction of COTP Philadelphia), west of a line from Montauk Point Light bearing 132° True to the outermost extent of the EEZ.	N/A	1) Waters as defined in the Pre-authorization zone between 1/2 and 3 nm from the Territorial Sea Baseline;[1] 2) Trial application may also be conducted in the following water bodies: Hudson River *south* of the George Washington Bridge; Upper New York Bay; The Narrows; Lower New York Bay; Raritan Bay, excluding Spermaceti Cove and not within 1/2 nm of Sandy Hook, NJ;	1) Waters under jurisdiction of COTP-NY and COTP-Long Island Sound that lie within 1/2 nm of the Territorial Sea Baseline, including all bays and coves. 2) Also includes the Hudson River *north* of the Tappan Zee Bridge and Long Island Sound, with the exception of the COTP-NY AOR falling into the Trial Application Zone.	N/A

		Arthur Kill; Newark Bay up to mouths of the Passaic and Hackensack Rivers; Kill Van Kull; East River south of Throgs Neck Bridge; Long Island Sound within COTP-NY AOR only, excluding Little Bay, Little Neck Bay, Manhasset Bay, Hempstead Harbor, Eastchester Bay, Pelham Bay and not within 1/2 nm of the northern shore of Long Island.

Region 2	Pre-authorization	Trial Application
New York & New Jersey	1) Waters under the jurisdiction of COTP-NY and COTP-Long Island Sound at least 3 nm of the Territorial Sea Baselineı along the coast of New Jersey (north of the demarcation of the jurisdiction of COTP-Philadelphia) and along the south shore of Long Island, NY west of a line from Montauk Point Light bearing 132° True to the outermost extent of the EEZ.	1) Waters under the jurisdiction of COTP-NY and COTP-Long Island Sound that lie between 1/2 and 3 nm from the Territorial Sea Baseline. In addition, specific water bodies are also included in Zone 2, and are as follows: Hudson River *south* of George Washington Bridge Upper New York Bay The Narrows Lower New York Bay Raritan Bay excluding Spermaceti Cove and not within 1/2 nautical miles of Sandy Hook, New Jersey Arthur Kill Newark Bay up to mouths of Passaic and Hackensack Rivers Kill Van Kull East River south of Throgs Neck Bridge Long Island Sound within COTP-NY area of

continues

Region 2	Pre-authorization	Trial Application
New York & New Jersey (cont'd.)	2) Including Ambrose Channel south of a line drawn between East Rockaway Inlet Breakwater Light and Sandy Hook Light and seaward of a line connecting the 10-meter soundings off the coasts of NJ and NY.	responsibility only, excluding Little Bay, Little Neck Bay, Manhasset Bay, Hempstead Harbor, Eastchester Bay, Pelham Bay and not within 1/2 nm of the northern shore of Long Island.

Region 3	Pre-authorization	Trial Application Zones	Restricted/Exclusion Areas
Coastal Delaware, Maryland, Virginia	1) Waters greater than 3 nm seaward of shoreline to the outermost extent of the EEZ.	1) Trial application may be conducted at distances 1/2 to 3 nm seaward of the shoreline or in depths greater than 40 feet, excluding bays and coves; 2) Without concurrence, the FOSC may authorize trial application on spills of only 50 bbls or less, or on portions 50 bbls or less of larger spills; 3) Concurrence/non-concurrence decision is limited to within 4 hours after agency communication has been established.	1) Limited pre-authorization granted for trial use only on spills 50 bbls or less, or on portions 50 bbls or less of larger spills, on waters within Big Stone Beach Anchorage in the Delaware Bay area; 2) Waters within 1/2 nm of shoreline; 3) Water depth less than 40 feet; 4) Limited to concurrence obtained within 4 hours after agency communication has been established.

Region 4	Pre-authorization	Case-by-Case Authorization	Special Considerations
North Carolina, South Carolina, Georgia, Florida, Alabama, Mississippi	1) Waters at least 3 nm seaward of shoreline extending to the EEZ; 2) Water depth is at least 10 meters.	1) Waters within 3 nm of shoreline; 2) Water depth less than 10 meters; 3) Waters that are under State or federal special management jurisdiction, which i ncludes anywaters designated as marine reserves, National Marine Sanctuaries, National or State Wildlife Refuges, units of	**Special Case for West Coast of FL: 1) State waters extend 9 nm seaward into the Gulf of Mexico; 2) No case-by-case authorization will be required or considered necessary from EPA, DOI, DOC, or the State of Florida for waters greater than 10 meters in depth that extend

		the National Park Service, or proposed or designated Critical Habitats; 4) Waters are in mangrove and coastal wetland ecosystems (which includes submerged algal beds and submerged sea grass beds), or directly over living coral communities, which are in less than 10 meters deep.	more than 3 nm seaward on FL's West Coast unless designated as meeting case-by-case criteria.

Region 6	Pre-authorization	Special Considerations
Texas & Louisiana	1) Waters farther. than 3 nm seaward of shoreline; *or* 2) Water depth greater than 10 meters, *whichever is farthest from shore*, extending to the EEZ	1) All dispersant spray operations are conducted during daylight hours only; 2) Flower Gardens National Marine Sanctuary: A. Decision to apply dispersants must be based upon the weather, sea state, water temperature, oil characteristics, history of spill, and risk of spill contact for particular life forms; B. All efforts must be made to apply them in water as deep as possible and as far from the Sanctuary as possible.

Region 9	Pre-authorization	Case-by-Case Authorization/RRT Approval Required	Special Considerations
California (North Coast Region: From the California-Oregon border to the southern edge of Sonoma County)	1) All federal waters off the north coast region 3–200 nm seaward.	N/A	N/A
San Diego Area	1) All federal waters 3–200 nm seaward	1) Federal waters of the Channel Islands National Marine Sanctuary (CINMS), including those waters 3 nm seaward of the baseline to the	1) Pre-authorization criteria do not

continues

Region 9	Pre-authorization	Case-by-Case Authorization/RRT Approval Required	Special Considerations
San Diego Area (cont'd.)	within the COTP-San Diego AOR excluding an area 3 nm from the U.S./Mexico border.	Sanctuary's outer boundary at 6 nm from the islands; 2) All remaining federal waters within COTP-San Diego area of responsibility not covered by the above zones remain under the current "Quick Approval Process" and are designated, "RRT Approval Required" zones.	apply to Group I oils (gasoline, diesel, and jet fuels).
San Francisco, Central Coast (Santa Cruz & Monterey counties), and San Francisco Bay & Delta	1) All federal waters 3–200 nm with the Central Coast region, the San Francisco Bay, and Delta AOR. 2) All federal waters 6–200 nm seaward of the Monterrey Bay National Marine Sanctuary (MBNMS).	1) Federal waters of the Gulf of the Farallones, Cordell Bank, and the portion of the Monterey Bay National Marine Sanctuary within the San Francisco Bay and Delta AOR are designated as "RRT Approval Required." 2) All remaining waters within the San Francisco Bay and Delta AOR not covered by the above zones, remain under the current "Quick Approval Process" and are designated, "RRT Approval Required" zones.	N/A
Los Angeles-Long Beach	1) All federal waters 3–200 nm with the COTP LA/LB.	1) The Federal waters of the Monterey Bay National Marine Sanctuary (MBNMS) within the Los Angeles-Long Beach region, and up to 3 nautical miles beyond the MBNMS as "RRT Approval Required;" 2) The Federal waters of the Channel Islands National Marine Sanctuary (CINMS) as "RRT Approval Required." This include those waters from 3 nautical miles seaward of the baseline to the Sanctuary's outer boundary at 6 nautical miles from the islands. 3) All remaining waters within our area of responsibility not covered by the above zones remain under the current "Quick Approval Process" and are designated "RRT Approval Required."	N/A

Region 10	Pre-authorization	Case-by-Case Authorization	Restricted/Exclusion Areas	Special Considerations
Pacific Northwest	N/A	1) All requests for dispersant application in the States of Washington and Oregon are subject to review on a case-by-case basis.	N/A	N/A
Alaska	Please refer to Alaska Dispersant link for detailed information.	Please refer to Alaska Dispersant link for detailed information.	Dispersant use is not recommended in the following zones: 1) *Upper Cook Inlet:* Inshore of the 5-fathom isobath, during the first 3 hours of an ebb tide, and for all periods outside of that, an area north of a line extending from Point Possession to the North Forelands; 2) *Middle Cook Inlet:* Inshore of the 5-fathom isobath near the northeast shoreline and inshore of the 10-fathom isobath along the southeast and west shorelines; 3) *Lower Cook Inlet:* Inshore of the 10-fathom isobath along the east and west shorelines, inshore of the 5-fathom isobath around Kalgin Island, and inshore of a 1-mile buffer along the extreme southern portions of Cook Inlet, where the 10-fathom isobath drops off rapidly near the shore; 4) *Port of Valdez and Valdez Arm:* Tatitlek Narrows and Columbia Bay; 5) *Main Body of Prince William Sound:* The	1) *Cook Inlet:* Due to the large numbers of commercially valuable adult salmon, the section of Cook Inlet north of a line drawn along the latitude at Anchor Point north of Kachemak Bay is considered to be "Case-by-Case" during the period from July 1st–August 15th 2) *Prince William Sound:* The tanker lanes have a pre-authorization designation, while most of the

continues

Region 10	Pre-authorization	Case-by-Case Authorization	Restricted/Exclusion Areas	Special Considerations
Alaska (cont'd.)			majority of the waters within this section, with the exception of tanker lanes and an appropriate buffer zone to either side of these lanes; 6) *Hinchinbrook Entrance:* The area around Seal Rocks; 7) *Copper River Delta:* Inshore of the 3-mile territorial limit along the coast from Cape Hinchinbrook to Kayak Island; 8) *Montague Island:* Inshore of a line drawn approximately 1 nm off the outside coasts of Montague and Elrington Islands and extending east to Cape Junken.	remaining area is "Case-by-Case" 3) Please refer to Alaska's Dispersant Use Policy for more in-depth guidelines concerning Cook Inlet and Prince William Sound.
Oceania	Pre-authorization		Case-by-Case Authorization	
Guam	N/A		N/A	
Hawaii	1) Water depths greater than 10 fathoms, with the exception of the Maui County four-island area bounded by La'au Point, Molokai to Kaena Point, Lanai; Kamaiki Point, Lanai to Cape Kuikui, Kahoolawe; Cape Kuikui, Kahoolawe to Cape Hanamanioa, Maui; and Lipoa Point, Maui to Cape Halawa, Molokai.		1) In any case where circumstances do not meet the guidelines, use of dispersants is subject to case-by-case authorization.	
American Samoa	N/A		1) Any applications of dispersants within the American Samoa AOR must be authorized by the Oceania Regional Response Team.	

[1]Reference 33 CFR 2.05-10 for the definition of Territorial Sea Baseline.

SOURCE: Categorized by National Response Team Region; U.S. Coast Guard, *http://www.uscg.mil/vrp/maps/popups/Dispersant%20Table%20Aug%2026.doc.*

C

Acronyms

ADEC	Alaska Department of Environmental Conservation
ADIOS	Automatic Data Inquiry for Oil Spills
ALC	Arabian light crude
ANS	Alaska North Slope
API	American Petroleum Institute
ASTM	American Society for Testing and Materials
BAF	bioaccumulation factors
BIOS	Baffin Island Oil Spill
BSC	Bass Strait crude oil
BTEX	benzene, toluene, ethylbenzene, and xylenes
CCD	charge-coupled device
CEWAF	chemically enhanced water accommodated fractions
CMC	critical micelle concentration
CROSERF	Chemical Response to Oil Spills Environmental Research Forum
cP	centipoise
cSt	centistokes
DFO	Department of Fisheries and Oceans, Canada
DOC	U.S. Department of Commerce
DOI	U.S. Department of the Interior
DOR	dispersant-to-oil ratio

EPA	Environmental Protection Agency
ERA	Ecological Risk Assessment
EVOS	*Exxon Valdez* oil spill
FDA	fluorescein diacetate
FOSC	Federal On-Scene Coordinator
GC	gas chromatography
GC-FID	gas chromatography-flame ionization detector
GC-MS	gas chromatography-mass spectroscopy
GNOME	GNU Network Object Model Environment
HLB	hydrophilic-lipophilic balance
IFO 300	Intermediate Fuel Oil 300
IMMSP	Institute of Mathematical Machines and Systems Problems
IR	infrared
LURSOT	Laser Ultrasonic Remote Sensing of Oil Thickness
MAH	mono-aromatic hydrocarbon
MMS	Minerals Management Service
MNS	Mackay-Nadeau-Steelman
NCP	National Contingency Plan
NOAA	National Oceanic and Atmospheric Administration
NOEC	no observed effect concentration
NRC	National Research Council
NRDAM	Natural Resource Damage Assessment Model
NRT	National Response Team
OHMSETT	Oil and Hazardous Materials Simulated Environmental Test Tank
OPA 90	Oil Pollution Act of 1990
OWR	oil-to-water ratio
PAH	polynuclear aromatic hydrocarbons
PANH	polynuclear aromatic nitrogen heterocycle
PBCO	Prudhoe Bay crude oil
PIV	particle image velocimetry
PWSRCAC	Prince William Sound Regional Citizens' Advisory Council

ROSS	River Oil Spill Simulation
RRT	Regional Response Team
SERF	Shoreline Environmental Research Facility
SETAC	Society of Environmental Toxicology and Chemistry
SIMAP	Spill Impact Model Application Package
SLAR	side-looking airborne radar
SMART	Specialized Monitoring of Advanced Response Technologies
SOP	Standard Operating Procedure
SOR	surfactant-to-oil ratio
SPC	simplified pseudo-component
SPM	suspended particulate material
TBP	true boiling point
TEM	total extractable material
THC	total hydrocarbon concentration
TI	Toxicity Index
TOC	total organic carbon
TPH	total petroleum hydrocarbons
TROPICS	Tropical Oil Pollution Investigations in Coastal Systems
T/V	Tanker/Vessel
UCM	unobserved component model
USCG	U.S. Coast Guard
USGS	U.S. Geological Survey
UV	ultraviolet
VCO	Venezuelan medium crude oil
VOC	volatile organic compound
WAF	water-accommodated fraction
WPMB	Water Planning and Management Branch
WSF	water-soluble fraction
WSL	Warren Springs Laboratory

D

Definitions and Unit Conversions

Conversions reported in the text conserve the number of significant figures of the original reported value using rules consistent with the NRC report on *Oil in the Sea III: Inputs, Fates and Effects* (NRC, 2003) and available on the following Massachusetts Institute of Technology website: http://web.mit.edu/10.001/Web/Course_Notes/Statistics_Notes/Significant_Figures.html.

We are reporting everything in metric units except where common or regular usage requires that values be reported in English units. In these cases, metric equivalents are provided in parenthesis.

barrels × 42 = US gallons
liters × 0.264 = US gallons
cubic meters × 264.2 = US gallons
cubic feet × 7.481 = US gallons
liters × 0.0009 = tonnes*
(note tonnes = metric tons)
tonnes × 294 = US gallons*
tonnes × 7.33 = barrels
US gallons × 0.0034 = tonnes*
US gallons × 3.785 = liters

---

*NOTE: The gallon is a volume measurement. The tonne is a weight measurement. For truly precise conversions between gallons and tonnes, it is important to take into account that equal volumes of different types of oil differ in their densities. The specific gravity (sp gr), or density in relation to pure water is generally less than 1.0. Specific gravity of petroleum products varies from about 0.735 for gasoline to about 0.90 for heavy crude to 0.95 for Bunker C (No. 6 fuel). In some cases the oil is even heavier than water, especially with some of the heavy No. 6 fuels. These oils can sink. The volume that a particular weight of oil takes up varies with temperature and atmospheric pressure.

Some common metric unites and their english equivalents:

miles × 1.609 = kilometers
miles × 1.1 = nautical miles
nautical miles × 1.852 = kilometers
feet × 0.304 = meters
nautical miles per hour (knots) × 1.852 = kilometers per hour
miles per hour × 1.609 = kilometers per hour
gallons/acre × 9.35 = liters/hectare
acre × 0.404 = hectare
inches × 25.4 = millimeters
fathom × 1.8288 = meters

The conversion factor of 294 gallons per tonne is derived from an average specific gravity of 0.83, which corresponds to an API gravity or degree API of 39. Note that API gravity and specific gravity are inversely proportional as per the formulae below. The 294 gallons/tonne conversion unit is also convenient because it happens that 294 gallons = 7 barrels.
API = (141.5/sp gr) – 131.5
sp gr = 141.5/(API + 131.5)

E

Analysis of the Sensitivity of Dispersed Oil Behavior to Various Processes

There are many complex mechanisms interacting to control oil transport and fate. They include oil surface spreading, evaporation, entrainment, emulsification, horizontal/vertical advection and natural diffusion (affected by current, wind, wave), sedimentation, oil droplet rising (due to buoyancy), biodegradation, dissolution, and chemical dispersion and associated oil droplet-size changes.

It is very difficult to quantify the importance of these mechanisms on oil concentrations changing with space and time. Decisionmakers contemplating potential use of dispersants must know where and how fast spilled oil is migrating. It is, therefore, important to evaluate how these mechanisms affect oil transport. As discussed in Chapter 4, oil transport and fate models integrate major controlling physical, chemical, and biological processes into one system to identify cause-effect relationships and to evaluate effects of these mechanisms on oil concentrations. Therefore, it was decided to conduct a sensitivity analysis as part of the committee's work on evaluating the existing literature.

The objective of the sensitivity analysis conducted here was to understand the role that various processes have in the transport and fate of spilled oil, both with and without the use of chemical dispersants, and to determine the sensitivity of this behavior to these processes. While some sensitivity was discussed analytically in earlier parts of Chapter 4, an integrated sensitivity analysis can only be performed with a comprehensive computer model that includes all of the relevant processes. The sensitivity modeling analysis discussed here was conducted for 14 cases with various oil types (crude oil and light and heavy refined oil), environmental conditions (wind speed, waves, diffusion), chemical dispersion effectiveness, and oil droplet sizes in nearshore and offshore Florida Coast. Biodegradation was not modeled in the sensitivity analysis, because models were run for only one or two simulation days.

Although the sensitivity results are somewhat dependent on the test conditions and simulation codes selected for this evaluation, they are good indicators of the importance of the model parameters tested. The sensitivity analysis helps identify knowledge gaps, future research needs, and a new approach (potential of using models to assist on-scene decision-makers for the possible dispersant use) to assess dispersant use. The analysis also can enhance our knowledge and understanding of the combined effects of these mechanisms on oil transport and fate, and beneficial and adverse impacts of using chemical dispersants.

MODEL SETUP

Code Selection

The two codes that were considered are SIMAP (French-McCay, 2003, 2004), and the combined use of ADIOS2 (Lehr et al., 2002) and Lagrangian 3-D GNOME (Simecek-Beatty et al., 2002); both simulate most mechanisms needed to assess dispersant use, except dispersant effectiveness itself and changes in the oil droplet sizes (which are user inputs). The latter codes are used by NOAA for their real-time response to an actual oil spill. The flow field is not simulated by 3-D GNOME, but it is supplied to the code as a model input. Although the code can accept a three-dimensional velocity distribution, NOAA usually uses a simpler two-dimensional flow field, balancing pressure forces, bottom friction, Coriolis force, and water density variation, adjusted by tide and wind for a real-time emergency response. Their capability to reflect a three-dimensional flow field in a real-time emergency response needs to be improved, especially in complex nearshore areas. Oil trajectory predictions during a spill are constantly adjusted to match observed trajectories by re-adjusting the model input, including the velocity and wind fields.

Because of the availability of the codes within the public domain, their ease of use, limited requirements to operate the models, and their potential use for determining dispersant use during an oil spill event, ADIOS2 and 3-D GNOME were used for the sensitivity analysis. However, this investigation was not intended to evaluate these codes, but rather to use their simulation results as indicators for sensitivity of oil concentrations to various transport and fate processes. Because the sensitivity results may reflect specifics of these two codes, additional sensitivity analysis is recommended with other codes.

Water Body

There are many different types of water bodies: nearshore, offshore, semi-confined water, estuaries, lakes, and rivers. Because nearshore environments may impose more difficulties for decisionmakers regarding the appropriateness of dispersant use (Reed et al. 2004), this evaluation focused more on nearshore waters with few additional offshore model runs. However, other water bodies should also be evaluated. A somewhat schematic representation of the southern Florida coast was selected to represent a relatively simple nearshore region. A coastal flow moving northeast along the Florida Keys from Key West toward Key Largo was imposed as the geostrophic flow without accounting for wind and tide. The oil spill location (25°01′N, 80°23′W) was selected to be in 10-m-deep water about 9 km offshore. For some sensitivity test cases, the oil was assumed to be spilled further offshore in 200-m-deep water (25°01′N, 80°11′W).

Oil Types

Three types of oil were selected: Alaska North Slope crude oil, Intermediate Fuel Oil 300 (IFO 300), and marine diesel primarily used in the southern United States for commercial marine operations. It is useful to evaluate other crude oils, including heavy California crudes, light and heavy crudes imported from Africa, Mexico, Venezuela, and elsewhere. But the Alaskan Crude, IFO 300, and diesel were selected, in part, due to the availability of data on their physical and chemical properties. Their physical properties built into the ADIOS2 code (Lehr et al., 2002) are shown in Table E-1. These oils also represent widely varying chemical components, as indicated in their distillation cuts (also built into the code)

TABLE E-1 Densities and Viscosities of Three Oils

Characteristics of Oil		Alaska North Slope Crude Oil	IFO 300	Diesel Oil
Density, g/mL		0.8936 at 15°C	0.9859 at 15°C	0.8362 at 15°C
Viscosity, cP		23.0 at 15°C	14,470 at 15°C	4.0 at 15°C
Surface Tension, dynes/cm	Oil-water	26.1 at 0°C	—	24.9 at 0°C
	Oil–seawater	23.8 at 0°C	37.3 at 15°C	24.6 at 0°C
Percent Oil Evaporation to Initiate Emulsification		18	100	100

shown in Table E-2. Intermediate fuel oil is heavy and viscous, and marine diesel is light and much less viscous. Alaska North Slope crude falls between them in density and viscosity. IFO 300 and this particular marine diesel have more higher-molecular-weight components than Alaska North Slope crude oil, as indicated by higher distillation temperatures assigned in ADIOS2. For the sensitivity analysis, 10,000 barrels (roughly 1,500 tonnes of crude; 1,600 tonnes of IFO 300; 1,400 tonnes of diesel) of oil were assumed to be released to the surface of this water body over one hour.

Environmental Conditions

Selected environmental variables were wind speed, wave height, horizontal diffusion, and vertical diffusion. These variables can either be entered independently, or the latter three can be computed by the model as a function of the wind speed and current, using relationships such as Morales et al. (1997) described in Lehr et al. (2002) and Simecek-Beatty et al. (2002). The latter option was used here, resulting in the following values:

TABLE E-2 Distillation Cuts of the Three Oils Used in the Modeling Sensitivity Analysis

	Alaska North Slope Crude Oil		Intermediate Fuel Oil 300		Diesel Fuel Oil	
Oil Cut Number	Weight Fraction, wt percent	Temperature, °C	Weight Fraction, wt percent	Temperature, °C	Weight Fraction, wt percent	Temperature, °C
1	1.0	42	1.1	180	1.1	120
2	4.0	98	1.1	200	1.1	140
3	5.0	127	6.4	250	1.1	160
4	5.0	147	9.4	300	3.2	180
5	5.0	172	7.2	350	5.2	200
6	10.0	216	8.1	400	20.4	250
7	10.0	238	6.0	450	31.9	300
8	5.0	247	3.0	500	25.5	350
9	5.0	258	4.9	550	9.7	400
10	5.0	265	9.8	600	1.0	450
11	5.0	272	14.7	650	—	—
12	10.0	282	10.7	700	—	—
13	30.0	>282	17.4	>700	—	—

SOURCE: Data from Environment Canada, 2005.

- 2, 10, and 25 m/s for the speed of south-southwesterly wind
- 0, 0.9, and 5.5 m for the breaking wave height
- 160, 350, and 660 cm^2/s for the horizontal diffusion coefficient
- 23, 51, and 97 cm^2/s for the vertical diffusion coefficient.

The mixing depth was selected to be 10 m for the oil releases on the 10-m-deep water, while it was assigned to be 200 m for the oil releases in deep water. The mixing depth can represent the maximum depth of surface mixing (e.g., by Langmuir circulation) or a diffusion floor imposed by a pycnocline or thermocline, though neither Langmuir circulation nor density stratification was simulated in the modeling. The vertical diffusion coefficient through and below the mixing depths was assigned as 0.11 cm^2/s. The diffusion coefficients calculated by ADIOS2 and 3-D GNOME above the mixing depth may be greater than realistic values, especially in offshore deep water. Moreover they are temporally and spatially constant. While it is possible to estimate surface diffusivities using real-time surface current data (e.g., Ojo and Bonner, 2002), for simplicity and comparative purposes, the above values were used for both the nearshore and offshore modeling cases.

Dispersant Application

The sensitivity analysis was performed with and without dispersant applications. It was assumed that dispersant application began six hours after the release and it took six hours to treat the entire slick, even though in a real spill case, there may be some operational limitations that prevent complete areal coverage. Because the dispersant effectiveness is the measure of how much oil on the water surface is entrained into the water column below, the oil concentration in the water column is almost directly proportional to the dispersant effectiveness. Thus, the sensitivity analysis did not vary the dispersant effectiveness. It was arbitrarily assumed that 50 percent of the oil on the water surface would be "dispersed" (entrained into the water column from the water surface), and that the oil droplet diameters would be the same as, or reduced by a factor of five from the baseline sizes, resulting in slower rise toward the water surface. The baseline oil droplet sizes were assigned to be between 10 and 70 μm. Although the dispersant does not increase the oil droplet sizes, two additional distributions of larger droplet sizes (two and four times the baseline sizes) were also selected to evaluate an effect of oil droplet sizes on the vertical oil distribution. These four oil droplet-size distributions, each including seven droplet sizes, are shown in Figure E-1. Figure E-2 shows the representative rising velocity of an oil droplet of Alaska North Slope crude oil in 20° C (roughly 68° F) sea water (kinematic viscosity

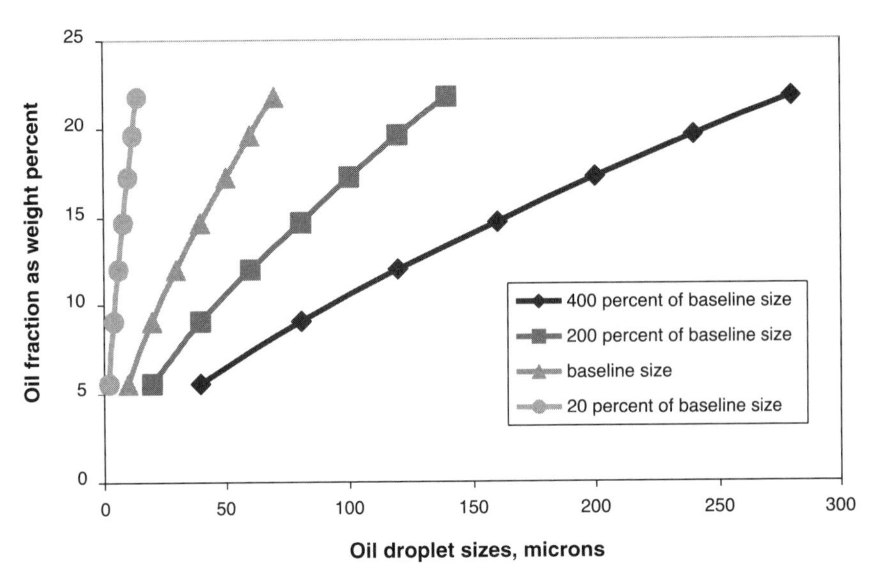

FIGURE E-1 Oil droplet-size distributions.

of 1.1×10^{-6} m^2/s). NOAA's ADIOS2 code includes oil and sediment interaction. For this modeling the suspended sediment concentration was assigned to be 5 mg/L. However, these models do not include oil-particle interaction, which may be important in estuaries and high energy coastal waters with varying salinity and high concentrations of suspended organic and inorganic matter.

Modeling Results and Evaluation

Sensitivity analyses were performed by combining the three oil types, three wind speeds, three wave heights, three horizontal diffusion coefficients, and three vertical diffusion coefficients with and without dispersant application, as well as four oil droplet-size distributions with dispersant applications. Table E-3 shows the 14 test cases for the sensitivity analysis. Cases 1 and 2 are baseline cases for the crude oil without and with a dispersant. Note that oil droplet sizes in Case 2 were assigned to be 20 percent of those of Case 1. Cases 3 through 6 are for environmental changes without and with a dispersant application. Cases 7 though 10 are for three oil types without and with a dispersant. Cases 1 through 10 are those whose oil releases are in 10-m-deep water (nearshore), while Cases 11 through 14 have 200 m of water depth (offshore) at the release point with four sets of oil droplet-size distributions.

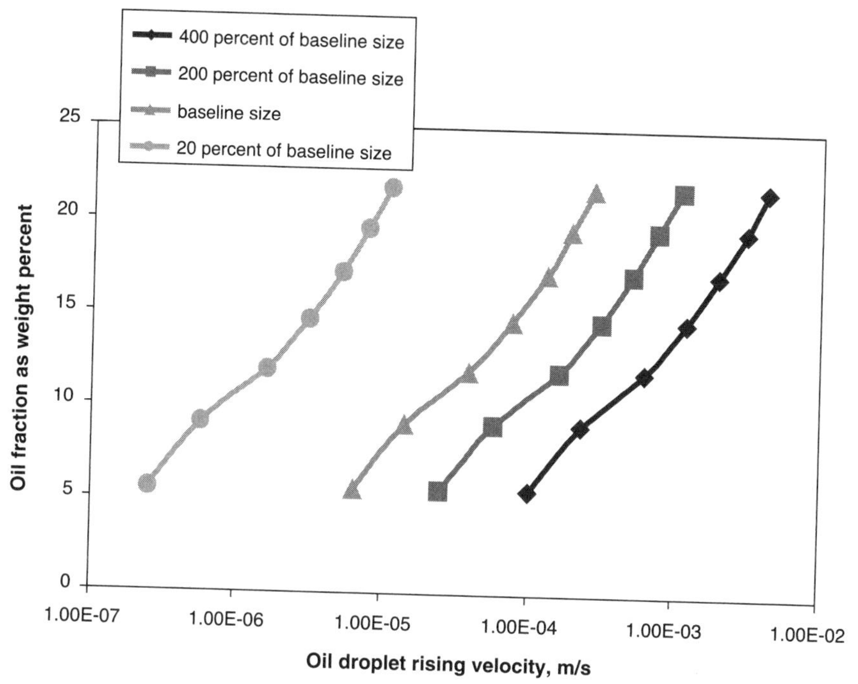

FIGURE E-2 Oil droplet rising velocity. Note: Oil viscosity in a surface slick changes as the oil weathers due mainly to emulsification. Thus, no single value for oil viscosity was applied to calculate the rising velocity. The rising velocities shown here were estimated with the viscosity (1.07×10^{-6} m^2/s) of sea water at 20°C.

The sensitivity modeling was performed for 24 simulation hours for the nearshore cases and 48 hours for the offshore cases. Simulation results for evaporation, natural dispersion, chemical dispersion, amount floating, and amount beached at 14, 24, and 48 hours are summarized in Table E-4. Oil volume percent may be converted to weight percent using oil density. However, as will be discussed later, oil density changes with time due to selective evaporation of lighter-molecular-weight components and emulsification. These sensitivity analysis results are discussed below under four categories: (i) Main characteristics of oil transport and fate with and without a dispersant (Cases 1 and 2), (ii) Environmental conditions (Cases 3-6), (iii) Oil types (Cases 7-10), and (iv) Oil droplet sizes (Cases 11-14). Model results of Case 1 (without a dispersant) and Case 2 (with a dispersant) are shown in Figure E-3, as an example of predicted oil distributions shown in Table E-4.

TABLE E-3 Model Input for Fourteen Sensitivity Test Cases

Case No.	Disp. Use	Oil Type	Wind Speed (m/s)	Wave Height (m)	Hori. Diff. Coeff. (cm^2/s)	Vert. Diff. Coeff. (cm^2/s)	Amount Evapor. to Start Emulsif. (percent)	Oil Dropl. Size as Percent of Baseline Sizes (percent)	Water Depth at Release Point (m)
1	No	Crude	10	0.9	350	51	18	100	10
2	Yes	Crude	10	0.9	350	51	18	20	10
3	No	Crude	2	0	160	23	18	100	10
4	No	Crude	25	5.5	660	97	18	100	10
5	Yes	Crude	2	0	160	23	18	20	10
6	Yes	Crude	25	5.5	660	97	18	20	10
7	No	IFO300	10	0.9	350	51	100	100	10
8	No	Diesel	10	0.9	350	51	100	100	10
9	Yes	IFO300	10	0.9	350	51	100	20	10
10	Yes	Diesel	10	0.9	350	51	100	20	10
11	Yes	Crude	10	0.9	350	51	18	20	200
12	Yes	Crude	10	0.9	350	51	18	100	200
13	Yes	Crude	10	0.9	350	51	18	200	200
14	Yes	Crude	10	0.9	350	51	18	400	200

Main Characteristics of Oil Transport and Fate With and Without a Dispersant

Cases 1 and 2 have a wind speed of 10 m/s with corresponding wave and diffusion coefficients (see Table E-3). The 3-D GNOME code prescribes that the oil slick is advected with a speed equal to the underlying current velocity plus 1 ~ 4 percent of the wind speed in the direction of the wind, while oil in the subsurface water is carried by the underlying current. For Case 1 without a dispersant application, Figure E-4 shows the location of the predicted oil plume 24 hours after the spill and the oil spill location marked by "+." Black spots represent oil floating on the water surface, that traveled about 50 km from the spill site over 24 hours. The colored areas show different ranges of oil concentrations in the top 1 m of the water column; note that oil plume in the water column is following a different trajectory than the oil on the surface. The figure also indicates the area of the top 1 m of water column containing oil to be roughly 25 km^2. Evaporation accounted for the loss of 34 volume percent of 10,000 barrels of the spilled Alaska North Slope crude oil over 24 hours, while only 3 volume percent was naturally entrained (dispersed) into the water column, as shown in Figure E-3. The remaining 63 percent was floating on

TABLE E-4 Summary of the Simulation Results at 14, 24, or 48 Hours after the Oil Spill

	Oil Distribution				
Case No.	Evaporation (volume percent)	Natural Dispersion (volume percent)	Chemical Dispersion (volume percent)	Floating (volume percent)	On Beach (volume percent)
1[a]	34	3	0	63	0
2[a]	31	3	37	29	0
3[a]	36	0	0	64	0
4[a]	30	31	0	38	1
5[a]	30	0	42	28	0
6[a]	28	31	24	17	0
7[a]	10	0	0	90	0
8[b]	18	73	0	9	0
9[a]	8	0	49	43	0
10[b]	21	44	34	1	0
11[c]	35	3	37	25	0
12[c]	35	3	37	25	0
13[c]	35	3	37	25	0
14[c]	35	3	37	25	0

[a]Values are at 24th simulation hour.
[b]Values are at 14th simulation hour because no floating oil existed after 16 hours.
[c]Values are at 48th simulation hour.

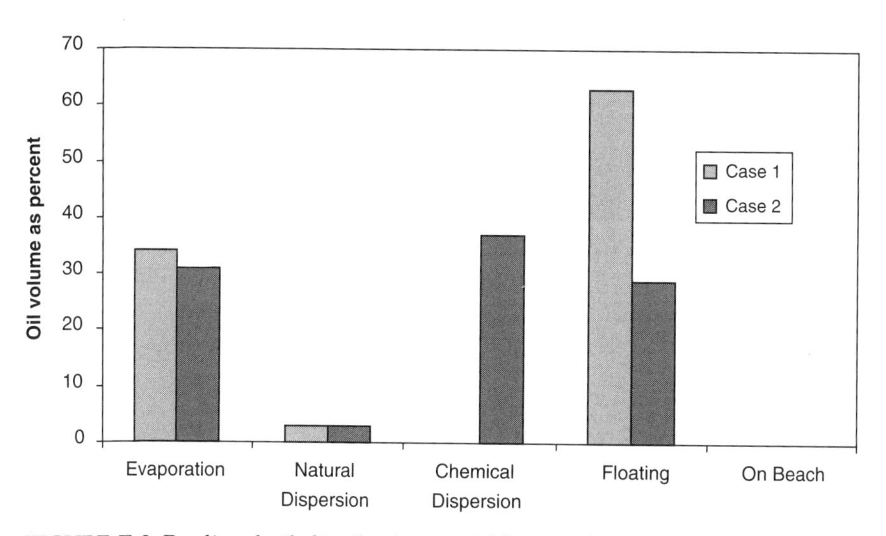

FIGURE E-3 Predicted oil distributions at 24 hours after the release of Alaskan North Slope crude oil with (Case 2) and without a dispersant (Case 1).

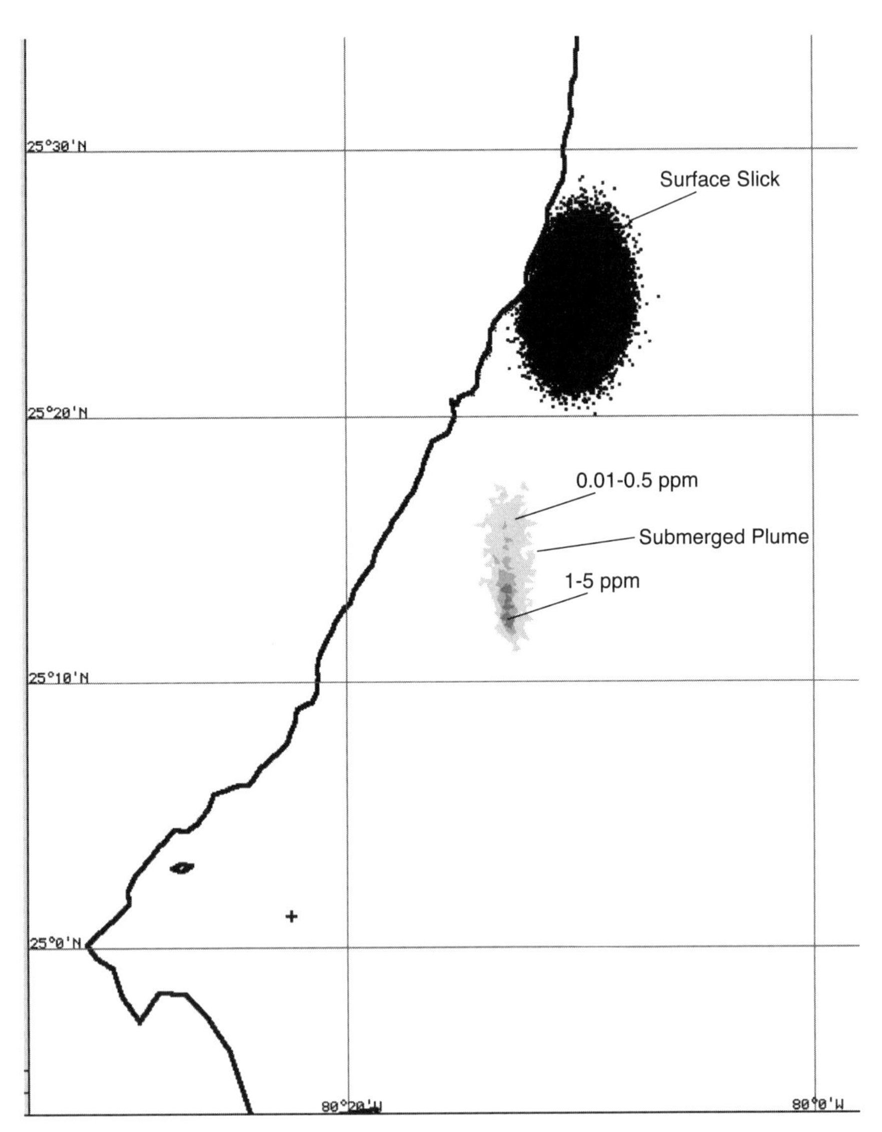

FIGURE E-4 Predicted oil movement at 24 hours after the release at point + for Case 1.

the water. Natural dispersion caused by wind, waves, and current is the only mechanism in Case 1 to disperse oil into the water column.

One aspect of complexity comes from the fact that oil consists of a wide range of hydrocarbons (see Table E-2). Although oil toxicity comes from the cumulative impacts of multiple hydrocarbon components, low and intermediate-molecular-weight components such as BTEX and PAH tend to cause more acute risks to aquatic biota, as discussed in Chapter 5. These components usually evaporate faster and to a greater extent than higher-molecular-weight components such as wax, resins, and asphaltenes. The latter are contributing components in the formation of mousse, which makes it more difficult for a dispersant to work effectively (see Chapter 3).

Figure E-5 shows the predicted composition (a relative volume fraction of each distillation cut) of the Alaska North Slope crude oil floating on the water surface at different times after the spill for Case 1. Because of evaporation, the oil composition after the release is changed from its initial composition. This figure shows that cuts with lower distillation temperatures evaporate faster and more completely than those with higher distillation temperatures, thus increasing the relative percentage of the latter components with time. For example, the heavy distillation cut #13 increased its relative volume fraction from the initial 30 percent of the oil to 46 percent under 10 m/s wind speed (Case 1) after 24 hours on the water surface. At that time, 63 volume percent of the oil was floating on the surface, thus potentially available to reach shorelines. As shown in Figure E-5, all of the first three cuts and most of cut #4 (those distilled at 147° C [roughly 296° F] or lower) evaporated within six hours. Compounds present in cuts 1–3 (e.g., below 127° C [roughly 260° F]) would include alkanes with fewer than 8 carbons and the monocyclic aromatics, benzene and toluene. Additional compounds present in cuts 1–5 (e.g., be-

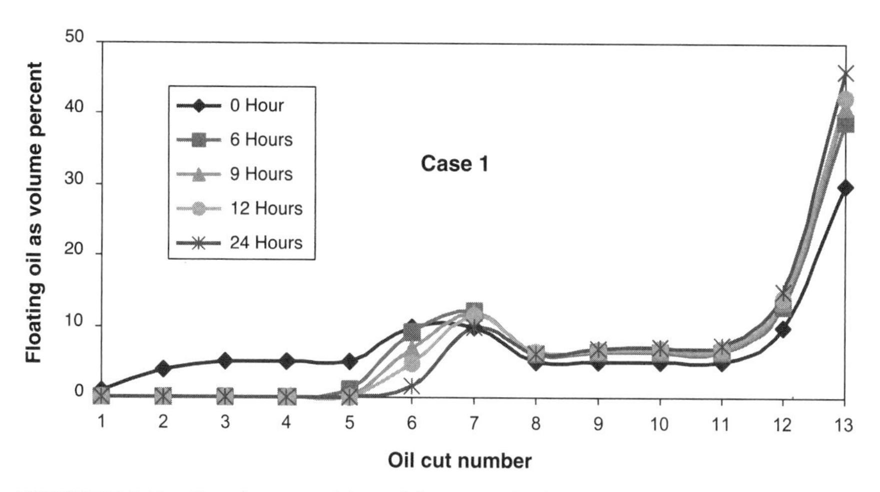

FIGURE E-5 Predicted composition of floating Alaskan North Slope crude oil with a dispersant under 10-m/s wind at various times (Case 1).

low 172° C [roughly 341° F]) would include alkanes with fewer than 10 carbons, ethylbenzene, o-, m-, and p-xylene, and several C3-benzene isomers. Thus, toxic components in these cuts are no longer on the water surface, nor would they be dispersed into the water column after 6 hours. For oil still floating on the water surface after 24 hours, those cuts distilled at 216° C (roughly 420° F) or lower (1 through 6) are mostly evaporated, and the surface oil is mostly composed of heavier-molecular-weight components. Oil with these compositions would be dispersed from the water surface to the water column below. Some of the oil in the water column would be dissolved, although ADIOS2 and 3-D GNOME do not simulate this process, as previously indicated. Ideally thermodynamic principles would be used for dissolution, as a part of the transport modeling of reactive chemicals by simulating chemical reactions (e.g., aqueous reactions, solid dissolution/precipitation, and adsorption/desorption), associated rheology and chemical property changes, and transport, coupled together (Onishi et al., 1999). However, due to a large number of oil chemical components and emulsion complexity, it is difficult to predict oil dissolution. As discussed in Chapter 4, Raoult's Law is sometimes used to estimate the dissolution (Page et al., 2000a; Sterling et al., 2003). Thus, dissolved and particulate PAH could be explicitly evaluated with this modeling approach. These results, for example, could be compared with PAH thresholds measured in toxicity tests. This has significant implications for assessing the effect of toxicity of oil on aquatic biota. However, additional data are required in order to be able to use this assessment approach.

When the dispersant was applied, the model predicted that significant changes occurred, as shown in Figure E-3. Because the dispersant application area (reflecting the operational effectiveness), dispersant effectiveness, and oil droplet-size changes are input to the ADIOS2 and 3-D GNOME codes, it is critical to know their values accurately before assessing the potential use of dispersants. There is currently no simulation code this committee is aware of that predicts dispersant effectiveness and oil droplet-size changes as a function of the controlling physical and chemical processes. This predictive capability should be developed. Furthermore, the operational effectiveness of a dispersant application to hit a targeted oil spill area should also be considered in setting the final dispersant effectiveness value.

Predicted oil movements with a dispersant application are shown in Figure E-6 after 24 simulation hours. Compared with Case 1 (without dispersant), this case has a large amount of oil in the water column (see Figure E-3 and compare Figures E-4 and E-6). With the dispersant application, 40 percent of the oil ends up in the water column (37 percent by chemical entrainment and 3 percent by natural entrainment). This figure also indicates that the area of the top 1 m of water column containing oil is

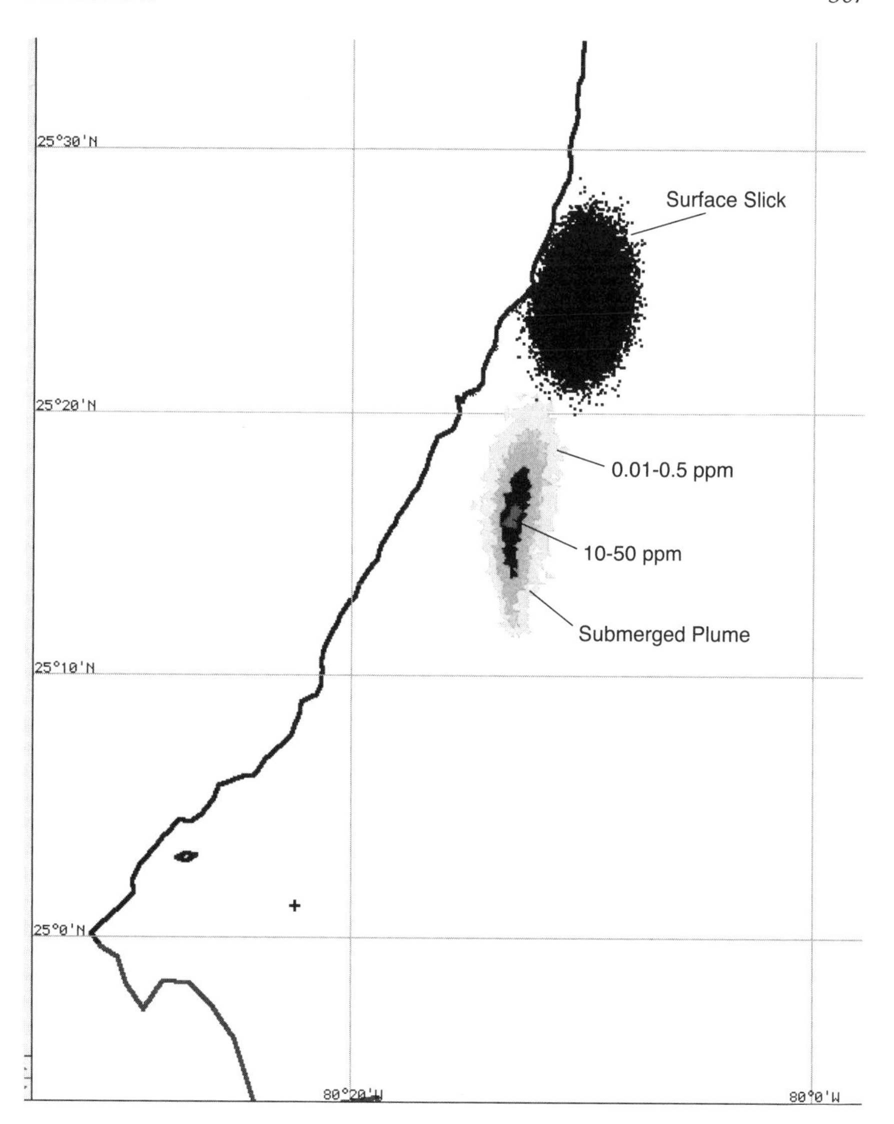

FIGURE E-6 Predicted oil movement at 24 hours after the release at point + for Case 2.

about 64 km^2, 2.5 times more than the contaminated top 1-m water area without dispersant application. Although these results are understandable, it is difficult to quantitatively estimate the time-varying oil concentration at the point of interest to assess the potential environmental impact of the oil spill without modeling it.

The predicted volume fraction of each distillation cut and the total volume of the floating Alaska North Slope crude oil over the first 24 hours after the oil spill for Case 2 were very similar to the undispersed oil slick, as shown in Figure E-7. As discussed above, most of the first five cuts (those distilled at 172° C [roughly 341° F] or lower—including alkanes with <10 carbons plus benzene and toluene, and most of the C2- and C3-substituted benzenes) evaporated within six hours, before the dispersant was applied. Thus, the dispersant applied from 6 to 12 hours after the oil spill does not introduce these chemicals to the water column. The remaining oil in both Cases 1 and 2 is mostly composed of heavier molecular-weight cuts.

Environmental Conditions

Wind and currents affect waves and diffusion in horizontal and vertical directions. As the wind becomes stronger, more oil is entrained into the water column. Consequently, less oil floats on the water surface and less oil is available for evaporation, although it is somewhat counter-intuitive for less evaporation with stronger wind. Table E-4 and Figure E-8 show the strong effect of wind on oil entrainment, with the percentage entrainment varying from 0 percent at 2 m/s (Case 3), to 3 percent at 10 m/s (Case 1) to 31 percent at 25 m/s (Case 4). Oil concentrations in the water column vary depending on the amount of oil naturally dispersed (entrained), but they also reflect the diffusivity (which increases at higher

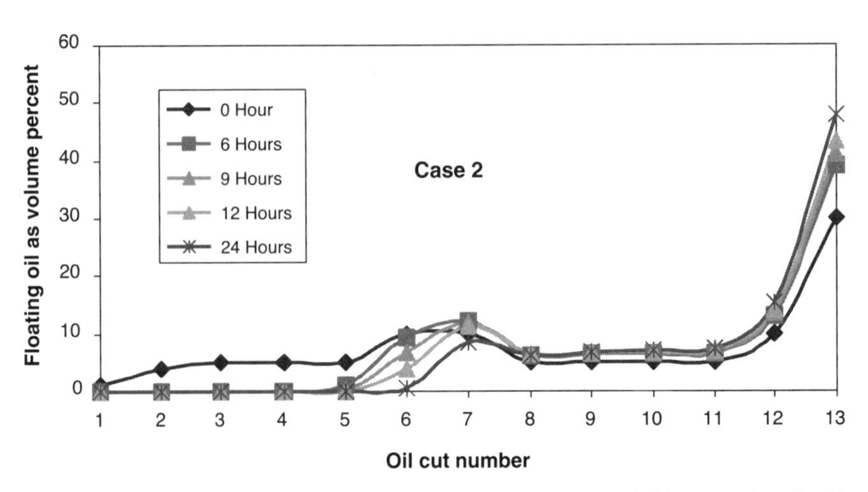

FIGURE E-7 Predicted composition of floating Alaskan North Slope crude oil with a dispersant under 10-m/s wind at various times (Case 2).

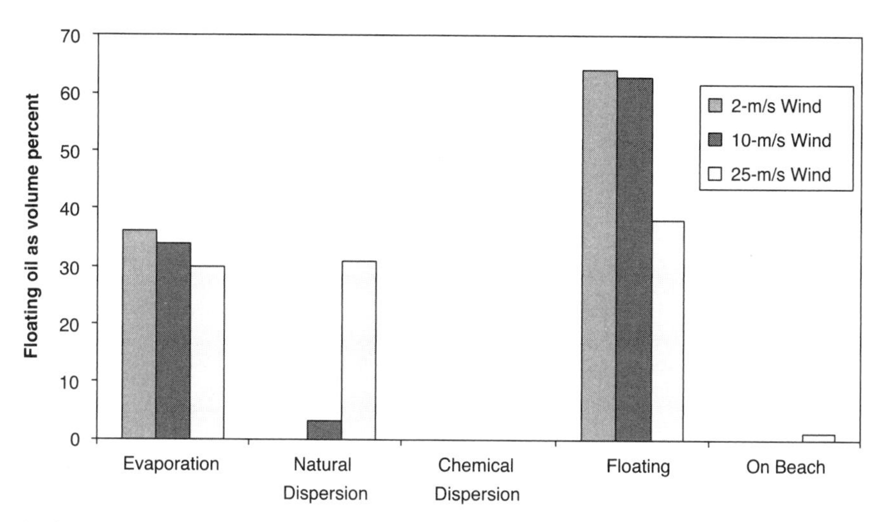

FIGURE E-8 Predicted oil distributions 24 hours after the release of Alaskan North Slope Crude (no dispersant applied) under 2-, 10-, and 25-m/s wind. There is no oil dispersed by a chemical dispersant for these three cases.

wind speed) in the water column. As discussed in Chapter 4, there is significant uncertainty in the structure of vertical diffusion, and this is manifest in uncertainty in the subsurface concentration of dispersed oil. This example reveals complexity of wind and currents controlling waves and diffusion in horizontal and vertical directions, affecting oil movement. Similar to Case 1 (10-m/s wind), Case 4 (25-m/s wind) caused all or most of the first five cuts to be evaporated, as shown in Figure E-9, which presents the predicted composition (relative volume fraction of each distillation cut) of the Alaska crude oil floating on the water surface.

The combined effects of wind (thus wave energy and diffusion) and the use of dispersants are examined by comparing results of Cases 2, 5 and 6 (wind speeds of 10, 2, and 25 m/s with dispersants) with Cases of 1, 3, and 4 (without chemical dispersants). An interesting aspect of these cases is that as the wind increases from 2 m/s to 10 m/s to 25 m/s, more natural dispersion occurs, resulting in less available oil on the water surface to be dispersed by the chemical dispersant (see Table E-4). However, these model results may be artifacts of the dispersion modeling, because the dispersant effectiveness is the measure of how much oil on the water surface is entrained into the water column below and the same 50 percent dispersant efficiencies were imposed in these three cases regardless of the wind conditions. In reality, stronger wind tends to increase dispersant effectiveness, at least up to a certain wind speed. As stated previously, the

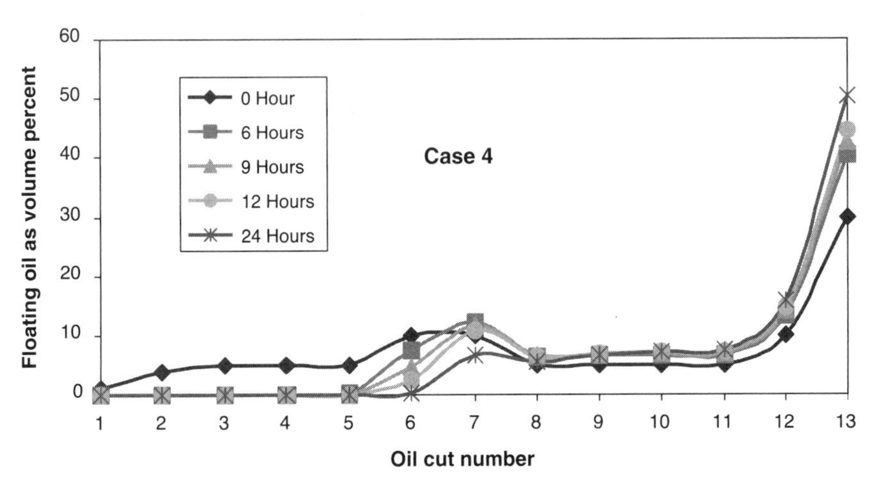

FIGURE E-9 Predicted composition of floating Alaskan North Slope crude oil without a dispersant under 25-m/s wind at various times (Case 4).

need to specify dispersant effectiveness as a model input is the weakest part of the dispersant application assessment, yet unfortunately the dispersant effectiveness is probably the most important parameter.

Effect of Oil Type

Cases 7 and 8 have the same conditions as Case 1 except for the oil type: Alaska North Slope crude oil in Case 1, heavier IFO 300 in Case 7, and lighter diesel in Case 8. The Alaska North Slope crude oil started to emulsify when 18 percent of the oil was evaporated. On the other hand, the IFO 300 and diesel are not expected to emulsify. The IFO 300 and diesel evaporated 10 volume percent and 18 volume percent, respectively, over 24 and 14 hours (see Figures E-10 and E-11 with and without a dispersant), which was less than the 34 volume percent for Alaska North Slope crude oil. This may be expected from Table E-2, which shows that these refined oil products have a low percentage of low-temperature distillation cuts. Moreover, the IFO 300 did not disperse into water due to its high viscosity (~ 15,000 cP). Diesel with very low viscosity (~4 cP), on the other hand, was greatly dispersed (73 percent), and after 16 simulation hours, there was no oil floating on the surface. Note that oil viscosity varies as oil weathers. The simulation results without dispersant application indicate that the viscosity of Alaskan North Slope crude oil floating on the water surface (Case 1) changed from its original (un-weathered) value of about 20 cP to about 200,000 cP (and still increasing) 24 hours after the spill. For the IFO 300 (Case 7), the floating oil viscosity changed from over

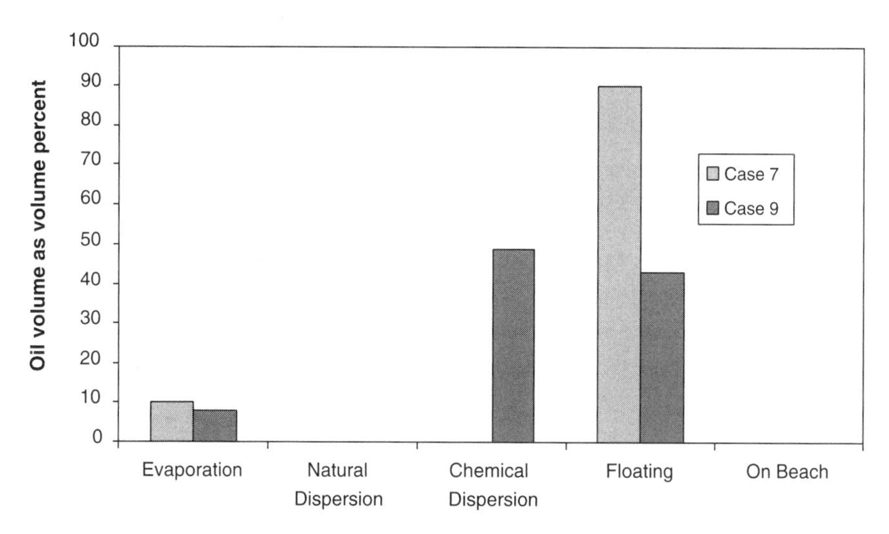

FIGURE E-10 Predicted oil distributions at 24th hour after the release of Intermediate Fuel Oil 300 with (Case 9) and without a dispersant (Case 7).

10,000 cP initially to 40,000 cP over 24 hours. For diesel (Case 8), it changed only from 4 cP to 8 cP over 16 hours. These oil viscosity changes have a significant effect on how oil spreads and on how effective a dispersant would be in dispersing it.

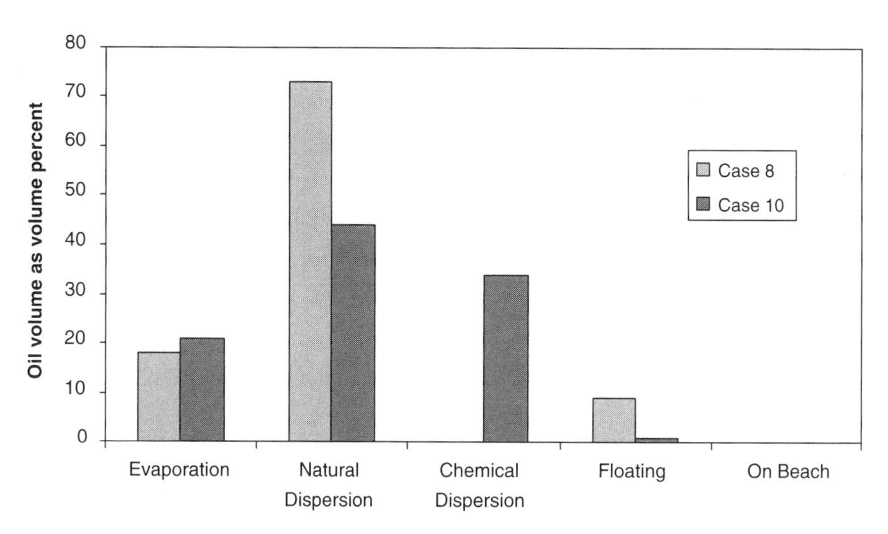

FIGURE E-11 Predicted oil distributions at 14th hour after the release of diesel oil with (Case 10) and without a dispersant (Case 8).

When it was assumed that the Alaska North Slope crude oil did not form an emulsion, the model predicted that 18 percent of the oil was naturally dispersed into the water column, compared to 3 percent when an emulsion was allowed to form. On the other hand, when mousse was assumed to be formed after 1 percent evaporation, only 1 percent of the oil dispersed into the water column. These results reveal the importance of oil type, oil properties, and emulsification on oil dispersion.

Cases 2, 9, and 10 with dispersant use correspond to Cases 1, 7, and 8 without dispersant use (Cases 1 and 2 are with Alaska North Slope crude oil, Cases 7 and 9 are with IFO 300, and Cases 8 and 10 are with marine diesel). In spite of the very high viscosity, the models predict that 49 percent of the IFO 300 would be in the water column within 24 hours as a result of dispersant, due to the assigned 50 percent dispersant effectiveness as model input (see Figure E-10 with and without a dispersant application). By contrast, Figure E-11 shows that 78 percent of the diesel was dispersed into the water by a combination of natural and chemical dispersal. This amount is essentially the same without the use of the chemical dispersant (Case 8), implying that if this diesel is spilled, there is no need to use the dispersant under a 10-m/s wind. Again, because these results are based on using dispersant effectiveness as a model input, the fate of real oil may be somewhat different.

Figures E-12 and E-13 present the predicted volume fraction of each distillation cut of floating IFO 300 and marine diesel, respectively, after the spill for Cases 9 and 10. As shown in Figure E-12, only the first cut and

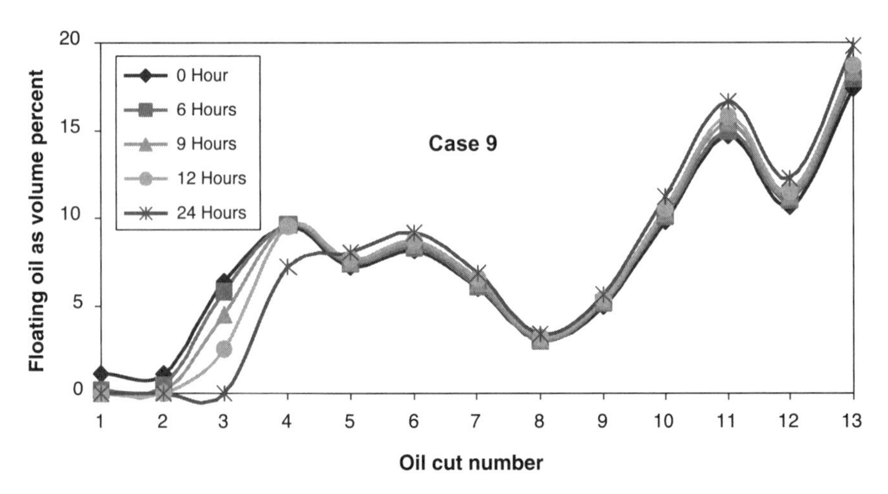

FIGURE E-12 Predicted composition of floating Intermediate Fuel Oil 300 with a dispersant at various times (Case 9).

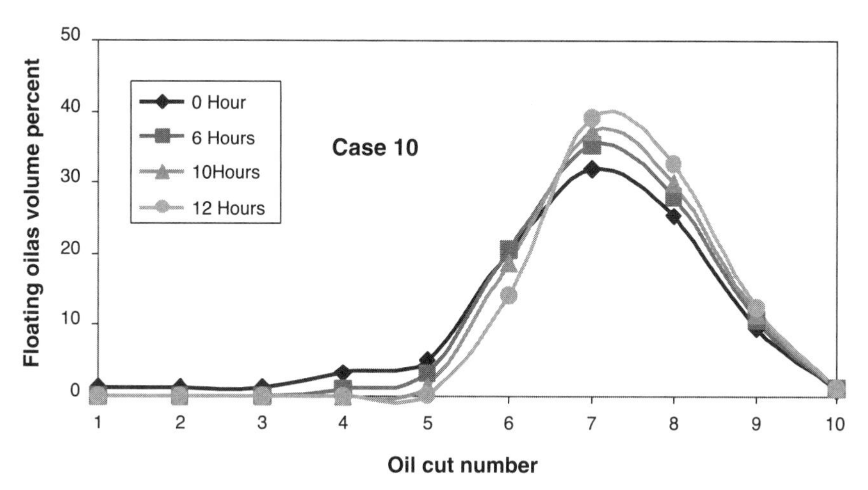

FIGURE E-13 Predicted composition of floating diesel oil with a dispersant at various times (Case 10).

some of the second cut (those distilled at 200° C [roughly 392° F] or lower) of the IFO 300 were evaporated within 6 hours; thus application of a dispersant after 6 hours would not entrain much of these components into the water column. For diesel, all or most of the first four cuts (those distilled at 180°C [roughly 356° F] or lower) were evaporated before the dispersant was applied (Figure E-13). For both refined oils, these cuts consist of only small portions (2 and 6 percent) of the refined fuels, and the majority of the oils are mostly composed of heavier-molecular-weight cuts (containing parent- and alkyl-substituted PAH).

These results indicate that without models it is very difficult to integrate all interacting and sometimes competing transport and fate processes, oil types/properties, and dispersant use to predict how much oil will be found in specific areas during an actual oil spill. Thus, transport and fate models should be used to assist decisionmakers to take appropriate remedial actions during an actual oil spill.

Effect of Oil Droplet Sizes

A dispersant application is expected to result in entrained oil composed of many more small droplets, which rise more slowly, as discussed previously (see Figure E-2). To evaluate this effect, four oil droplet size distributions (see Figure E-1) were simulated. Case 12 has the baseline droplet sizes (diameters varying from 10 μm to 70 μm), whereas Case 11 has an 80 percent reduction in droplet size due to the dispersant application. Because of uncertainty in oil droplet sizes, simulations with droplet sizes larger than base case (Case 13 with twofold increase in diameter, and Case 14 with 4-fold increase in diameter) were also evaluated.

To isolate the effect of droplet-size distribution (uninfluenced by bathymetry, and uncertainty in vertical diffusivity), these four cases were run with oil spilled on the surface of 200-m deep water, further offshore along the Florida coast, and constant diffusivities were assigned over a mixing depth of 200 m.

Figure E-14 presents predicted oil migration on the water surface and in the top 1-m water column 48 hours after the oil spill for Case 11. This figure also indicates the oil spill location by "+." Unlike the shallow water applications (Cases 1 through 10), oil in this case traveled through deep water, ranging in depth from 200 m to over 350 m. Predicted time-varying average and maximum oil concentrations in the top 1-m water column are shown in Figure E-15, indicating the increase of oil concentrations during the first 6 and 12 hours after dispersant application. At 24 hours, the average and maximum concentrations were 0.7 and 3.4 mg/L, respectively, while at 48 hours, they were reduced to 0.4 and 1.7 mg/L. These concentrations are much lower than those appearing in the nearshore case (Case 2) due to unrestricted vertical diffusion in the offshore case, and because the same diffusion coefficients were used over the entire 200-m mixing depth as were used in the shallow water. This latter assumption was used for simplicity and comparative purposes; in reality diffusion in the deeper water is expected to be less than that in the shallow water.

As indicated in Figure E-2, the rise velocity of oil droplets ranges from about 2.5×10^{-7} m/s for a diameter of 2 μm to 4.3×10^{-3} m/s for a diameter of 260 μm. Droplets moving at 2.5×10^{-7} m/s will rise only 0.001 m and 0.02 m, over periods of 1 hour and 24 hours, while over the same periods, droplets rising at 4.3×10^{-3} m/s will rise 15 m and 370 m. Meanwhile, a vertical diffusivity of 51 cm^2/s will spread oil droplets (both upward and downward) about 6 m and 30 m over the same periods. Thus, the smallest oil droplets behave as if they are neutrally buoyant—transported only by diffusion—while the largest droplets are advected mainly by their buoyancy.

Predicted vertical distributions of oil at 48 hours are shown in Figure E-16 for four size distribution cases (Cases 11 through 14). Each vertical distribution is plotted at the horizontal location where the oil concentration within the top 1-m layer is the highest over the area contaminated by oil after 48 hours. Although four cases are plotted in the same figure, they are at slightly different locations. As expected, when the oil droplet sizes increase, more oil is found near the water surface. Thus Case 14 (droplet diameters of 40 to 280 μm) has the highest concentration of 18 mg/L in the top 1-m water column, while Case 11 (diameters of 2 to 14 μm) has a

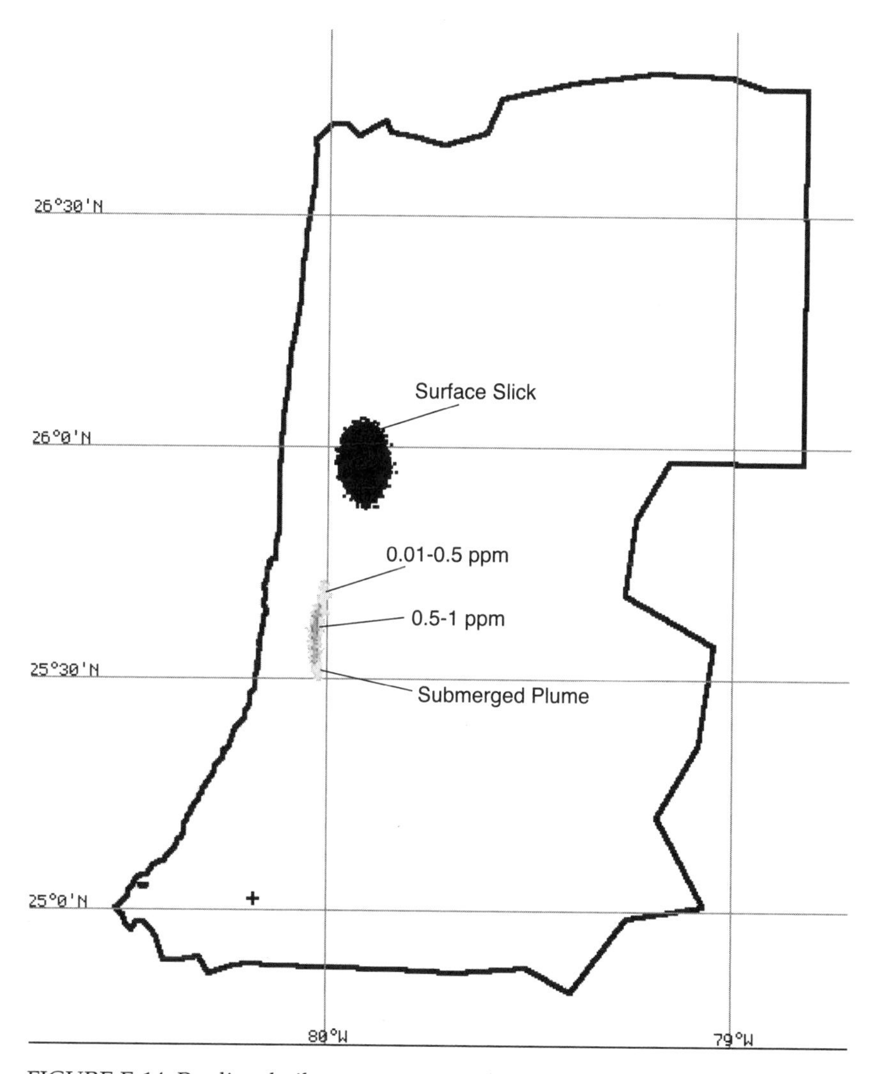

FIGURE E-14 Predicted oil movement at 48 hours after release of Alaskan North Slope crude at point + for Case 11 (80 percent reduction in oil droplet size due to dispersant application).

corresponding concentration of 1.7 μg/L. Currently the effect of chemical dispersant application on oil droplet sizes is a model input, but the ability to predict droplet size should be developed and incorporated into oil transport and fate codes.

These results again indicate that it is very difficult to integrate all of the interacting and sometimes competing transport and fate processes and

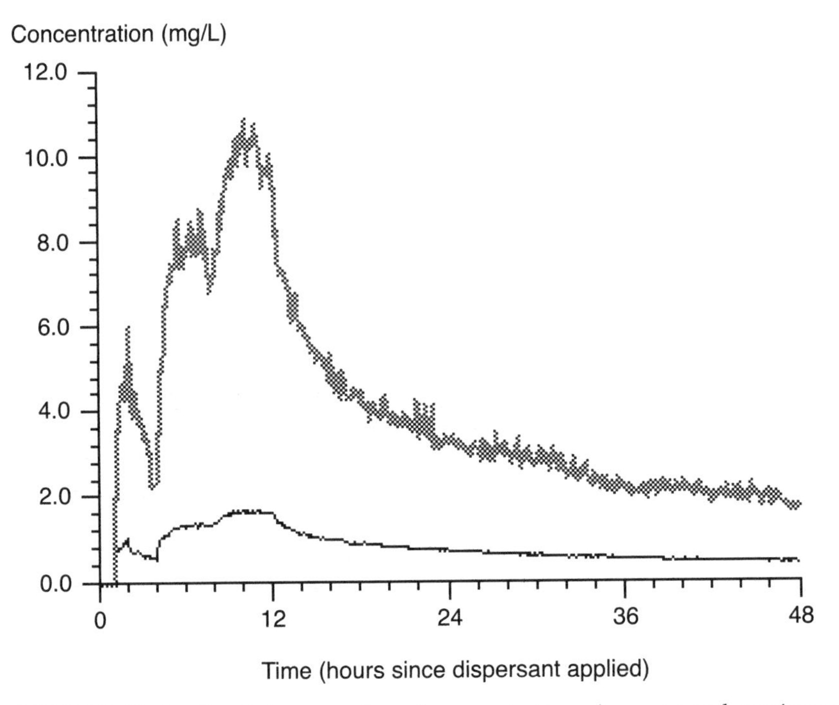

FIGURE E-15 Predicted time-varying oil concentrations (average and maximum concentration following plume versus time) in top 1-m water column for Case 11. Note: There was an 80 percent reduction in oil droplet size due to dispersant application.

oil types/properties to predict when and how much oil will move to specific areas with and without dispersant application during an actual oil spill event. Thus, transport and fate models should be used to assist decisionmakers in choosing appropriate remedial actions during an oil spill by providing quantitative estimates of oil distributions that change with time and space. This is especially important in nearshore areas, which might experience the greatest environmental sensitivity. Yet these same areas are likely to have the most complex flow fields. Limitations on computer speed and human resources will likely limit, for some time, the accuracy of numerical models to simulate advection and diffusion in near real time, especially considering that spill locations are unpredictable, and multiple "what if" scenarios runs must be run. A consensus regarding "how good is good enough" needs to be developed among decisionmakers and model developers, and used to guide the future development of models and to optimize their use in real time.

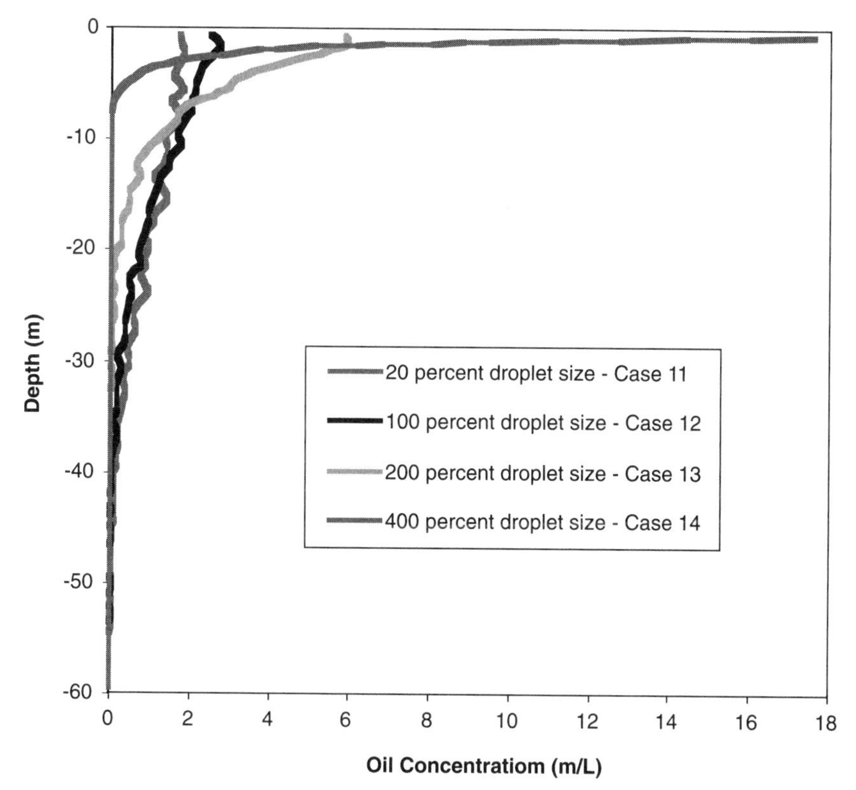

FIGURE E-16 Predicted vertical distribution of oil concentrations at the horizontal location where the oil concentration within the top 1-m layer is the highest over the area contaminated by oil after 48 hours for Cases 11 through 14.

In the meantime, efforts should be made to improve and validate models. This effort should include undertaking research at laboratory and mesoscales to define parameters that control oil dispersibility. The improved models should be used to assist on-scene decisionmakers to determine whether to use dispersants during an actual spill, and feedback should be sought from these decisionmakers as to the utility of the models in this regard. The ADIOS2 and 3-D GNOME codes, and possibly other codes, may support emergency response to provide "rough-cut" predictions within hours of oil spills.